Vienna Circle Institute Yearbook

Institute Vienna Circle, University of Vienna
Vienna Circle Society, Society for the Advancement of Scientific World Conceptions

Volume 25

Editorial Address
Institute Vienna Circle and Vienna Circle Society
Wiener Kreis Gesellschaft
Universitätscampus, Hof 1
Spitalgasse 2-4, A-1090 Wien, Austria
Tel.: +431/4277 46501 (international) or 01/4277 46501 (national)
Email: vcs@univie.ac.at
Homepage: http://univie.ac.at/vcs/

More information about this series at http://www.springer.com/series/6669

Esther Ramharter
Editor

The Vienna Circle and Religion

Springer

Editor
Esther Ramharter
Institute for Philosophy and
Institute Vienna Circle
University of Vienna
Vienna, Austria

ISSN 0929-6328 ISSN 2215-1818 (electronic)
Vienna Circle Institute Yearbook
ISBN 978-3-030-76150-9 ISBN 978-3-030-76151-6 (eBook)
https://doi.org/10.1007/978-3-030-76151-6

This Springer imprint is published by the registered company Springer Nature Switzerland AG
The registered company address is: Gewerbestrasse 11, 6330 Cham, Switzerland

Editorial

The well-known rejection of metaphysics by the Vienna Circle can be understood with good reason as also implying a rejection of the claim that there can be (scientific) knowledge in religion or theology. However, specifics of the Vienna Circle's engagement with religion, both in political and in theoretical-philosophical terms, have rarely been studied so far. This volume is now intended to spark interest in this direction.

An anti-metaphysical and anti-religious or at least anti-clerical stance of the Vienna Circle can be found not only in the 1929 manifesto *Wissenschaftliche Weltanschauung*, but also throughout the writings of the members of the Vienna Circle. There are, however, rare statements or considerations by members of the Circle who concede a positive role of religion. As an aside, one should also mention rather peculiar mediations between empiricism and religious topics such as Moritz Schlick's argument that the existence of life after death is an empirical question (see the paper by Massimo Ferrari, Chap. 3, in this volume).

This volume begins with contributions on the treatment of religion by the three central figures of the Vienna Circle Carnap, Schlick, and Neurath. A. W. Carus and Christian Damböck both dedicate their papers to Rudolf Carnap's non-cognitivism, but Carus follows Carnap's development (in particular, the emergence of what Alan Richardson calls Carnap's "fundamental philosophical commitment") from its religious origins to the sort of Kantian non-cognitivism that was the result of a shift from a Romantic and Idealist understanding of Kant to a more scientific and Enlightenment-oriented one, whereas Damböck focuses on the influence of the Herbartian philosophy of Carnap's grandfather Friedrich Wilhelm Dörpfeld on Carnap. Moritz Schlick has one of the most complex relationships with religion. Like the other members of the Vienna Circle, Schlick certainly did not consider religion as a core issue of his philosophical inquiry. But he was, as Massimo Ferrari shows in his contribution, "sensitive to a monistic worldview, according to which religion could be at least conceived as the feeling of a deep harmony with the universe" and he "regarded religion [...] as part of emotional experience able to enhance our faith in future". Marco Brusotti's contribution deals, in his own words, "with Neurath's quest for a non-metaphysical sociology of magic and religion and takes into account a broader spectrum of anthropological positions known to him in the late twenties and early thirties".

Hans Hahn, a very active member of the Vienna Circle, was an outstanding mathematician, his philosophical publications, however, are comparably less well-known, and he wrote almost nothing about religion in particular. In her paper in this volume, Julia Schäfer collects the scattered remarks on religion and God in his writings and identifies some elements of "faith" in his philosophical considerations. Malachi Hacohen's paper and mine are dedicated to Josef Schächter. Schächter, an ordained rabbi, is the only member of the Vienna Circle who could probably be called a religious person. Hacohen presents Schächter as a "religious Jewish maskil" who was central to the group in one respect, but in another was "an outlier in the Vienna Circle. He shared with them only part of his world". Schächter was present as a person, but only partly as a thinker; Philipp Frank, on the other hand, was a member who, having left Vienna in 1912 to accept a chair in Prague, was present in Vienna for only some of the meetings, but remained a "full-blooded" logical-empiricist. In the 1940s, then living in the USA, he participated in the annual interdisciplinary "Conference on Science, Philosophy and Religion" to defend modern science and its philosophy against metaphysics, ethics, and theology. Friedrich Stadler, in his contribution to this volume, explains Frank's role in these meetings.

The last two papers of this volume, one by Tim Lethen and the other by Lethen and Annika Kanckos, address the work of a person who was not a member of the Circle in the narrower sense, but was nevertheless very important for the Circle: Kurt Gödel. The first of these papers deals with Gödel's dogmatic theology and the second with Gödel's reception of Charles Hartshorne's ontological proof.

In this volume, there are at least two apparent gaps, which should be addressed: The first consists in the absence of an essay on Wittgenstein. The decision not to include such a paper is due to the fact that, in contrast to the other members of the Circle, a lot has already been published about Wittgenstein and religion. The second gap that this volume unfortunately has will soon be filled: A volume on Edgar Zilsel and his conception of a *Geniereligion*, edited by Donata Romizi, Monika Wulz, and Elisabeth Nemeth will be published in the Vienna Circle Institute Yearbook Series (vol. 27, *Edgar Zilsel: Philosopher, Historian, Sociologist*).

The book at hand offers obviously a person-centred approach.[1] I think that this is appropriate for a first publication on the subject, as the members of the Circle held very diverse views and attitudes towards religion. Malachi Hacohen and Friedrich Stadler provide, however, some synthesising perspectives.

Finally, I want to make explicit one editorial decision: the specification "name year" may stand for publications, but it may also stand for the persons who have written the publications. The contributors of this book did not treat this issue uniformly, and I decided to allow both.

---

[1] I would like to thank Sami Pihlström for pointing out in his report that this is something worth to be explained.

My thanks go to Elisabeth Nemeth, with whom I organized the conference "The Vienna Circle and Religion", held on October 26, 2019, at the Department of Philosophy of the University of Vienna. The conference was financed by the Institute Vienna Circle, the Vienna Circle Society, and the Research Centre "Religion and Transformation of Contemporary Society". My thanks go also to Friedrich Stadler and Martin Kusch for including this volume in the "Yearbook" Series, and to Solomon George at Springer, who was very helpful throughout the publication process. Last but not least, I thank all the contributors, who made it possible to give such a broad picture of the topic in this volume.

Vienna, Austria
February 21, 2021

Esther Ramharter

Contents

Part I
The Vienna Circle and Religion

Chapter 1
Carnap's Fundamental Philosophical Commitment: From Religious Origins to Kantian Non-cognitivism, 1911–21

A.W. Carus

In his excellent paper on "Carnapian Pragmatism," Alan Richardson (2007, 297–303) summarizes what he calls Carnap's "fundamental philosophical commitment" as an insistence on *both* scientific standards for philosophy on the one hand *and* practical freedom in the choice of linguistic framework, on the other.[1] The origins of this dual commitment (which also underlies Carnap's later non-cognitivism) he traces to the fundamental Kantian distinction between the theoretical and the practical uses of reason. "The former," says Richardson, "is the role of reason in constructing systems of representations of the world – the use of reason in constructing theories of how the world works. The latter is the role of reason in explaining, justifying, and motivating action – the use of reason in deciding what to do." (Richardson 2007, 298).

This not only seems entirely consistent with what we know about Carnap's early neo-Kantianism, but would appear to be corroborated by a passage in a document of 1916, when Carnap was 25 and a soldier on the eastern front. In this document, an open letter to Eduard Le Seur, a protestant pastor from Berlin, Carnap attacks Le Seur's call for a return to traditional religion in view of the abject failure of "modern culture" to provide a moral basis for coping with the catastrophe of the war. Among other arguments, Le Seur claims that only Christianity can provide a way of escaping from the deterministic world-view of modern science; "a grasp of this unconditional unfreedom leads to despair," says Le Seur, and Jesus is our only way out. Carnap objects:

[1]A characterization of Carnap's fundamental position that is consistent with Richardson's but approaches it from a somewhat different angle is given in my paper on "Carnapian Rationality" (Carus 2017); see also Carus (forthcoming).

A.W. Carus (✉)
MCMP, Ludwig-Maximilians-Universität München, Munich, Germany

E. Ramharter (ed.), *The Vienna Circle and Religion*, Vienna Circle Institute Yearbook 25, https://doi.org/10.1007/978-3-030-76151-6_1

> But only as a natural being am I situated within the causal nexus. As an ethical being I am myself, on the contrary, the freely deciding subject of my actions. (Carnap 1916, 17–18, 2018, 485)

Carnap would evidently appear to have had Kant's distinction between the theoretical and the practical in mind as early as 1916. So is Richardson vindicated? This paper will show that, concerning the ultimate *origins* of Carnap's non-cognitivism, his hypothesis fails. But regarding the subsequent history of Carnap's non-cognitivism, Richardson's hypothesis is even more interestingly suggestive than his original formulation would have led us to suspect.

Section 1.1 of the paper takes us back another 5 years, to a document from 1911, in which we discover a quite different source of Carnap's non-cognitivism, in his efforts to find a way of understanding religion and the religious discourse he had grown up with. Section 1.2 looks at the context of this 1911 document, as far as it can be reconstructed from other evidence about the pre-war Carnap. Section 1.3 returns to the Le Seur letter and considers the changes between 1911 and 1916. Section 1.4 looks more closely at Carnap's transition, apparently via Goethe, from religious to Kantian ways of framing his non-cognitivism, and Sect. 1.5 concludes by considering the significance of Carnap's further transition, apparently before 1916, from one version of Kantianism to a quite different one.

1.1 Carnap's Earliest Non-cognitivism (1911)

Carnap's noncognitivism is first clearly articulated in a set of notes for a talk to the *Freischar* (a student organization independent of the traditional fraternities) in Freiburg, where Carnap had gone to study for three semesters in 1911–12; he also played a role in founding and organizing the *Freischar*. The title of the talk is "Religion and the Church." In contrast to his later talk notes, these are fully written out in complete sentences, though with many corrections, deletions, and additions crowded into the margins, making his shorthand sometimes a challenge to read. Carnap's main point in this talk is to parse all statements concerning religion or commitment [*Gesinnung*] into two basic components, which almost always occur together, but have completely different logical functions and must therefore be strictly distinguished: knowledge statements [*Wissenssätze*] and faith statements [*Glaubenssätze*]. The talk first seeks to make this distinction clear, and then illustrates it with a number of historical examples. Carnap begins by distinguishing three aspects of religion as it manifests itself empirically:

1. *Ritual*, or actions that express a stance [*Gesinnung*] toward higher things or aspirations [*gegenüber dem Höheren*].
2. *Ethics*, or the stance toward others and toward society that results from the stance toward higher aspirations.
3. *Doctrine*, which is supposed to articulate such a stance in words. Carnap is decidedly skeptical toward doctrine: "religion not only does not consist in

doctrine – as everyone will admit – but can neither be supported nor undermined by it, as it has no contact with it whatever."

One might object, Carnap continues, that "every religion is passed from person to person *by means of words*." True, he says, but there are two fundamentally different kinds of "words" involved here, two different kinds of statements, and this brings him straight to his main distinction between knowledge statements and faith statements:

> Among the statements, though, that have to be made for this purpose, I want to distinguish between those that are about things that can be grasped by the understanding and those that express ethical requirements or the subjective awareness of the world and human life as a whole. The first kind of claim, concerning theses that are by nature objective (if perhaps not yet known or knowable) I will call "knowledge statements." ... In contrast to them there are statements about our stance toward this world picture that give expression to our pessimism, say, or idealism or something of that sort; these statements cannot be shown to be true purely by the understanding, nor refuted, therefore they cannot be discussed. I will call them "faith statements." I want to propose the expressions "knowledge statements" and "faith statements" for the purposes of our discussion today, without any further presumption, so that we have a common terminology. The expressions "knowledge" and "faith" by themselves, on the other hand, I would ask you to avoid, since prior to a precise definition they will certainly not be understood in any uniform sense. (Carnap 1911, 5, 2018, 477)

This is followed by an historical excursus to show that no "religion" in Carnap's sense (which comprises not only religions in the ordinary sense, but any ethical community, including e.g. the ancient philosophical schools) actually depended on knowledge statements, or was regarded as depending on such statements by its founders and prophets.

Carnap's principle concern in this talk was to distinguish sharply and radically between any form of religion or highest *value* and any form of *knowledge*: "religion not only does not consist in doctrine – as everyone will admit – but can neither be supported nor undermined by it, *as it has no contact with it whatever* [my emphasis]." To appreciate the full scope of this distinction it is important to take Carnap's very broad conception of "religion" in this talk into account:

> I am conceiving of religion here more broadly than is usual. I am considering it as something universal to humans, which depends neither on belief in a god (I have so far not even mentioned any sort of faith in this sense) nor on any particular sort of ideal. According to my way of looking at it here, then, patriotism is a religion, too, for instance, and acting in accordance with it is the practice of a religion, specifically for the person who puts his country at the highest level of his value system. *What* occupies this level in someone is in principle of no relevance to the question whether his relationship to it is religion or not, even if where other people are concerned we may sometimes value one kind of religion in this sense more highly than another. For one person a personal god may occupy this highest spot, or the whole of the world as an organism in the pantheistic sense, while for another it may be art in general, or a particular art, or science; for others again it may be family, country, race, or humanity as a whole – (Carnap 1911, 3–4, 2018, 476)

What he would later call descriptive or "cognitive" statements could not be derived from statements expressing whatever occupies the highest level in a person's scale of values. And conversely such a value statement could not be derived from cognitive statements. This is precisely what Carnap (1963, 1000) would define as "non-cognitivism" half a century later. Nor is this idea merely implicit in the 20-year-old Carnap's lecture; it is the main point. Evidently this question had been very much on his mind during these years, as we can also document from other sources.

1.2 The Background to "Religion and the Church"

The young Carnap's intense interest in the question of highest values and what to base them on (or not) is hardly surprising, given his upbringing. In the original draft of his autobiography he not only describes the utopian religious ideas of his ancestors, but even suggests a direct connection between his own prophetic enunciation of new ideas and the religious family background:

> The vivid interest which these people took in their religion and their way of life could sometimes not be satisfied by merely accepting the Lutheran doctrine. The Reformed Church, which was strongly influenced by Calvinist ideas, had many adherents. About the year 1742 some members of this church, among them two ancestors of my father, found the worldly life in the great city of Elberfeld too sinful and intolerable. Eventually a group comprising about fifty families emigrated to the other side of the mountain and founded the town of Ronsdorf, which they called their "Zion," devoted to a new and better life... I believe there is still a trace [of this impulse] in me, derived from the strivings of these people for the realisation of a visionary aim, and from their missionary spirit... (UCLA Box 2, CM3, folder M-A5, A6-7)

But in apparent contrast to the somewhat fanatical religiosity of these ancestors, the emphasis in Carnap's description of his own upbringing is on the fact that for his parents what mattered was not doctrine but ethical *practice*. The important thing about religion "was not so much the acceptance of a creed, but the living of a good life" (Carnap 1963, 3). As an example of this attitude Carnap cites his mother's interest in the pantheistic philosophical writings of Gustav Theodor Fechner, the founder of psychophysics:

> My mother liked to read at that time his little-known religious-philosophical works; it was characteristic for her attitude toward religious questions that she did not find any difficulty in reconciling Fechner's pantheism (or panentheism) with her own Lutheran-Protestant faith. (UCLA Box 2, CM3, folder M-A5, B20)

Furthermore, Carnap traces the easy and untraumatic nature of his disengagement from religion in his youth to this very subordination of doctrine to ethical practice:

> I relinquished my religious beliefs by a gradual development, a continual transformation... It was not, as I had often seen it with others, a matter of a sudden and violent rebellion with vehement emotional upheavals, where love is transformed into hatred and reverence into contempt and derision. I think that this is chiefly due to the influence of my mother's attitude. Since childhood I had learned from her not to regard changes of convictions as moral problems, and to regard the doctrinal side of religion as much less important than the ethical side. (UCLA Box 2, CM3, folder M-A5, B23)

Anna Carnap was interested not only in Fechner, but also in many other less orthodox religious writers, among them especially Heinrich Lhotzky and Johannes Müller, who had since 1897 been bringing out their widely-read *Blätter zur Pflege persönlichen Lebens* [*Leaflets for the fostering of personal life*], known as the *grüne Blätter* [green leaflets] for their pale-green covers. She had presumably first encountered Müller in his lectures, since Wuppertal was one of the first places outside of Berlin where Müller's events had attracted great interest and a substantial following after 1898 (Haury 2005, 70–71).[2] And it was a local business family, inspired by these lectures, who had financed Müller's und Lhotzky's "convalescent home for adults [*Erholungsheim für Erwachsene*]" in the Palace of Mainberg in a reclusive corner of Franconia (Haury 2005, 79), which began receiving guests in 1903, and soon became popular among the troubled religious consciences of German (and even foreign) high society. We don't know when Anna Carnap first went to Mainberg, but it (as well as the *grüne Blätter*) often came up in the prewar (and even in the wartime) correpondence between Carnap and his mother. Carnap himself spent a month or two there in the autumn of 1911, and as he wrote his Swedish friend Tilly Neovius (whom he had evidently met there), it had left a strong impression:

> But I must get back to Mainberg again this year. The thought won't let me go. Since I left, it exercises such a strong attractive force on me. When I was there, you probably didn't notice that so much; I think I hardly noticed it myself. But now I'm always thinking: when can I get back? (ASP 025-71-01, 13)

In the idyllic rural setting overlooking the river Main there was not only lively dancing every evening (which would have appealed to Carnap's youth-movement enthusiasm for the Sera dances; see Werner 2003, 124–30), but Johannes Müller personally engaged each individual guest in private conversations about their personal concerns and spiritual well-being (Haury 2005, 154–162). Carnap's talk to the Freiburg *Freischar* on "Religion and the Church" was composed in the period immediately following this visit, under the strong impression left by his Mainberg experience.

But what exactly about Müller could or would have appealed to Carnap at this juncture? To appreciate that, one needs to understand where Müller's standpoint fit into the overall spectrum of religious and spiritual offerings in the first decade of the twentieth century. For from today's standpoint, nothing very exceptional about the *content* of his message seems to set him apart from the many competing spiritual and religious subcults of the day – such as the "movement of awakening" [*Erweckungsbewegung*] within the established Lutheran churches (Benrath 2000;

[2]The Wupper valley (the present city of Wuppertal was incorporated as a consolidated entity only in 1929) and the surrounding area ("Bergisches Land") were, as Anna Carnap writes in the introductory chapter to her biography of her father – and as Carnap himself, in his original (unpublished) autobiography recapitulates, largely drawing on his mother's account – were already heavily pietist and Calvinist in the seventeenth and eighteenth century. This region is also regarded as a center of the movement of "awakening" within German protestantism of the nineteenth century (Benrath 2000; Müller-Späth 1984).

Beyreuther 1977), the many "back to nature" movements, including those within the Youth Movement (Hepp 1987), Monism (Hübinger 1997; Carus 2007b), or the cults surrounding various magnetic personalities such as Stefan George, Rudolf Steiner, Ludwig Klages, or C.G. Jung (Noll 1994). Carnap's environment in Jena was a paradigmatic example of the extensive syncretism among these intellectual and spiritual movements; the Sera-Circle around the publisher Eugen Diederichs, in which Carnap was such an enthusiastic participant in 1908–14, was a typical product of its time, as Meike Werner (2003, 130–156) shows in convincing detail, not least in its attitude(s) toward religion. So what role could Müller have played for someone like Carnap?

This question why Carnap might have been interested in Müller, of all the possible choices out there, is of particular interest since Diederichs himself, the moving spirit of the Sera circle, kept Müller at a distance. Despite repeated contacts between them – Diederichs came to Mainberg more than once – Müller evidently did not in the end qualify as one of the "modern spirits" that Diederichs wanted to cultivate and bring together in his publishing house – to Müller's great disappointment (Haury 2005, 175). The many intellectuals who wanted to leave Christianity behind and find a "new religion" (among whom Diederichs counted himself; see Werner 2003, 130–56) missed in Müller the sort of radicalism they admired (in very different forms) among the Monists, say, or in the writings of Nietzsche, or in the workers' movement.

But it was precisely this moderation that made Müller's standpoint unique – that in the spectrum of (free) religious movements around 1900 he represented a *middle way*, so to speak, between the radical rejection of Christianity at the one end and the "movement of awakening" that remained safely within the protestant church, at the other end. Intellectually he could hardly be considered Lutheran, in any serious sense; he had left too many of the church's doctrines behind. He denied original sin, for instance, and played everything supernatural in the New Testament down so far that Jesus became essentially just an ethical paradigm. God, for Müller, was to be understood in his natural manifestations, even if the conception of nature involved was vitalistic (Haury 2005, 102–111); this was all pretty far removed from any remotely acceptable Lutheran doctrine. But nonetheless Müller had grown up within the church, and identified as a Christian, in the sense, minimally, of the "modern theology" of Ritschl and Harnack (Nipperdey 1988; Haury 2005, 28–33), which enabled him to take account of certain recent developments in society and in the sciences, which he did not (like some other unorthodox Lutherans) regard as an obstacle to the continuation of a living protestant tradition. Precisely this careful walking of a fine line between the extremes was what made him, then, an attractive mediator, especially to educated people who had grown up within Lutheranism and felt comfortable in it, but who also saw the contradictions between such a traditional faith and the new developments in science. That was presumably also the attraction for Anna Carnap, who certainly belonged to the *Bildungsbürgertum* but whose neo-pietist upbringing made more radical alternatives outside the church unpalatable.

In his precarious balancing act between the extremes it was essential for Müller to avoid any trace of doctrine and to stay as far away as possible from any explicit confession of faith. In fact he completely rejected all theology as a "cult of consciousness [*Bewußtseinskultur*]" – as an articifial and harmful intellectualization of inner experiences or affective insights that can and should only emerge from individual intuition against the background of a particular biography (Haury 2005, 93–95). In his 1911 talk Carnap merely systematizes and generalizes what he had read in Müller and heard in Mainberg. Carnap's radical dissociation of "faith statements" and "knowledge statements" would have been unequivocally endorsed by Müller. So noncognitivism was not much of a stretch for Carnap; in fact, given his upbringing and his family-mediated experience of Mainberg, it is hard to see how he could have avoided it.

1.3 Carnap's Non-cognitivism After 1911

The European upheavals of the following years dragged Carnap into their midst; he was involved in some of the bloodiest fighting on both the Eastern and the Western fronts and emerged a pacifist, at latest by the time he was reassigned to a research post in Berlin to develop wireless telegraphy in late 1917. He remained there for over a year, got involved in left-wing politics, and joined the USPD (the breakaway anti-war wing of the social democrats) from August 1918, gradually developing his own perspective on what was going on around him, which he summarized in a wide-ranging reflection on "Germany's Defeat" written just a few days before the Revolution that ended the Hohenzollern monarchy. These events, and Carnap's direct involvement in them, obviously influenced his thinking very deeply.

This is evident in the next document (after the 1911 talk) we have of Carnap's overall perspective, from early 1916, when he was still on the Eastern Front. This is the already-mentioned open letter to Le Seur, where he rejects not only the doctrinal Christian "knowledge statements" but now also the corresponding "faith statements":

> The expression "acknowledgement of sin" derives from a conceptual world that has grown alien to me. It was not always so; as a child I plagued myself a great deal with guilt feelings and the search for forgivenness by other people and by God. Such experiences are only good for keeping one's conscience awake, though, and barren beyond that. In the same barren conceptual world I would put: the bondage of sin, grace, forgivenness, redemption. I do not at all deny the realities underlying these terms. I know my errors and know that I have more of them than I can see. I know that my capabilities are limited though the task is limitless. (Carnap 1916, 13–14, 2018, 485)

And for the 1916 Carnap, Jesus no longer qualifies even as an ethical paradigm, the function he had served for Johannes Müller:

> For me he can't be that. The entire context of his life is too different from ours for that. The community he lived in, the way he arranged his life, how he treats people, his views on family, state, employment, and mental life of the people, nearly every single one of his actions, especially the ones emphasized in the reports about him, are so totally alien to our

> life and our goals that if I even wanted to search out a particular person as the "image before me," I could not possibly choose him. (Carnap 1916, 18–19, 2018, 485–6)

And in place of the Christian acquiescence in divine grace, which remained of critical importance to Müller despite his attempted modernization of Christianity in other respects – we see unmistakable first signs of Carnap's later "voluntarism" (Jeffrey 1994; Carus 2018); "Don't demand acknowledgement of sin from people today, or an appreciation of their need for redemption, but rather grab them from the precisely opposite, positive side: strengthen their self-confidence," Carnap suggests to Le Seur, "Show them the tasks in the service of a spirit that goes beyond the individual person but is recognizeable by him — tasks that correspond to their capabilities." And "Teach them on the one hand the humility in the valuation of their own person in comparison to the overriding idea whose servants they are," However, "On the other hand teach them the pride that accepts no grace or mercy."

> If a person who has gone astray suddenly realizes his fault and tortures himself with the burden of his guilt, don't call out to him: do you now understand your hideous sin? – now you are ripe for our gospel! But tell him that he not only has the right to absolve himself of his past, but rather the duty, even, to ward off all excruciating thoughts about his guilty life, so as to work with full attention on his present task. Don't talk to him about a redeemer; don't tell him that someone died for him once long ago. If you want to give him a good example, show him people of his own kind, from our own time. Those he can relate to; he can identify with their experiences. In their work, which he understands because it has some relation to his own, he discerns the larger aspiration, to which even the greatest men subordinated themselves as servants. In this way he can arrive at clarity about his own task and the desire to join its service. (Carnap 1916, 21–2, 2018, 486)

Carnap fully acknowledges that Christianity, like many other religions, expresses something fundamental about the human condition; it articulates the question we all have about the big picture, the destiny of the human species, and the place of each individual within that larger pattern. He admits that he has no answer to this question; "I have no idea where the eternal unchanging comes from that breathes life into us for a short time and uses us as its organs and then lets us sink into nothingness again, like a tree that pulls up a lifeless atom from the ground and uses it to make leaves." And he acknowledges not only the legitimacy of this question, but admits that we can't help asking it: "where does the great unchanging come from that forces us into its service, and where is it going? Whence does it derive the power and the right to decide about our lives?" Luckily, he says, we don't need an *answer* to this question to go about our daily tasks, and to make a contribution to the great collective project "in the service of that spirit" – which he translates into more secular terms as "in the service of cultural development." Nonetheless, regarding Le Seur's religious articulation of that great question, Carnap admits that he is "in the unfortunate position, without knowing any solution to it myself, of having to reject yours unequivocally." (Carnap 1916, 11, 2018, 484)

> When you describe our searching and inquiring with the words of the Psalm, "My soul thirsts for God, the living God," I can't reject that. For perhaps you understand something by these words that I would express in cruder terms. There is a thirst, I feel it; what God is I have no idea. (Carnap 1916, 11–12, 2018, 484)

Nonetheless, despite this repudiation of the "faith statements" of Christianity (and thus of Johannes Müller's middle way) – along with the "knowledge statements" he had already rejected in 1911 – the previous noncognitivism survives intact. Though Carnap completely rejects Le Seur's religious path ("that is not a path, it is a way for you and people of your faith to bring the inner quest for an answer to that question [the above question] to a halt."), he admits that it is a normative choice (a matter of "faith statements"), which descriptive or cognitive (factual or theoretical) considerations ("knowledge statements") cannot decide: "At the level of theory, of course, nothing can be said about this." Still, he does not think reasoning about it is impossible, since he goes on to offer reasons against Le Seur's normative choice:

> But it is enough to undermine the general validity of the path by showing that I, for one, cannot take it, even if it weren't for all the others who are like me. Yes, you will think, that is the 'iron cage.' But our directions are perhaps not as diametrically opposed as you might have expected. I will try to approach as closely as possible to your standpoint. That will make the unbridgeable gap that still separates us most precisely apparent. (Carnap 1916, 12–13, 2018, 484–5)

What Carnap appears to be saying here is that while fact and theory have no bearing on the normative question, Le Seur's answer can nonetheless be challenged, e.g. on the grounds that alternative answers are possible. To Le Seur's anticipated possible response that no dialogue is possible between such radically different standpoints (this is presumably the point of the "iron cage" allusion to Le Seur's text), Carnap responds that the jury is out; we shouldn't give up too soon, and he for one will try to articulate his own viewpoint as far as possible in the terms of Le Seur's, so as to pinpoint exactly where the remaining differences are.

The strict distinction of 1911 between "faith statements" and "knowledge statements" remains intact, but there is evidently a difference regarding the possibility of bringing (some form of) "reason" to bear on "faith sentences." In 1911, we are told simply and baldly that "these statements cannot be shown to be true purely by the understanding, nor refuted, therefore they cannot be discussed." In 1916, there is obviously an echo of this radical impulse in his "At the level of theory, of course, nothing can be said about this." But the fact that he *does* then say something about it nonetheless indicates that the influence of Kant (which we noted in this letter to Le Seur at the beginning of this paper) had by 1916 been deeply internalized, and that the idea of a practical reason (*praktische Vernunft*) distinct from, or reason (*Vernunft*) extending beyond, empirical or theoretical reason (*Verstand*) had become a mental habit even if Carnap's first impulse, rhetorically at least, was to exaggerate the extent of the dissociation between "faith statements" and "knowledge statements."

My interpretation of this rhetorical impulse is the same as my interpretation of the *very same* rhetorical impulse half a century later. The older Carnap, like the younger Carnap, had a tendency to overstate a point (exhibited most notoriously in the polemical writings of the Vienna Circle years); and in the particular case we are discussing here, his way of overstating the point remained almost exactly the same between 1911 and 1958. To stress that factual or descriptive knowledge cannot *determine* normative choices, he says (in 1958):

> Although all logic, including inductive logic, and factual knowledge are irrelevant, it nonetheless seems to me that there are other, purely valuational criteria by which to judge a [comprehensive] value function as more or less rational than another. (Carnap 2017, 193)

As I have argued elsewhere in more detail (e.g. Carus 2017, 176), this sweeping "irrelevant" is clearly not to be taken altogether literally; in fact, Carnap explicitly says elsewhere that factual knowledge has an important role in guiding choices of values (e.g. Carnap 1963, 83). What he meant to say is that it can't by itself *determine* the choice of values. These must be, as he put it in 1958, decided on by "other, purely valuational criteria by which to judge a value function as more or less rational than another." Given his invocation of Kant elsewhere in the 1916 Le Seur letter and his explicit reasoning against Le Seur's normative choices,[3] it seems reasonable to read something like the 1958 attitude back into Carnap's earlier discussions of values.

In any case, the important point to keep in view here is that Carnap's noncognitivism survived the abandonment of its original motivation, i.e. of Johannes Müller's strategy (adopted by Carnap in 1911) for remaining nominally within the church while rejecting its official doctrines. Kant appears to have been drafted in to provide the basis for noncognitivism that Müller had previously supplied; we will now look at this transition – in which Goethe played a key role – in more detail.

1.4 From Müller to Kant via Goethe

In Carnap's 1911 talk to the Freiburg *Freischar*, religion is in the foreground and Kant is absent; he is not even in the background (Rickert is mentioned, but only parenthetically). Carnap's original non-cognitivism, it seems, can be explained directly from his upbringing and his family – together with Johannes Müller, to whom (and to whose writings) his family had introduced him. In the Le Seur letter of 1916 it is precisely the other way round. Here, the original religious motivation is invisible (from this letter by itself it could not have been inferred), and Kantian ways of thinking have become practically second nature. So although Richardson's hypothesis about the *origin* of Carnap's noncognitivism appears to be mistaken, it seems that the Kantian distinction between theoretical and practical had a critical role in the *preservation* of Carnap's noncognitivism after its religious basis crumbled away. This would if anything make Richardson's hypothesis more interesting, since then Kant's philosophy would not just have been imbibed passively with Carnap's upbringing; it would have been something he consciously *chose* to assimilate, and to integrate into an existing mental world.

---

[3] As well as his equally explicit reasoning, a couple of years later, about other (in this case political) normative choices in his unpublished "Deutschlands Niederlage" of 1918, mentioned above; on this document and the discussions surrounding it, see Carus 2020c.

Carnap had certainly encountered Kant's works in some form in Jena before his three Freiburg semesters in 1911–12. But his description of Rickert's lectures in Freiburg in the winter semester 1911/12 sound very much as if they had revealed to him for the first time what the *point* of Kant was. In January of 1912 he wrote to Tilly Neovius:

> And then philosophy. (You're not allowed to stop reading now; you have to pay attention! As a punishment, you know.) Rickert is talking now, one of our most significant philosophers right now in Germany, about philosophy from Kant to Nietzsche (so basically about the 19th century). When he talked about Kant, that was terrific. I don't think anyone else could get that across so well. He didn't take Kant and go through it book by book. He did it the other way around; he started from the problems, made *those* completely clear to us, and then showed us Kant's solution, as far as possible without Kant's terminology that's so hard for us to understand. This way we could appreciate why it even mattered that someone in Königsberg sat there 100 years ago and spent his whole life thinking about questions whose point and importance are hard to figure out at first glance. (ASP 025-71-01, 6)

But what mainly impressed Carnap at the time he was writing this letter was not Kant's epistemology or even his ethics, but rather the kinship Rickert portrayed between Kant and – yes – Goethe. The idea was by no means unique to Rickert; it had become something of a preoccupation among Neo-Kantians at around this time. Among many others, Georg Simmel and above all Ernst Cassirer made some form of complementarity between Kant and Goethe into a major theme of their work. In 1916, Cassirer would publish *Freedom and Form: Studies in German Intellectual History* [*Freiheit und Form: Studien zur deutschen Geistesgeschichte*], in which the long central chapter on Goethe not only integrates Kant with German classicism as its philosophical basis, but also portrays Goethe as a "liberator" of subjectivity. And just this aspect appears also to have been Rickert's theme in the lectures Carnap describes; Kant and Goethe, according to this description, support each other in their rejection of the "artificial culture of intellect [*gekünstelte Verstandeskultur*]" of the Enlightenment, in favor of an "pristine, genuine naturalness [*einer ursprünglichen, echten Natürlichkeit*]."

> And despite all that I wouldn't have dragged you into that lecture, if you'd been here. But to this most recent lecture I would have, in which he was talking about Goethe. Not about the poet Goethe; you've heard enough about him. But rather about the philosopher. And nonetheless you'd have found it interesting! He shows us Goethe during his "*Sturm und Drang* [Storm and Stress]" period via the *Urfaust* (the oldest version of the tragedy). Just as the Renaissance man Faust struggles against scholasticism, so Goethe against the movement of the "Enlightenment" (which you with your studies of French probably know better than I do), just as Kant does, but by different pathways. And goes on to show us how Goethe finds his ideal of original, undistorted *nature* – in contrast to the artificial *culture* of the Enlightenment – in Shakespeare (who was looked down on at the time): "Nature, nature!" (Goethe cries) "or rather Shakespearean people! For how should we know nature, after all? Since here everything is so fake and so suffocating." (ASP 025-71-01, 6-7)

But when we read on in this same letter, we are suddenly reminded, with a jolt, that it stems from exactly the same period as Carnap's 1911 lecture to the *Freischar*:

> So expressed bilaterally, in short: negatively – against the intellectual *culture* of the Enlightenment; positively – for original, authentic *nature* and the natural. All that is missing

is the use of the words "cult of consciousness" or "cult of knowledge [*Bewußtseins- und Wissenskultur*]" to show clearly how close these ideas are to those of Johannes Müller." (ASP 025-71-01, 7)

It really does look as if Carnap's first real access to Kant ran via Johannes Müller – and Goethe! To our amazement (given our acquaintance with the later Carnap), we find that this even extends to a certain sympathy for Goethe's conception of natural science:

> Then there is Goethe's scientific position with respect to nature, in contrast to Kant's, whose writings were therefore indigestible for him, but nonetheless influenced him indirectly via Schiller. Goethe never took an analytical stance toward his scientific object, as something concrete, but always a comprehensively intuitive one. Thence also this conception of nature that he didn't regard like Kant as an object opposed to the cognizant person, but as something that he himself was entirely involved in. Nature was for him the creator and basis for both everything material and everything spiritual. (ASP 025-71-01, 7)

Of course that doesn't mean that Carnap wholly endorsed the Goethean conception of nature at this time. In his physics courses we can safely assume that he appreciated the advantages of mathematical physics in the tradition of Newton. Even in passages of his correspondence with Tilly Neovius he extols the virtues of theoretical physics – especially in contrast to experimental physics. But during this period before 1914, the precarious middle way of Johannes Müller spared him the necessity of choosing between these different options he found himself able to sympathize with in different contexts. It spared him the effort, in any case, of having to weigh them against each other and undertake attempts to reconcile them.

Only later, in the trenches of the war, did he get around to confronting these various incompatible components of his pre-war mental life with each other. In 1916, the year of his open letter to Le Seur, his friend Wilhelm Flitner challenged him with something a mutual friend had said, regarding Goethe's critique in the *Theory of Colors* of Newton's "epistemologically monstrous notion ... that a mechanical process, ether waves, should be more real to the person experiencing them than the color he sees – that being the most real of all, to which everything else has to be reduced in order to be graspable, explainable." (ASP/WF 12) Carnap rejected this. "Such a reduction does not seem an 'epistemological monstrosity' to me, for it doesn't claim that the mechanical phenomena are *more real* than the non-mechanical ones derived from them."

> For the *contents of sensation*, all physical qualities have equal rights (weight, tone, color). But since *physics* has the task of describing natural processes as simply as possible, it is undeniably a big step forward for it when it has e.g. succeeded in portraying acoustical phenomena as pure processes of motion (specifically as vibrations of bodies) that are knowable, measurable, and countable, and with which, while we abstract from physiological effects, *no* (that is the important thing!) other qualities are intermixed than those of bodily motion (mass, spatial coordinates, temporal coordinates). A longitudinal vibration in the air of frequency 435 [cycles per second], regardless of how it is induced, and the sound wave of the pitch 'a', don't just match, they are identical. The fact that motion processes and pitches affect different sense modalities is no business of physics. (ASP/WF 12)

Here[4] he is well on the way to the conception expounded in more detail to Flitner in a later letter (1921), in response to a book Flitner had sent, *The Religious Decision* by Friedrich Gogarten, with the remark that it contradicts something Carnap had evidently said in conversation; in contrast to that view, Flitner says, Gogarten portrays a "gaping rift between cultural occurrence and divine occurrence" in the vivid terms "of a person strongly gripped by certain experiences." (ASP 081-48-04, 2) But having now read the book, Carnap denies that it reveals any new "region of reality [*Wirklichkeitsbereich*]" that would show his "idealistic conception" to be one-sided or incomplete. Of course, he admits, Gogarten could respond that Carnap is simply blind to the realms of which he writes: "So to him, my failure to see would not be an objection, but would merely show that I (so far) don't belong to the elect." So there is a stand-off between these positions; how should one go about resolving it?

> Initially, only the relativist view is possible: he admits that he can't make the other reality visible to me, I have to admit that I can't refute it; so each of us can initially maintain his position. But surely there has to be a way of getting beyond such relativism (which my belief in rationality can only allow as a temporary expedient) and arriving at an objective, generally binding knowledge. (ASP 081-48-04, 2)

Carnap suggests a parable to show how that might come about. Imagine most people were blind, and only a few could see. There would be a similar standoff regarding the existence of a visual world. But the blind could impose a standard for recognizing the existence of such a world; they could say "if you perceive more than we do, then you must discern more connections in things *that reach over into our world*, i.e. you must sometimes notice a connection as necessary that seems coincidental to us and surprises us." The seeing person can do that; he can predict things about their commonly perceived (heard and felt) surroundings that the blind person can't; so the latter "must acknowledge the peculiar perceptive powers of the seeing person" (ASP 081-48-04, 3). The same sort of standard can be generalized to cases that involve not a specific sense modality or immediate perception, but different learned interpretative capabilities, as Carnap illustrates with another parable. Take a completely unmusical person U, to whom a musical person M is incapable of explaining or showing directly what more there is to a melody than just a temporal series of pitches, or to harmony than combinations of simultaneous pitches.

> But as soon as they both direct their attention to the consideration of history as a whole, it becomes clear that M can notice connections where U sees none. If they are presented with a musical work, say, M is capable of specifying its period and cultural context from the musical character he notices. These are connections that U can check. I'm not saying that for M and for us the point of musical activity lies in this sort of historical analysis. Certainly not. But the point is that M is *also* capable of making connections among things *in U's world* that remain unconnected for U. (ASP 081-48-04, 4)

[4]This later response to Goethe's scientific views closely echoes that of Helmholtz (1853), which Carnap had indeed read, and still referred to half a century later (1966, 111); more details in Carus 2007a, 95.

This is the same demand, Carnap says, that he would make of Gogarten: show me what you can predict, by virtue of your vision, in the subset of your world that is accessible to both of us, that I am unable to make sense of without your vision. "For me it is a kind of postulate or article of faith that everything that is to be called real in some sense has ultimately to stand in a determinate relation to my (unique) reality." (ASP 081-48-04, 4)

When Carnap wrote these lines, he had just finished writing his dissertation. What he says in this letter to Flitner, then, indicates his philosophical starting point and embraces also the stance that resulted in – and emerged from – his political "awakening" of 1918.[5] What characterizes this starting point is the drive to *unity*; everything should be relatable to everything else within a single framework or system ('[. . .] surely there has to be a way of getting beyond such relativism [. . .] and arriving at an objective, generally binding knowledge')." (From which also proceeded, evidently, the impulse to construct a "system of all concepts," first sketched out in the following year.) In "Germany's Defeat," Carnap's wide-ranging political manifesto of 1918, the rational collaboration of all the different capabilities within a society requires a system of human knowledge that can ground a *Gemeinschaftsgestalt* [form of community] that is capable of coordinating the kaleidoscopic variety of human activity into a coherent pattern – "to remove it from the realm of chaotic whim and subordinate it to goal-oriented reason [*der chaotischen Willkür zu entziehen und der zielbewußten Vernunft zu unterwerfen*]" (ASP 089-72-04, 18).[6]

So the chaotic assortment of ideas that had cohabited in Carnap's mental life before 1914 had been resolved, by 1918–21, in favor of a stance in which all the parts were subordinated to the "goal-oriented reason" of the Enlightenment. This would become progressively clearer over the following years; perhaps its clearest articulation is in Carnap's 1929 talk to the Ernst Mach Society, motivating the constitution system of the *Aufbau*, called "On God and the Soul [*Über Gott und Seele*]" (Carnap 1929/2004) – which colleagues advised him not to publish if he ever wanted a university job in Germany. But we also find many other, smaller-scale articulations of this Enlightenment orientation, such as the remarks Carnap made at a friend's wedding in 1924, which he jotted down in shorthand in the woods just before the celebration; here he even picks up on Kant's famous definition of Enlightenment as the "emergence of humankind from its self-imposed immaturity."

After a few personal remarks he goes on to address the felt need, on such occasions, to think about the larger context, the big picture. Especially, he says, because that has now become harder for us. "The generations before us had well-worn traditions of casting their thoughts to the transpersonal [*Überpersonliches*] at

---

[5]See Werner 2015 and Carus 2020c on more details about Carnap's political evolution during this period. Also, the social context was obviously of great importance; Carnap's new postwar stance was supported and amplified as it developed in the following years by the modernist movements of the Weimar Republic and especially by the *Neue Sachlichkeit* [new objectivity] (Dahms 2004).

[6]These exact words recur in the concluding paragraph of one of Carnap's first publications (Carnap 1923, 240/241).

these occasions, via religion. We have lost the ability to speak of a God. For us, the meaning of life no longer resides in the beyond [*in einem Jenseits*]; so we are in danger of not finding it at all." (ASP 021-74-05, 1) That makes it a lot harder to reflect on the big picture, since we can never entertain such very definite ideas about it as our forefathers, but are rather permanently searching. We are a bit like children, Carnap says, who have no perspective of their own on the larger context into which they are growing up. Only unfortunately there are no adults who could explain this context to us. "Humanity [...] is a large family of nothing but children, all carried onwards by forces that stem not from any insight into the path or goals of humanity, but rather from immediate advantages of the moment in a narrower or wider sense." In that sort of short-sighted, limited way of operating we can survive, more or less:

> But now the time has come for human kind to make an effort to start growing up – that is, not to do anything very differently from what happens anyway, but to do it consciously and thereby gradually to learn better and better how to avoid detours. But those fruits come later. For the time being, the task is to become more conscious, even if nothing that happened were actually very different in substance. We need eventually to arrive at the feeling that it is beneath us just to proceed on our way instinctively, unconsciously, when we are actually capable of making quite a bit of it conscious. Philosophers have sometimes, it is true, thought that they were the adults, that they have the larger perspective, and the others should pay more attention to them, so that things would go better. (Plato: philosophers should be kings). I think though, that philosophers (I count myself among the members of this guild) should be more modest. If our work consists in achieving some degree of consciousness about what is going on, we are certainly still very much at the beginning of this task. We are somewhat more reflective children, but nonetheless children like the others, only that because of this reflectiveness we screw things up in practical life more than the others. (ASP 021-74-05, 2)

By this period, the first few years after the war, Carnap had arrived at something very like his later "fundamental philosophical commitment" and was writing his first philosophical works.

1.5 From Kant II to Kant I

We have seen (in Sects. 1.1 and 1.2 above) how Carnap's early non-cognitivism was a natural outgrowth of his upbringing and, specifically in the form it had been given by Johannes Müller, enabled Carnap to remain within the church, avoiding too strenuous a commitment (and preserving some level of family harmony), while rejecting or ignoring its doctrines. In Sect. 1.3 we saw (from the letter to Le Seur) how Carnap had left Müller behind, but had brought in Kant to supply the basis for his noncognitivism. In Sect. 1.4 we saw a hint of how this transition had begun, by way of Rickert's assimilation of Goethe and Kant, which again reminded Carnap of Müller, thus affording the young Carnap a first access route to Kant.

Then we also saw how a very different, more Enlightenment-oriented Carnap emerged from his 1918 political awakening, in which Kant once again played a (perhaps regulative) role; it turns out that one of Carnap's home-made "political circulars" of excerpts from various texts (news reports, political speeches, pacificist polemics on which he and various friends commented) took Kant's "On Eternal Peace" as its sole text (Werner 2015; Carus 2020c). More generally, Kant clearly served as the overall framework within which Carnap's thinking developed over the decade from Rickert's lectures during the winter semester 1911–12 to his dissertation in 1921 (published as Carnap (1922)). It is also clear that over the course of this development, especially over the first few years of it, apparently, Carnap shifted from one conception of Kant to a very different one.

For even then, Kant was contested terrain. In fact, he was contested terrain from the day he died, or perhaps even before. There were two main camps, right from the beginning: on the one hand was German idealism, on the other hand were more scientifically-inclined post-Kantians such as Fries and Herbart, leading to physiological investigations of the human neurological and perceptual apparatus to subject Kant's hypotheses about the mental synthesis of the world of objects in space and time to experimental tests. The leading figure among these early Kant-inspired neurophysiologists was one Johannes Müller (1801–1858 – i.e. a different Johannes Müller!), whose student Hermann von Helmholtz initiated the movement of scientific neo-Kantianism in the second half of the nineteenth century (and whose own path-breaking physiological studies of perception were the basis for his philosophical reflections). This was a Kantianism even positivists and monists could endorse; the founder of the Monist Society, Ernst Haeckel, was a great admirer of Kant – *this* Kant, anyway. In his famous book *Die Welträtsel* [*The World-Riddles*] (1899) Haeckel warned, though, that where Kant was concerned, you always had to check to be sure *which* Kant someone was talking about – the scientific monist Kant I or the anti-scientific dualist Kant II. To tell them apart, he helpfully supplied a full-page table in two columns, contrasting these two very different philosophers point for point (Haeckel 1899, 415).[7]

The Kant presented by Rickert in Freiburg (in 1911–12) was clearly a species of Kant II; Rickert had, after all, been one of the philosophical champions of German historicism and of the methodological incommensurability between the human (social) and natural sciences (Rickert 1913; Beiser 2011, Ch. 10). It is no accident that Rickert's lectures during this period also inspired the young Heidegger (who may even have attended the very lectures Carnap was so enthusiastic about), whose own later portrayal of Kant – in Heidegger (1929) and in his face-off against

[7]Obviously, Haeckel oversimplifies for rhetorical effect; neither his remarks nor his table in *Die Welträtsel* can be taken seriously as Kant interpretation. There were some (such as Cassirer) whose Kant was either in between the two extremes or an attempted synthesis of them. But Haeckel's rhetorical device wouldn't have been effective if the wide disparity in Kant interpretation hadn't been such a commonplace around 1900; for a more detailed discussion of the state of this discussion, and the perception of Kant as Janus-faced, see Carus 2007a, 68–77. I draw on Haeckel's device here simply to indicate that Carnap's initial attraction to Kant (in 1911–12) involved a very different understanding of Kant from the overall framework visible in his remarks to Flitner in 1916 and 1921 discussed above. And that the "fundamental philosophical commitment" at which he had arrived by around 1920 was based on something closer to Haeckel's Kant I than to the Kant who had so enthralled him in Rickert's lectures in 1911–12.

Cassirer at Davos – was an extreme version of Kant II (Friedman 2000, 59–61). In any case, the Kant Carnap began with in 1911–12 was a version of Kant II, and where he ended the decade (in the quotations we saw in Sect. 1.4 above about Goethe and Gogarten, and in his dissertation *Der Raum*) was with a version of Kant I. It would appear, from the little we know, that this transition was already well under way by 1916; the "parting of the ways" between Carnap and Heidegger that Friedman (2000) recounts the final chapter of – Davos in 1929 – seems to have been largely complete by the time Carnap was writing to Le Seur – despite the common starting point in Rickert's lecture-hall 5 years earlier! How and why Carnap underwent this transition, though, is something we still know too little about.[8]

In any case, by 1918–21 or thereabouts this transition from Kant II to Kant I had, against the background of Carnap's political awakening in 1918, had led to Carnap's "fundamental philosophical commitment" – which then remained essentially stable throughout his philosophical career. Of course it continued to develop, and it was only in late 1932 that Carnap explicitly formulated his principle of tolerance, which finally gave adequate expression to the "practical freedom" part of this fundamental commitment, which had been largely occluded in his published works up to that time. Only on the Kantian basis of practical freedom does the principle of tolerance make sense, as Richardson points out. And only on the basis of the "scientific monist" Kant I is the principle of tolerance even interesting; there is no *point* to a principle of tolerance without some regulative principle of unity in the cognitive realm[9] – without intended unity, there is plurality (and thus tolerance of a sort) by default. The critical difference between Kant I and Kant II, for these purposes, revolves around the question whether the distinction between the theoretical and the practical is a difference of *function* (Kant I)[10] or a difference *within* the cognitive realm (Kant II) – the latter interpretation[11] regards the theoretical-practical distinction as the foundation for the distinction between the *Geisteswissenschaften* and the *Naturwissenschaften*. Kant II, as Haeckel complained, was a dualist who thought that the social and human sciences could not be made commensurable with the natural sciences. Without the aspiration to unity, tolerance has no bite.

[8]For some tentative thoughts on the final stages of this development *after* 1916, see Carus (2020c); one obvious difference between Carnap and Heidegger that may have influenced their opposite developments was that while Carnap volunteered for the war and was in the thick of some of the deadliest conflicts, at the front, Heidegger was drafted, and manned a weather station far from any actual fighting.

[9]In one of Carnap's last pronouncements on the unity of science, his reply to Feigl (Carnap 1963), he makes very clear that he is not claiming unity as any sort of fact and calls it a "distant goal." If it hadn't been too much of a concession to traditional philosophy of the kind he was attempting to grow out of, he might have called it a "regulative principle."

[10]In the sense of what has come to be called "non-cognitivism" (a term introduced by Carnap in 1963), which has now come to be widely regarded as a special case of a certain "functionalism" – as applied to Carnap, see the discussions by Richard Creath (1994) and Huw Price (1997), as well as the survey of Price's functionalism in relation to Carnap in Carus 2018 and Price's (2018) reply.

[11]For which Dilthey, but especially Windelband and his student Rickert, are best known; see Beiser 2011, Ch. 8–10.

What makes Carnap's fundamental philosophical commitment interesting and original is the *combination* of a striving for scientific standards in philosophy (his inheritance from scientific neo-Kantianism in the Helmholtz tradition) – and hence an aspiration to unity – with his insistence on practical freedom in the choice of conceptual or linguistic framework (which included an insistence that such choices, like all practical choices among courses of action, are not themselves part of, or governed by, the cognitive realm). Much more typical among scientific philosophers – philosophers who try to apply scientific standards to philosophy – is Quine, for instance, who also aspired to unity, and was even a noncognitivist about normative statements, but for whom practical freedom did not extend to the choice among conceptual frameworks (cf. Friedman 2006; Carus 2017). Much more typical among philosophers who are tolerant among conceptual frameworks is the rejection of *any* unifying framework for the whole of knowledge (their tolerance comes cheap, you might say). Carnap combined these rarely-conjoined commitments to scientific standards and to tolerance, was aware that it was a delicate balancing act, but stuck with it to the end. He did not work out all the details, of course; that is left to us.

1.6 Unpublished Sources

Documents from the following collections are cited: Archives of Scientific Philosophy, Hillman Library, University of Pittsburgh, Carnap Papers (abbreviated ASP); the Carnap papers (manuscript collection 1029) in the Special Collections Department, Young Research Library, University of California at Los Angeles (abbreviated UCLA); the Wilhelm Flitner papers in the manuscript collection of the Staats- und Universitätsbibliothek Hamburg, of which a few photocopies are available in the ASP (abbreviated ASP/WF). Documents from these archives are cited by the abbreviation given above followed by the catalog number of the document in its respective archive (and if applicable the page number within the document). Several of the documents quoted in this paper were previously discussed in Carus (2020a, 2020b), where some of them are quoted in their original German.

References

Beiser, Frederick C. 2011. *The German Historicist Tradition*. Oxford: Oxford University Press.
Benrath, Gustav A. 2000. Die Erweckung innerhalb der deutschen Landeskirchen 1815-1888: Ein Überblick. In *Der Pietismus im neunzehnten und zwanzigsten Jahrhundert*, ed. Ulrich Gäbler, 150–271. Göttingen: Vandenhoeck & Ruprecht.
Beyreuther, Erich. 1977. *Die Erweckungsbewegung*. Göttingen: Vandenhoeck & Ruprecht.
Carnap, Rudolf. 1911. Religion und Kirche, notes for a talk to the *Freischar* in Freiburg. In ASP; English translation in Carnap 2018, 475–482.

———. 1929/2004. In *Über Gott und Seele; Scheinfragen in Metaphysik und Theologie, in Rudolf Carnap, Scheinprobleme in der Philosophie und andere metaphysikkritische Schriften*, ed. Thomas Mormann, 49–62. Hamburg: Meiner.

———. 1916. Open letter to Pastor Eduard Le Seur, in ASP; English translation in Carnap 2018, 482–486.

———. 1922. *Der Raum: Ein Beitrag zur Wissenschaftslehre, Kant-Studien* Ergänzungshefte, Nr. 56. Berlin: Reuther & Reichard; reprinted with facing-page English translation and commentary in Carnap 2019, 21–208.

———. 1923. Über die Aufgabe der Physik und die Anwendung des Grundsatzes der Einfachstheit. *Kant-Studien* 28: 90–107, reprinted with facing-page English translation Carnap 2019, 209–45.

———. 1963. Intellectual Autobiography and Replies and Systematic Expositions. In Schilpp 1963, 1–84 and 859–1013.

———. 1966. *Philosophical Foundations of Physics. An Introduction to the Philosophy of Science*. New York: Basic Books.

———. 2017. Value Concepts (1958). *Synthese* 194 (1): 185–194.

———. 2018. Two Early Metaphilosophical Texts. *The Monist* 100: 475–486.

———. 2019. In *Rudolf Carnap: Early Writings*, The Collected Works of Rudolf Carnap, vol. 1, ed. A.W. Carus, Michael Friedman, Wolfgang Kienzler, Alan Richardson, and Sven Schlotter. Oxford: Oxford University Press.

Carus, A.W. 2007a. *Carnap and Twentieth-Century Thought: Explication as Enlightenment*. Cambridge: Cambridge University Press.

———. 2007b. Gedanken zum amerikanischen Monismus. *Jahrbuch für europäische Wissenschaftskultur* 3: 205–228.

———. 2017. Carnapian Rationality. *Synthese* 194: 163–184.

———. 2018. Going Global: Carnap's Voluntarism and Price's Expressivism. *The Monist* 101: 441–467.

———. 2020a. Werte beim jungen Carnap. In *Der junge Carnap im historischen Kontext 1918–35*, ed. Christian Damböck and Gereon Wolters. Wien: Springer (to appear).

———. 2020b. Die religiösen Ursprünge des Nonkognitivismus bei Carnap. In *Logical Empiricism, Life Reform, and the German Youth Movement/Logischer Empirismus, Lebensreform, und die deutsche Jugendbewegung*, ed. Christian Damböck, Günther Sandner, and Meike Werner. Dordrecht: Springer (to appear).

———. 2020c. Carnap's politische Entwicklung 1917–18. *Carnap Blog*, April 2020.

Creath, Richard. 1994. Functionalist theories of meaning and the defense of analyticity. In *Logic, language, and the structure of scientific theories*, ed. Wesley Salmon and Gereon Wolters, 287–304. Pittsburgh: University of Pittsburgh Press.

Dahms, Hans-Joachim. 2004. *Neue Sachlichkeit* in the Architecture and Philosophy of the 1920s. In *Carnap Brought Home: The View from Jena*, ed. Steve Awodey and Carsten Klein, 353–372. LaSalle: Open Court.

Friedman, Michael. 2000. *A Parting of the Ways: Carnap, Cassirer, and Heidegger*. LaSalle: Open Court.

———. 2006. Carnap and Quine: Twentieth-Century Echoes of Kant and Hume. *Philosophical Topics* 34: 35–58.

Haeckel, Ernst. 1899. Die Welträtsel: Gemeinverständliche Studien über monistische Philosophie (Bonn; Strauß).

Haury, Harald. 2005. *Von Riesa nach Schloß Elmau: Johannes Müller (1864–1949) als Prophet, Unternehmer und Seelenführer eines völkisch naturfrommen Protestantismus*. Gütersloh: Gütersloher Verlagshaus.

Heidegger, Martin. 1929. *Kant und das Problem der Metaphysik*. Tübingen: Klostermann.

Helmholtz, Hermann von. 1853. Über Goethes naturwissenschaftliche Arbeiten. Repr. In Hermann von Helmholtz: *Vorträge und Reden*, vol. 1 (Braunschweig 1903: Vieweg), 1–24; translated as On Goethe's Scientific Researches. In Hermann von Helmholtz: *Science and Culture: Popular and Philosophical Essays*, ed. David Cahan, 1–17. Chicago: University of Chicago Press 1995.

Hepp, Corona. 1987. *Avantgarde: Moderne Kunst, Kulturkritik und Reformbewegung nach der Jahrhundertwende*. München: dtv.
Hübinger, Gangolf. 1997. In *Die monistische Bewegung. Sozialingenieure und Kulturprediger, in Kultur und Kulturwissenschaften um 1900*, ed. Gangolf Hübinger, Rüdiger vom Bruch, and Friedrich Wilhelm Graf, 246–259. Stuttgart: Steiner.
Jeffrey, Richard C. 1994. Carnap's Voluntarism. In *Logic, Methodology, and Philosophy of Science IX*, ed. Dag Prawitz, Brian Skyrms, and Dag Westerståhl, 847–866. Amsterdam: Elsevier.
Müller-Späth, Jürgen. 1984. Protestantismus und Gründerzeit im Wuppertal. In *Gründerzeit: Versuch einer Grenzbestimmung im Wuppertal*, ed. Karl-Hermann Beeck, 360–419. Köln: Rheinlang-Verlag.
Nipperdey, Thomas. 1988. *Religion im Umbruch. Deutschland 1870–1918*. München: C. H. Beck.
Noll, Richard. 1994. *The Jung Cult: Origins of a Charismatic Movement*. Princeton: Princeton University Press.
Price, Huw. 1997. Naturalism and the Fate of the M-Worlds. Repr. In Huw Price: *Naturalism Without Mirrors*, 132–147. Oxford: Oxford University Press 2011.
———. 2018. Carnapian Voluntarism and Global Expressivism: Reply to Carus. *The Monist* 101: 468–475.
Richardson, Alan. 2007. Carnapian Pragmatism. In *The Cambridge Companion to Carnap*, ed. Michael Friedman and Richard Creath, 295–315. Cambridge: Cambridge University Press.
Rickert, Heinrich. 1913. *Die Grenzen der naturwissenschaftlichen Begriffsbildung: Eine logische Einleitung in die historischen Wissenschaften*. 2nd ed. Tübingen: Mohr Siebeck.
Werner, Meike. 2003. *Moderne in der Provinz: Kulturelle Experimente im Fin de Siècle Jena*. Göttingen: Wallstein.
———. 2015. Freideutsche Jugend und Politik; Rudolf Carnaps *Politische Rundbriefe* 1918. In *Geschichte intellektuell: theoriegeschichtliche Perspektiven*, ed. Friedrich Wilhelm Graf, Edith Hanke, and Barbara Picht, 465–486. Tübingen: Mohr Siebeck.

Chapter 2
Carnap's Non-cognitivism and His Views on Religion, Against the Background of the Herbartian Philosophy of His Grandfather Friedrich Wilhelm Dörpfeld

Christian Damböck

The importance of protestant theology for Carnap's early intellectual development was recently highlighted by A.W. Carus by means of Carnap's relationship with the German Youth Movement and several unknown writings of Carnap from the 1910s.[1] Michael Heidelberger recently pointed to the role that Carnap's grandfather – the pedagogue Friedrich Wilhelm Dörpfeld (1824–1893) – played in his intellectual development.[2] In this paper, I will elaborate on this and highlight two aspects of Dörpfeld's work: (1) the striking similarities between Dörpfeld's views on ethics and Carnap's non-cognitivism; and (2) Dörpfeld's enlightened approach toward religion, which might also have influenced Carnap's world views. Both aspects of Dörpfeld's work might allow us to add further clarification to Carus's account, which has already pointed to the importance that Carnap's early intellectual development had for the development of his non-cognitivism, but without considering Dörpfeld in that regard.

The main hypothesis of this paper is that Carnap's non-cognitivism can be interpreted as a secularized version of Dörpfeld's meta-ethical views, whereas Carnap's views on religion share certain basic features with Dörpfeld's conception. For this purpose, I first provide an outline of Dörpfeld's system of the sciences (Sect. 2.1), which then becomes confronted with Carnap's meta-ethical views (Sect. 2.2). After a longer digression where I sum up all crucial theoretical and political

[1] See (Carus 2021a, b) as well as the appendix of the forthcoming volume (Damböck et al. 2021) which will contain two early texts on religious matters by Carnap, together with an introduction by A.W. Carus.

[2] See (Heidelberger forthcoming) which provides a comprehensive analysis of Dörpfeld's philosophical background.

C. Damböck (✉)
Institut Wiener Kreis, Universität Wien, Vienna, Austria
e-mail: christian.damboeck@univie.ac.at

E. Ramharter (ed.), *The Vienna Circle and Religion*, Vienna Circle Institute Yearbook 25, https://doi.org/10.1007/978-3-030-76151-6_2

aspects of Carnap's mature views on religion (Sect. 2.3), I end with some facts and speculations about the way in which both Carnap's non-cognitivism and his views on religion might have been inspired by Dörpfeld (Sect. 2.4).

2.1 Dörpfeld's System of the Sciences and Its Meta-ethical Aspects

Although Friedrich Wilhelm Dörpfeld never worked at a university – he spent his entire professional career as a teacher and director of an elementary school – he was in his time considered an important and influential pedagogue who stood out with a remarkable list of publications on pedagogics, psychology, and philosophy. His work is documented by an edition of his writings that appeared posthumously in ten volumes in the renowned publishing house of C. Bertelsmann. His daughter Anna Carnap, Rudolf Carnap's mother, wrote a voluminous account of his life and work that appeared in two editions (Carnap 1897, 1903). But even in the twentieth century, the pedagogue Dörpfeld was not completely forgotten, as is documented by the monograph (Beeck 1975). Though Dörpfeld was mainly influential as a pedagogue in the tradition of Herbart, in his main philosophical work *On Ethics [Zur Ethik]* (Dörpfeld 1895), he also outlined a striking philosophical account which contextualizes ethics and religion in the context of the sciences, viz. represents the then influential approach toward religion that restricts itself to positive empirical knowledge. Dörpfeld's account, which was well-known to Carnap (see Sect. 2.4), is a rather characteristic product of the empirically minded philosophy that was widespread in the German-speaking world during the nineteenth century (Damböck 2017). As revealed in the programmatic preface to his psychology (Dörpfeld 1894, IX-XXVI), Dörpfeld was mainly influenced by Herbart but also committed himself to the Völkerpsychologie of Moritz Lazarus (XV) and William Whewell's approach to the inductive sciences (XXIII). To get a grasp of Dörpfeld's views, we will now review the main aspects of his "system of the sciences" as it becomes developed in (Dörpfeld 1895).

There are two varieties of evidence: psychological self-evidence and formal-logical evidence (pp. 1–9)

Evidence in science, for Dörpfeld, is possible on two different grounds. First, there is the phenomenon of self-evidence [Selbstgewissheit], which is something that occurs only, rather than to follow from a doctrine, an axiom, or the adoption of a method. Self-evidence is the point, and the only point, where the voice of God becomes heard by humans. "*God's* revelations are not words, not a doctrine but *deeds, works, events*" [Dörpfeld 1895, 7: *Gottes* Offenbarungen sind nicht Worte, nicht Doktrin, sondern *Thaten, Werke, Geschehnisse*]. God reveals herself not at the level of anything that could be justified via an axiom, doctrine, or theorem; there is no objective but only subjective evidence for the voice of God. Self-evidence is a psychological phenomenon, a "fact of consciousness" [Bewusstseinstatsache] that always affects only the one concrete case, the "thought process" in which it occurs. Therefore, self-evidence does not establish anything that could be systematically reproduced. The reproduction of evidence, in turn, becomes possible only based on certain formal rules that we establish in science and logic. Logical axioms allow us to establish a method that adds to the unique cases of self-evidence, a procedure through which evidence becomes reproducible in a step by step process. Thus, the second form of evidence that exists, for Dörpfeld, is logical evidence that establishes truth via logical axioms and rules. Only here we obtain objective truth via reproducible deductions. The first evidence of Dörpfeld's system – self-evidence – is subjective, psychological, and non-reproducible, whereas the second – logical evidence – is objective, formal, and reproducible. For reasons that will become clear in the following, however, Dörpfeld does not think that logical evidence is able to finally justify its results, while real justification is possible only at the level of self-evidence.

Mathematics is the only non-empirical science, plus Dörpfeld's system of the empirical sciences (pp. 10–18)

Though based on logic, mathematics differs from the former in that it develops a system of axioms and deductions as well as cases of self-evidence that correspond to them, which is entirely theoretical and detached from the empirical world. Every other science, for Dörpfeld, is empirical. This is a remarkable variety of empiricism, because it identifies not only the natural sciences, but also disciplines such as logic, ethics, and theology, as empirical sciences. More precisely, Dörpfeld identifies two types of empirical sciences:

(1) *Inductive or explanatory sciences that include (a) the whole of the natural sciences, and (b) psychology, linguistics, and theology (and possibly also metaphysics, but Dörpfeld leaves this open).*

The systematic key point here is that not only the natural sciences and the Geisteswissenschaften belong to the inductive sciences – which is something that would also hold for several other German philosophers of that time (e.g., Dilthey or Wundt) – but also theology. This implies that theological doctrines are relevant and significant only insofar as they can become subject to empirical testing and result from an inductive process of abstraction from the facts. Theology relies on the search for empirical evidence for the existence of God (via signs and wonders). What theology, for Dörpfeld, cannot provide though is any other kind of evidence. For Dörpfeld, there is no ontological proof in theology and also no emotional evidence that might establish the existence of God.

From the standpoint of the modern philosophy of science, the weak point of this conception is its empirical indifference. Though everything in the inductive sciences, for Dörpfeld, is significant only insofar as it becomes abstracted from the facts and becomes subject to further empirical testing, there is no way to further qualify theories once they are inductively found and empirically tested. Dörpfeld's conception shows traces of the hypothetico-deductive conception of the sciences because it grants that every claim that is formulated in the inductive sciences remains hypothetical, as soon as it is not to be verified or falsified by a singular event. However, Dörpfeld also goes on to say that these "hypothetical explanations" say "nothing else than that the matter is conceivable, i.e., logically possible" (Dörpfeld 1895, 11), whereas, at the same time, it is granted that the matter could also be entirely different. This is, to be sure, also what the hypothetico-deductive standpoint would imply. However, Dörpfeld also deduces from this that inductive science "only entails logical possibility but no reasons of probability". That is, it seems that for Dörpfeld, all empirical hypotheses that withstood empirical testing are equally "logically possible", viz. equally probable, no matter to which degree they are backed up by empirical evidence. An entirely vacuous hypothesis that is compatible with every possible empirical reality, for Dörpfeld, is epistemically equivalent to an extremely rich hypothesis being backed up by a wealth of successful and significant empirical tests. In other words, for Dörpfeld, falsifiability is no epistemic criterion at all. There is no genuine method of the inductive sciences that allows us to distinguish between rich and vacuous hypotheses, and there is also no possibility to identify degrees of truth, probability, or empirical significance for any hypothesis of the inductive sciences.

(2) *So-called norm searching [normsuchende] sciences which "ask for the signs or norms of something ideal" and include logic, aesthetics, and ethics.*

As a result of the somewhat vacuous status of inductive science – all empirically adequate theories are equally probable –, the other fields of empirical science as being outlined by Dörpfeld become extremely important. These so-called norm-searching sciences might seem to be inspired by Southwest German Neo-Kantianism and Hermann Lotze. However, unlike Lotze et al., who develop a strictly non-psychological brand of norms that leads to varieties of Platonism and a vacuous philosophy where norms become deduced from the scientific facts via synthetic judgments a priori, Dörpfeld's conception remains in the empirical realm, here following the psychologist tradition of Herbart and also the Völkerpsychologie of Moritz Lazarus. For Dörpfeld, there is no external (Platonist) realm in which norms can be fastened down, independently of their appearances in human beings. Rather, norms only show up in our mind, and there are only certain abilities of the mind (psychological abilities in the Herbartian sense) that allow us to identify the proper norms. There are feelings [Gefühle] of a certain kind that allow us to establish norms with "natural necessity" [naturgegebener Notwendigkeit] (Dörpfeld 1895, 14). The only way to proceed in our business of norm searching is to clean up the mind, make room for exactly those norms identifying feelings and follow their advice. This does not boil down norms to questions of mere taste and arbitrary choice, because "the observer" of norms needs to be "purely objective, i.e., his view may not be clouded by personal interests, partisan sympathy or antipathy etc." [Dörpfeld 1895, 14: Dabei muß aber der Beobachter rein *objektiv* schauen, d.h. sein Blick darf nicht durch irgend ein persönliches Interesse, durch parteiische Sympathie oder Antipathie u.s.w. getrübt und bestochen werden.] We need to realize the "involuntarily emerging characteristic feeling" [das unwillkürlich entstandene eigenartige Gefühl] that induces moral values and then will be able to inductively carry out the "formal properties of the good" in that we only need to find those attitudes that regularly accompany the feeling of the good. Such formal properties, according to Dörpfeld, are things like the feeling of justice, inner freedom, willpower, and the like.

It is not only ethics and aesthetics but also logic whose basic norms (viz. the logical axioms), according to Dörpfeld, can only be established via feeling based evidence. The idea is that whereas logic in itself forms the basic discipline for both mathematics and the inductive sciences, its construction takes place in the field of norm searching science, via feelings that establish self-evidence for the basic axioms and notions of logic.

Dörpfeld's system in comparison with the modern scientific world view

Dörpfeld's system of the sciences offers a remarkable gap because the fundamental principles (axioms) of the inductive sciences somewhat hang in the air, whereas only the axioms of logic, ethics, and aesthetics actually become justified in the process of the search for (self-)evidence via feelings. Dörpfeld seems to think that our feelings might establish trustworthy evidence only in those fields being intimately connected to the mind, which includes, in his view, ethics, aesthetics, and logic, whereas every other empirical discipline – the natural and human sciences, together with theology – is doomed to the business of induction which, for Dörpfeld, does not establish evidence at all but only uncovers the "logical possibility" of hypotheses.

This conception is far from a modern picture of the sciences insofar as it devalues inductive science. If there is no evidence in inductive science but only the weak outcome of "logical possibility" for empirically adequate hypotheses, there is no real possibility of fruitful empirical research at all. But there is also an aspect in Dörpfeld's conception where it appears to be rather modern, namely, the fact that he includes not only natural sciences and Geisteswissenschaften to the field of inductive sciences but also theology. This leads to a radically enlightened and disenchanted notion of theology, which leaves no room for speculations and ontological proofs. Theism, in other words, is not a miracle, for Dörpfeld, but rather only a bunch of weak and risky empirical hypotheses. Theism also is not based on self-evidence, because in this system of the sciences, it is only ethics and not theology that can establish evidence at all. Therefore, the weakness of Dörpfeld's notion of the inductive sciences seems to be hardly motivated by considerations on the natural sciences, which certainly were not Dörpfeld's business. Rather, the weakness of his system is a result of the radically enlightened and disenchanted world view that Dörpfeld adopts, here following the tradition of Herbart and Völkerpsychologie, and insofar appears to be a real strength, from the perspective of modern science.

Religious belief is a psychological quality (pp. 65–84)

Because science is concerned with evidence and the establishment of knowledge, for Dörpfeld, religious belief is nothing that belongs to science. Whereas in theology, we are only able to establish those aspects of theism that can be carried out in an inductive and empirical way (with the downside that these aspects hardly exist at all); everything that goes beyond this and actually connects the otherworld of heaven and God is a matter of belief rather than knowledge. Belief, in turn, is nothing that we gain from science but only develop in real life, because of certain cultural and psychological processes. Therefore, the study of belief is a matter of psychology which here, in Dörpfeld's view, includes both what even today is called psychology but also such disciplines as sociology and history. What science can do here is only to study under which historical, psychological, and sociological circumstances religious belief is present or absent. In other words, the study of belief belongs to the human sciences and has nothing to do either with theology or ethics or what Dörpfeld calls lore of religion [Religionslehre], since religious evidence (knowledge) is established only in the latter fields (viz. in a combination of both).

Religionslehre is an empirical science which is based on theology and ethics

Theology, for Dörpfeld, neither offers ontological proofs nor evidence via feelings: it offers no evidence at all but rather empirical hypotheses, which at most can establish "logical possibility" (if they stand empirical testing). Therefore, theology needs to be backed up by another discipline that only establishes evidence. Theology only adds to the whole discipline of a Religionslehre what can be carried out in an inductive way, for example, on the basis of empirical evidence on the world's being a product of God's will and creation (via signs and wonders). Theology, in other words, concerns only those aspects of religion that address the transcendental otherworldly reality of Heaven and God, whereas ethics is concerned with this world, establishing all these rules that guide our living and acting. Ethics, therefore, is not concerned with speculations about heaven and God but only considers practical decisions. These decisions, however, as soon as they are guided by pure feelings, are automatically guided by God, because pure feelings open our senses to hear the voice of God. The divine voice is in itself a pure voice, of course, and therefore every situation must be evaluated exactly in the same way by every human that puts herself in the position of being guided by purified feelings that open her senses for the voice of God. In other words, moral absolutism follows here only at a secondary level, because Dörpfeld thinks that if people only follow their pure feelings, they necessarily must all arrive at the same values and moral attitudes because these pure feelings open their minds to the unique voice of God, which is the same voice for everybody.

The key features of Dörpfeld's metaethics are, to conclude, that (1) there is no way to derive moral rules, either inductively or deductively; (2) moral rules become visible in practical situations only, by means of pure feelings of those human beings who made their mind ready for them; (3) every human being which is prepared for pure feelings necessarily arrives at the same moral attitude regarding a particular situation X.

2.2 A Comparison with Carnap's Non-cognitivism

Carnap's non-cognitivism was developed over decades (Reisch 2005, 47–53, 382–384; Richardson 2007; Mormann 2007; Uebel 2010; Siegetsleitner 2014, 89–162; Carus 2017, 2021b; Damböck 2017, 199–203, 2018, 2021a, c; Damböck et al. 2021). For the present purpose, we widely ignore changes in his position and rather focus on the most comprehensive account of his views in "Kaplan on value judgments" (Schilpp 1963, 999–1013).

Carnap first grants that there is a wealth of empirical and logical statements about values. Empirical statements may contain (a) psychological, sociological, and historical analyses of concrete evaluations in certain persons or groups; (b) considerations about means-end relationships; and (c) statements on "the utility of a possible event [...] for a person". Logical statements are either deductions from empirical statements about values or concern the semantics and the "explication" of value concepts. Any statement "on values and valuations", which is neither empirical nor logical, then, is called "non-cognitive" by Carnap (Schilpp 1963, 999).

Non-cognitive statements, on the other hand, need to be distinguished from "emotive statements" because a non-cognitive value statement expresses "more than merely a momentary feeling of desire, liking, being satisfied, or the like" but rather "satisfaction in the long run" (Schilpp 1963, 1009). Now, Carnap's overall meta-ethical claim is twofold:

(1) He claims that a good decision must be made in a sober and maximally informed and rational way. This means, on the one hand, that we may not let our actions be guided by momentary feelings and partial temptations but rather dig deeper to find out what our genuine attitudes are. On the other hand, we should always try to be on top of the available empirical and logical evidence and consider the entirety of our knowledge, before we go on to decide.
(2) He claims that even as soon as we find a proper way to decide in perfect accordance with the rules as resulting from point (1), it appears that every moral judgment is ultimately non-cognitive, even if it is also guided by a huge amount of empirical knowledge and logical considerations.

There is also an important corollary to point (2), namely, that non-cognitivism also implies that two persons may arrive at diverging decisions and incompatible value statements:

> It is logically possible that two persons A and B at a certain time agree in all beliefs, that their reasoning is in perfect accord with the deductive and inductive standards, and that they nevertheless differ in an optative attitude component. (Schilpp 1963, 1008)

In other words, Carnap's non-cognitivism involves a variety of moral relativism. Though empirical and logical evidence can and should have strong implications for our practical decisions, our decisions might still turn out to be different, even if we act rationally and take into consideration the available empirical evidence and let our feelings only guide us insofar as they are justified to be pure feelings that satisfy our intuition in the long run. Even if we follow the available evidence and pure rather than gut feelings, there still remains a non-cognitive basis to our decisions that can be different in different people with the implication that also the practical decisions may differ among different humans, even if they use exactly the same scientific basis for them.

Carnap's account is certainly much more refined than Dörpfeld's. From a Carnapian standpoint, one might object to Dörpfeld that his account is a little bit naïve in that he seems to think that it might be sufficient in a situation X to get yourself free of temptations and momentary emotions and automatically you will arrive at the correct decision. Carnap would object here that pure feelings alone are by no means sufficient; they also need to be accompanied by a rational process that considers the entirety of the available empirical and logical evidence; practical decisions involve theoretical questions that (a) show us the means toward an end and (b) make the consequences that we may expect from an action firstly visible (Carnap 1934). However, apart from this obvious criticism, there is also a key point where Carnap's and Dörpfeld's conceptions widely agree, namely, the overall non-cognitive attitude. Carnap and Dörpfeld agree in their conviction that moral judgments are, in the end, something that depends only on our feelings, although we first need to free these feelings from temptations and passing emotions. In this sense, it seems that Carnap's theory is ultimately identical to Dörpfeld's. They are both non-cognitive, in the same way. The only difference here is that Dörpfeld and Carnap have different judgments about the social status of non-cognitive statements. Whereas for Dörpfeld, two persons who agree in all their believes and logical considerations and also both base their decisions on purified feelings necessarily arrive at exactly the same decision because their attitudes are necessarily the same, for Carnap, this is simply not the case, because even here attitudes may diverge. The unifying voice of God, which leads to moral absolutism in Dörpfeld's variety of non-cognitivism, is absent in Carnap's account, and therefore moral relativism arises because of the absence of a guiding voice that is unequivocally hearable for everyone.

To conclude, there is one key point of agreement between the meta-ethical conceptions of Carnap and Dörpfeld, namely, the idea that moral judgments are ultimately based on purified emotions. Apart from this, however, there are also two crucial aspects where Carnap's and Dörpfeld's theories diverge. First, Dörpfeld's theory only provides a rather naïve conception of practical decision making, which, unlike Carnap's, fail to incorporate the relevant theoretical questions. Second, and even more fundamentally, Carnap and Dörpfeld disagree in that for Dörpfeld, two persons who make a value judgment against exactly the same epistemic background necessarily have to judge identically and will base their judgments on the same attitudes, whereas for Carnap, in cases of identical epistemic background, the attitudes may be different and therefore two persons could also arrive at different moral judgments.

2.3 Digression: Carnap's Views on Religion

Carnap's views on religion can be divided into a political and a more theoretical part. As for the theoretical part, one might further distinguish between the theoretical aspects of Carnap's own atheist attitude and his theoretical attitude towards believers. Carnap arrived at atheism around 1920. At this stage of his intellectual development, the commitment to atheism was no longer based on any considerations about lack of support of theism via empirical evidence or the rejection of ontological proofs and other justifying strategies for theism because all these epistemic possibilities were already rejected by Carnap in his youth. In his autobiography (Schilpp 1963, 3–84), this early turn toward a secular scientific worldview is attributed by Carnap to his adoption of the sober scientific and anti-theist attitude of the freethinkers' movement as being represented by Ernst Haeckel and Wilhelm Ostwald, whose writings Carnap already read in his early student years (1911–1914) (Schilpp 1963, 7–9, Dahms 2016). After he had rejected any traditional approach of epistemic justification of religion, with the inclusion of Spinozist pantheism, which happened to be the final theist strategy he adopted before moving forward to atheism (ibid.), the question of religion, for Carnap, ultimately became a mere psychological and cultural-historical one. This becomes obvious in two remarks that document his final abandonment of religion in 1920. In a discussion note on a meeting at the Wartburg in Mai 1920, Carnap says the following:

> First insight: the distinction between clerical and religious; but: even the frankly clericals are not all religious; I state two types: religious and irreligious predisposition, the latter not as a flaw (examples: classic human being, Goethe, many recent people (me)); the expression 'religiousness' should be saved for another type (gothic human being, possibly Egyptians [...]).[3]

This is clearly a cultural-historical notion of "religiousness". To be clerical only means that one is a member of a club that has no necessary implications for his feelings and does not say anything about the presence or absence of religiousness. The latter, on the other hand, as it is described here by Carnap, also has nothing to do with questions of (scientific) evidence. Religiousness – and the adoption of theism – is nothing that follows from any insight, either empirical or theoretical. Rather, to be a religious or irreligious person is a question of character, and it is also a cultural-historical question. There are periods (Carnap mentions the Gothic period and ancient Egypt) where people tend to develop religious characters, while in other periods (Carnap mentions ancient Greek and the present age), people

[3]In the following I quote from the Rudolf Carnap Papers at the Hillman Library, University of Pittsburgh, always using the format "RC XXX-XX-XX, page number" The present document is RC 091-18-02, 3. ‚Erste Erkenntnis: Unterscheidung von kirchlich und religiös; aber: auch die aufrichtig kirchlich[en] nicht alle religiös; ich stelle 2 Typen auf: rel[igiöse] und irrel[igiöse] Veranlagung, letzteres nicht als Fehler (Beispiel: klass[ischer] Mensch, Goethe, viele heutige Menschen (ich)); Ausdruck 'Religiosität' soll aufgespart werden für anderen Typ (gotischer Mensch, viell[eicht] Ägypter [...])"

mainly develop irreligious characters. So, for Carnap, to be irreligious is something that follows from someone's own personal disposition but also from the historical disposition of the present age. Carnap only realized this during the days at the Wartburg in May 1920, and his convictions became consolidated in the following weeks, as he reports in a circular letter from November 1920:

> In summer, I have been in Jena for several months; at Whitsun with Flitner, Räubers and Rugard von Rohden with Baußnerns in the Pfiffelbach parsonage. During our discussions on Christianity and church, my skepticism about what benedictory we might anticipate from the church for our time developed to such clarity that I turned my back to it.[4]

Here the motive is again outlined as an entirely cultural-historical one. Evidence (or doubt) does not play a role at all for Carnap's decision. This is, to be sure, also in accordance with the Herbartian understanding of religion as it was defended by Carnap's grandfather. Religious belief is something that exists only as a psychological quality, for Dörpfeld. The presence or absence of this quality is not a question of evidence at all. We may not change the religious or irreligious disposition of a person through the presentation of either empirical evidence or logical proof. Therefore, what happened with Carnap in 1920, viz., that he lost his religiosity, is a very simple psychological and cultural-historical process not only for him, Carnap, but also in the religious system of his grandfather: for both of them, the process might be something that one could either welcome or regret, whereas one could hardly do anything positive to prevent or support it.

That Carnap, in his mature views, felt no sympathy for scholasticism, and the search for ontological proof is obvious, for his anti-metaphysical brand of empiricism considers only empirical claims and logical tautologies to be meaningful. Therefore, any attempt to formulate meaningful statements about a transcendental being is certainly meaningless metaphysics for Carnap (Carnap 1932, 1934, 2004, 49–62, Damböck 2018). But Carnap also rejected alternative attempts towards providing evidential support for theism via emotive talk. This is well-documented by a discussion that Carnap had in the early 1920s with his friend Wilhelm Flitner about the theology of Friedrich Gogarten (Carus 2007, 95–97, 2021b, 37–38, Damböck et al. 2021). Carnap rejects the idea of a "religious seeing" that allows the believer to gain insights which are entirely inaccessible to the nonbeliever. Gogarten's approach, which was a key example for the new varieties of theology that repressed in the Weimar republic the traditional empirically minded accounts of religion, represented, for Carnap, only another brand of metaphysical reasoning that was for him probably even more dangerous than the scholastic attempts toward ontological proof, because the new irrationalists around Gogarten hid from the start every rational discourse about religion behind a wall of dubious emotions.

[4]RC 081-47-01. ‚Im Sommer war ich einige Monate in Jena; Pfingsten mit Flitner, Räubers und Rugard von Rohden bei Baußnerns im Pfiffelbacher Pfarrhaus. Bei unsern Gesprächen über Christentum und Kirche wurde mir meine Skepsis über das, was von der Kirche Segensreiches für unsre Zeit zu erwarten sei, zu solcher Klarheit, daß ich ihr den Rücken wandte."

The rejection of religion and the religious attitude by Carnap as it was described so far is twofold. On the one hand, Carnap rejects religion at a non-cognitive level, here based on a personal and a historical argument. The personal argument is that Carnap simply realized to be an irreligious character and, therefore, had to give up the religious attitude. The historical argument goes even further – and here also connects to the political argumentation that will be described below –, stating that it is not just a question of his own personality to reject the religious attitude but also that the entire present age – after the First World War and the German revolution – is one that tends to be irreligious. The second part of Carnap's theoretical attitude toward religion is that he, here again on a line with his grandfather, rejects all kinds of cognitive approaches toward justification of religiousness. This includes ontological proofs as well as emotive justifications in the sense of Gogarten.

These theoretical reasons for the rejection of religion did not hinder Carnap to respect religious persons.

> Since I experienced the positive effect of a living religion in the lives of my parents and in my own life during childhood, my respect for any man whose character I esteem highly is not diminished by the fact that he embraces some form of religion, traditional or otherwise. At the present stage of development of our culture, many people still need religious mythological symbols and images. It seems to me wrong to try to deprive them of the support they obtain from these ideas, let alone to ridicule them. (Schilpp 1963, 8)

This tolerance – which certainly was mainly due to the religious attitude of Carnap's mother that he admired for his whole life (Schilpp 1963, 3) – was practiced by Carnap even against fellow philosophers, such as the philosopher of religion Charles Hartshorne with whom he was a close friend during his time in Chicago, although he did not share any of Hartshorne's religious and metaphysical views (Carnap forthcoming). Religion, as Carnap realized, though it is a dead branch of cultural history that ultimately prevents progress, is still something that many people need to morally succeed. What the above statement implicitly says is that there are people who – possibly for a weakness in character – develop moral attitudes being favorable for Carnap only because they base their reasoning on religious believes. Therefore, religion is something that is still needed in our present culture, say, as a moral crutch for weak characters who might be endangered to become immoral without it, although in the future, religion might become more seamlessly replaced by the scientific world-conception and finally die out.

These considerations on Carnap's attitude toward religious people directly lead to an examination of the political aspects of Carnap's views on religion. As a social democrat and part-time communist, Carnap shared the alleged Marxist view that religion is a dangerous narcotic for the masses. He added to this a story about metaphysics, which was thought by him to be an equally dangerous narcotic for the educated: "philosophical and religious metaphysics is a potentially dangerous reason-damaging narcotic" [daß philosophische und religiöse Metaphysik ein unter Umständen gefährliches vernunftschädigendes Narkotikum ist] (Carnap 1934, 260). Whether this view is genuinely Marxist could be questioned, because Marx seems to defend the view that religion is taken by the people as a narcotic, in order to sustain otherwise unsustainable working conditions,[5] whereas, for Carnap, the point seems to be rather either religion or metaphysics is administered to the people by right-wing politicians. This view becomes obvious in a manuscript that Carnap wrote for a talk he gave in Brno on "Philosophy Opium for the Educated" (RC 110-08-17). There, Carnap unequivocally calls religion and philosophy "ideological weapons" [ideologische Kampfmittel] that are used by right-wing politicians, in order to "dissuade people from enlightenment, from a rational-scientific analysis of the situation, from the will to empower a change" [die Menschen *abzubringen* von der *Aufklärung, vom rational-wissenschaftlichen Analysieren* der Situation, *vom Selbst-ändern-Wollen*]. The narcotic, therefore, "serves the interests of certain groups" [dient dem Interesse bestimmter Gruppen] – Carnap mentions the "ruling classes" [herrschende Schichten] in Italy, Germany, and Austria – although these groups in themselves do not consciously use the narcotic: they do not follow a "calculating decision" [berechnende Überlegung] but rather act "unconsciously", driven by sociological causes they themselves do not understand.

At any rate, religion, and metaphysics (or bad philosophy) were intimately connected for Carnap and the left-wing of the Vienna circle. The "scientific world-conception" is directed against "metaphysical and theologizing thought" that show up as a couple, in the Vienna Circle's manifesto and several other texts by Carnap from that time (Carnap 1932, 1934, 2004, 49–62, Stadler and Uebel 2012).

> The representatives of the scientific world-conception resolutely stand on the ground of simple human experience. They confidently set to work on the task of removing the metaphysical and theological debris of millennia. [...] The increase of metaphysical and theologizing leanings which shows itself today in many associations and sects, in books and journals, in lectures and university courses, seems to be based on the fierce social and economic struggles of the present. One group of combatants, holding fast to traditional social forms, cultivates traditional attitudes of metaphysics and theology whose content has long since been superseded; while the other group, especially in central Europe, facing the new age, rejects these views and adopts empirical science as its basis. (Stadler and Uebel 2012, 89)

The scientific world-conception of the Vienna circle, therefore, in the words of Carnap and Neurath, is the view that adopts empirical science as a new basis of life, which is thought to replace metaphysics and theology. This is more than an inner-scientific attitude. Rather, the aim is a fundamental life reform (Damböck et al. 2021), which is thought to accompany a political reform in such a way that the "socialist attitudes" of "the masses" converge (no longer with the outdated brand of "materialism" but) with "modern empiricism", which "has left behind a number of inadequate forms in its development and has found a defensible form in the scientific world-conception" (Stadler and Uebel 2012, 90).

In Carnap's political conception, metaphysics and theology are two sides of the same coin. They function as tools of mass manipulation that obscure our view on the empirical and logical facts. Therefore, it is the same sin of meaningless metaphysics

[5]This I learned from a remark by Hans-Joachim Dahms to this manuscript.

that obscures scientific reasoning and confuses the intellectuals and is also in effect in the everyday world, here confusing the masses via theological narratives.

There is an obvious difference to be noted here, between the more moderate stance of Carnap in his time in the US as it became formulated in his autobiography and the more radical stance of the Vienna Circle manifesto and the anti-metaphysical writings of the early 1930s. Whereas in the earlier writings, Carnap ultimately aims at a total and immediate removal of religion and metaphysics from all manifestations of culture, in his later years, he grants that some varieties of theology (and probably also metaphysics) are still needed. However, these changes were not necessarily accompanied by any change of the fundamental political and theoretical views of Carnap's, for even in his later years, he obviously thought that metaphysics and theology are harmful and ultimately should become removed from everyday and scientific culture. The only difference is that in his later years, Carnap also became more realist and saw that for many people, it is at present simply not possible to live without these narcotics. The metaphysics and theology addicts are unable to live without their drugs. Doctor Carnap, though being a strict abstainer who propagates a life without cultural drugs, is also a realist who grants those being already addicted to taking their life-sustaining doses.

To sum up, Carnap's views were certainly anti-religious and anti-theist, both at a theoretical and a political level. The key arguments against religion that Carnap put forth are cultural-historical and cultural-philosophical. He thought that for our present culture, religion and theism are harmful because they attempt to lead culture in the wrong direction, namely, fascism instead of social democracy. This, on the other hand, also meant that Carnap did not reject all possible manifestation of religion and theism, for he indeed only took the reactionary and fascist religions and theisms as a threat to human culture. Only here, Carnap was seamlessly opposed to all manifestations of religion and theism. But there was also the level of – as one could put it in Neurathian terms – religion and theism as a cultural agglomeration [Ballung], which is sometimes used as an ideological crutch even by rather progressive and well-meaning people. This came to Carnap as a surprise, but he also was willing to accept it without developing aversive feelings against those good-natured theology addicts. Apart from this, Carnap also dealt with religion and theology at a mere theoretical level where the argumentation was not affected by any political or moral attitudes. At this neutral level, Carnap started to reason about theological topics already in his earliest youth, whereas the political and cultural-historical aspects only developed in the course of Carnap's politicization at the end of the First World War (Damböck et al. 2021).

What about Dörpfeld, in relation to Carnap's views on religion? In his youth, Carnap moved from an enlightened Protestantism toward atheism. The leaders in this development are mentioned in his autobiography, and included enlightened natural scientists and philosophers, such as Ernst Haeckel and Wilhelm Ostwald. But it also becomes obvious that Carnap's theoretical atheist stance is not far at all from the enlightened theism of his grandfather. Dörpfeld, to be sure, might have had absolutely nothing in common with Carnap's political and cultural-historical views on religion, for Dörpfeld hardly ever committed himself to socialism. However, at the theoretical level, Dörpfeld's enlightened and sober attitude toward religion and science is something that Carnap widely shared in all periods of his life. Dörpfeld comes even closer here to Carnap's overall views than Haeckel and Ostwald because he shares with Carnap the non-cognitivist attitude, which is something that hardly can be found in those more traditionally minded empiricists and positivists. Dörpfeld's theism was based on an inductive stance toward theology. This is something that Carnap could always adopt, for it leads to an adoption of atheism as soon as our empirical views become updated with evidence from modern science. The second element of Dörpfeld's system of Religionslehre was non-cognitivist ethics. Even this could be adopted by Carnap, after removing the theist point. Therefore, at the theoretical level, Carnap's views on religion and ethics are pretty similar to Dörpfeld's and share with the latter much more than they have in common with Haeckel, Ostwald, and other contemporary philosophers, such as the Neo-Kantians or Dilthey. This is not to claim that all those philosophers were not important for Carnap, for his views in the theory of knowledge and philosophy of science were certainly influenced by those and other contemporary philosophers. It is only Carnap's non-cognitivism and his related views on religion where we can find hardly any other contemporary match than the astonishing philosophy of his grandfather.

2.4 Some Facts and Speculations on the Influence That Dörpfeld Had on Carnap

In the previous sections, we uncovered certain striking similarities between Dörpfeld and Carnap. Are these only accidental family resemblances, or is there also a chance that Carnap might have been directly influenced by the philosophy of his grandfather? In Carnap's published work, we find almost no mentions of Dörpfeld at all. Dörpfeld is never cited, and even in the autobiography, he is mentioned only once as the person whose biography Carnap's mother wrote while he was a child (Schilpp 1963, 3). However, the published version of Carnap's autobiography is only a considerably shortened version that lacks everything that was considered not sufficiently relevant for the contemporary American reader of the *Schilpp volume*. For information on Carnap's early European years, therefore, one must look at the original version of the autobiography, which was kept by Carnap with the intention of a later publication. In this unabbreviated version, Dörpfeld becomes identified as a crucial figure:

> There were two relatives, both highly revered by my mother, whom I regarded from childhood as models of men, admirable for the fact that in their scholarly field they did not simply follow traditional ways but searched for their own new paths. One was my maternal grandfather, the other my uncle, the archeologist Wilhelm Dörpfeld (1853–

1940), the explorer of the remnants of Greek antiquity, especially at Olympia, Troy, and Pergamum.[6]

Wilhelm Dörpfeld, the brother of his mother, was important for Carnap probably as an example of a sober and incorruptible and, last not least also extraordinarily successful, scientist. When Carnap decided to become a university scholar, he certainly took Wilhelm Dörpfeld as his main role model. On the other hand, Wilhelm Dörpfeld was less important for the development of Carnap's philosophical views, since, successful as an excavator, architect, and real estate developer, he was much more a practitioner than a theoretician. Friedrich Wilhelm Dörpfeld, by contrast, though being "an influential pedagogical author", still "remained throughout his life in the modest position of principal of a grade school".[7] Whereas Carnap learned from Wilhelm Dörpfeld to be smart, ambitious and success-oriented, it was probably rather certain inner values that he took from his grandfather, because he hardly ever aimed at life as a philosophical eremite. Friedrich Wilhelm Dörpfeld also seems to have functioned as a philosophical role model for the entire family Carnap-Dörpfeld. His daughter Anna – Carnap's mother – worked for several years as his assistant and, after her late marriage – she was 37 when she married Johannes Sebulon Carnap in 1889 – and after the death of her father in 1893, she spent several years on his legacy, writing a voluminous biography that actually had the status of a compilation from his writings and correspondence. Anna Carnap certainly deeply influenced her family in spreading the philosophical and religious views of her father. So, when Carnap, in the unpublished parts of his autobiography, associates key aspects of his philosophy, including his tolerant attitude, his humanism and his non-cognitivism to the legacy of his mother,[8] he implicitly also commits himself to the influence that the family philosophy of Friedrich Wilhelm Dörpfeld had on him. He became familiar with this philosophy, first trough the instructions of his mother, who also taught the kids at ground school level at home.[9] Later, until the death of Anna Carnap in 1924, his mother also remained an important discussion partner for Carnap, which becomes visible in the voluminous correspondence and also in Carnap's diaries.[10] But Carnap also became acquainted with Dörpfeld's philosophy as a reader of his mother's biography of his grandfather – which he read several times, using both editions of the work (Carnap 1897, 1903) – and of Dörpfeld's ethics, which he

[6]Young Research Library, University of California at Los Angeles, Special Collections Department, Manuscript Collection No. 1029, Rudolf Carnap Papers, Box 2, CM3, M-A5, p. A7

[7]Young Research Library, University of California at Los Angeles, Special Collections Department, Manuscript Collection No. 1029, Rudolf Carnap Papers, Box 2, CM3, M-A5, p. A3.

[8]Young Research Library, University of California at Los Angeles, Special Collections Department, Manuscript Collection No. 1029, Rudolf Carnap Papers, Box 2, CM3, M-A5, pp. A8-A15.

[9]Young Research Library, University of California at Los Angeles, Special Collections Department, Manuscript Collection No. 1029, Rudolf Carnap Papers, Box 2, CM3, M-A5, p. A11.

[10]See RC 025, several folders, as well as (Carnap 2021a, b).

read at least once in the early 1920s.[11] The exact extent and the way in which Carnap became influenced by his grandfather remains unclear. However, it seems that there is at least a very plausible candidate here, for the tracking of a quite crucial line of influence on Carnap's thought, which addresses an important aspect of his philosophy, namely, non-cognitivism.

Acknowledgements For comments on an earlier version of this paper I am grateful to Hans-Joachim Dahms, Günther Sandner, and Anne Siegetsleitner. Regarding the general way of sensing Carnap's early intellectual sphere, I owe very much to discussions with Michael Heidelberger, A.W. Carus and, especially, Meike Werner. I apologize for the poor linguistic quality of the text.

References

Beeck, Karl-Hermann. 1975. *Friedrich Wilhelm Dörpfeld – Anpassung im Zwiespalt*. Neuwied: Luchterhand.

Carnap, Anna. 1897. *Friedrich Wilhelm Dörpfeld. Aus seinem Leben und Wirken*. Gütersloh: Druck und Verlag von C. Bertelsmann.

———. 1903. *Friedrich Wilhelm Dörpfeld. Aus seinem Leben und Wirken*. Zweite Auflage ed. Gütersloh: Druck und Verlag von C. Bertelsmann.

Carnap, Rudolf. 1932. Überwindung der Metaphysik durch logische Analyse der Sprache. *Erkenntnis* 2: 219–241.

———. 1934. Theoretische Fragen und praktische Entscheidungen. *Natur und Geist* 2: 257–260.

———. 2004. *Scheinprobleme in der Philosophie und andere metaphysikkritische Schriften*. In *Herausgegeben, eingeleitet und mit Anmerkungen versehen von Thomas Mormann*. Hamburg: Felix Meiner Verlag.

———. 2021a. *Tagebücher 1908–1919. Herausgegeben von Christian Damböck, unter Mitarbeit von Brigitta Arden, Brigitte Parakenings, Roman Jordan und Lois M. Rendl*. Hamburg: Felix Meiner Verlag.

———. 2021b. *Tagebücher 1920–1935. Herausgegeben von Christian Damböck, unter Mitarbeit von Brigitta Arden, Brigitte Parakenings, Roman Jordan und Lois M. Rendl*. Hamburg: Felix Meiner Verlag.

———. forthcoming. *Tagebücher 1935–1970*. Hamburg: Felix Meiner Verlag.

Carus, A.W. 2007. *Carnap and Twentieth-Century Thought. Explication as Enlightenment*. Cambridge: Cambridge University Press.

———. 2017. Carnapian Rationality. *Synthese* 194: 163–184.

———. 2021a. Die religiösen Ursprünge des Nonkognitivismus bei Carnap. In *Logical Empiricism, Life Reform, and the German Youth Movement/Logischer Empirismus, Lebensreform und die deutsche Jugendbewegung*, ed. Christian Damböck, Günther Sandner, and Meike Werner. Dordrecht: Springer.

———. 2021b. Werte beim frühen Carnap: Von den Anfängen bis zum *Aufbau*. In *Der junge Carnap in historischem Kontext 1918–1935/Young Carnap in an Historical Context 1918–1935*, eds. Christian Damböck and Gereon Wolters. Dordrecht: Springer, 1–18.

[11] Carnap requested a copy of Dörpfeld's *Ethics* in a letter to his mother from February 20, 1920 (RC 025-85-32). From the formulation it becomes likely that Carnap already knew the book at this stage. Acquaintance with his mother's biography of the grandfather is documented at several places of the Nachlass, with the inclusion of the before-quoted passages of the unabbreviated autobiography and the correspondence.

Dahms, Hans-Joachim. 2016. Carnap's Early Conception of a "System of the Sciences": The Importance of Wilhelm Ostwald. In *Influences on the Aufbau*, ed. Christian Damböck, 163–185. Dordrecht: Springer.

Damböck, Christian. 2017. *Deutscher Empirismus. Studien zur Philosophie im deutschsprachigen Raum 1830–1930*. Dordrecht: Springer.

———. 2018. Die Entwicklung von Carnaps Antimetaphysik, vor und nach der Emigration. In *Deutschsprachige Philosophie im amerikanischen Exil 1933–1945. Historische Erfahrung und begriffliche Transformation*, ed. Max Beck and Nicholas Coomann, 37–60. Vienna: LIT.

———. 2021a. Carnap, Reichenbach, Freyer. Non-cognitivist Ethics and Politics in the Spirit of the German Youth-Movement. In *Logical Empiricism, Life Reform, and the German Youth Movement/Logischer Empirismus, Lebensreform und die deutsche Jugendbewegung*, ed. Christian Damböck, Günther Sandner, and Meike Werner. Dordrecht: Springer.

———. 2021b. Die Entwicklung von Carnaps *Aufbau* 1920–1928. In *Der junge Carnap in historischem Kontext 1918–1935/Young Carnap in an Historical Context 1918–1935*, eds. Christian Damböck and Gereon Wolters. Dordrecht: Springer, 19–54.

———. 2021c. The Politics of Carnap's Non-Cognitivism and the Scientific World-Conception of Left-Wing Logical Empiricism, *Perspectives on Science*, https://doi.org/10.1162/posc_a_00372.

Damböck, Christian, Günther Sandner, and Meike Werner, eds. 2021. *Logical Empiricism, Life Reform, and the German Youth Movement/Logischer Empirismus, Lebensreform und die deutsche Jugendbewegung*. Dordrecht: Springer.

Dörpfeld, Friedrich Wilhelm. 1894. *Denken und Gedächtnis. Eine psychologische Monographie*. 5th ed. Gütersloh: Druck und Verlag von C. Bertelsmann.

———. 1895. *Zur Ethik*. Gütersloh: Druck und Verlag von C. Bertelsmann.

Heidelberger, Michael. forthcoming. Between Pietism and Herbartianism. Archaeological Vestiges in Carnap's Thought. *Scientitfic World Conceptions: Rudolf Carnap and Otto Neurath*. ed. Christian Damböck, Johannes Friedl, and Ulf Hoefer. In preparation.

Mormann, Thomas. 2007. Carnap's Logical Empiricism, Values, and American Pragmatism. *Journal for General Philosophy of Science* 38: 127–146.

Reisch, George A. 2005. *How the Cold War Transformed Philosophy of Science. To the Icy Slopes of Logic*. Cambridge: Cambridge University Press.

Richardson, Alan. 2007. Carnapian Pragmatism. In *The Cambridge Companion to Carnap*, ed. Michael Friedman and Richard Creath, 295–315. Cambridge: Cambridge University Press.

Schilpp, Paul Arthur, ed. 1963. *The Philosophy of Rudolf Carnap*. Chicago: Open Court.

Siegetsleitner, Anne. 2014. *Ethik und Moral im Wiener Kreis. Zur Geschichte eines engagierten Humanismus*. Vienna: Böhlau.

Stadler, Friedrich, and Thomas Uebel, eds. 2012. *Wissenschaftliche Weltauffassung. Der Wiener Kreis. Hrsg. vom Verein Ernst Mach. (1929)*. Vienna: Springer.

Uebel, Thomas. 2010. "BLUBO-Metaphysik": Die Verwerfung der Werttheorie des Südwestdeutschen Neukantianismus durch Carnap und Neurath. In *Logischer Empirismus, Werte und Moral: Eine Neubewertung*, ed. Anne Siegetsleitner, 103–130. Wien: Springer.

Chapter 3
The Place of Religion. An Open Question in Schlick's Philosophy of Culture

Massimo Ferrari

3.1 The Place of Religion in Schlick's Philosophy of Culture

Neither religion nor the philosophy of religion play any significant role within Moritz Schlick's work. By scrutinizing his papers and books, but also his unpublished lessons and his letters, it is quite difficult to find religion as a core issue or, at least, as an autonomous field of inquiry. Even in *Natur und Kultur* – a posthumous collection of manuscripts highlighting Schlick's never-accomplished book about philosophy of culture – we do read a wide treatment of history, politics, moral, economy and technology, but Schlick makes no reference to religion as an independent aspect of human culture (Schlick 1952). According to Schlick, the very pivotal point of any form of culture is rather the "moral consciousness (*Gesinnung*)", which makes possible the conciliation between culture and nature through the moral responsibility proper to reason (Schlick 1962, 17–18). Nonetheless, the question concerning the place of religion in Schlick's philosophy of culture cannot be simply disregarded, since it in any case represents a field of some interest if we take into account his thought beyond the usual (albeit *per se* perfectly correct) portrait of Schlick as outstanding philosopher of science.

Indeed, immediately after his tragic death, former students and colleagues of Schlick already emphasized how the wide frame of his philosophy has to be understood as a veritable philosophy of culture. Edgard Zilsel pointed out, for instance, that throughout his life, Schlick manifested an ongoing interest in the questions of both life and culture, far beyond the classical issues of philosophy of science (Zilsel 1937, 162). No differently, Herbert Feigl stressed that since his very beginnings Schlick had dealt with the "problems of life", which could

M. Ferrari (✉)
Dipartimento di Filosofia e scienze dell'educazione, Università degli studi di Torino, Turin, Italy
e-mail: massimo.ferrari@unito.it

E. Ramharter (ed.), *The Vienna Circle and Religion*, Vienna Circle Institute Yearbook 25, https://doi.org/10.1007/978-3-030-76151-6_3

be considered no less important than the theory of knowledge (Feigl 1937/1938, 395). Even a distinguished figure for the intellectual exchange between Schlick and Wittgenstein, namely Friedrich Waismann, would speak of a genuine "philosophy of life (*Lebensphilosophie*)" on which Schlick's philosophical work had rested since his early days (Waismann 1938, XII).

Similar assessments are undoubtedly correct. The first aspect to be recalled in this context is surely Schlick's juvenile book on the wisdom of life (*Lebensweisheit*), which was published in 1908 (Schlick 2006). Significantly, the main sources for Schlick here are Schopenhauer and Nietzsche, who inspire his diagnosis of modern culture in the age of both technical domination and the loss of unity between nature and culture.[1] In order to regain what seems to be in deep contrast with modernity (Mormann 2010), Schlick considers the achievement of *pleasure* as the goal of human life. By using Schopenhauer's term, but distancing himself from his pessimism, Schlick speaks of a "will for pleasure" dominating all the world and motivating human action (Schlick 2006, 71, 83). "The will to happiness of men mastering human life – Schlick argues – is only a small revelation of the greater will to pleasure in nature, which in turn domains the world" (Schlick 2006, 84). Accordingly, both instincts and feelings, but in no way the rigorous law of duty in Kantian sense, are the roots of Schlick's account of moral experience as a mundane form of life.[2]

Given this conception of ethics as ongoing search for pleasure and happiness, what is now important for our question is that in Schlick's juvenile book we also find a chapter devoted to religion. To be sure, these pages are not enough to gain a comprehensive view of his conception of religion, but they are still worth considering. Schlick claims that what we call religion can be identified with the emotional state we experience as we are in intimate contact with the universe. This is, properly speaking, and unlike the "worldview (*Weltanschauung*)", the "feeling of the world" (*Weltgefühl)*", which coincides with the feeling of pleasure arousing from the effects of nature on man's impulses (Schlick 2006, 185). Only in this way can our soul actually achieve felicity.[3] But what remains, thus, of religion? Schlick believes that every religion rests just upon the "feeling of the world", which is strictly connected with the "faith in the future", with the open sea of possibilities: "the boundless sea is the symbol of future" (Schlick 2006, 191). And future signifies

[1] Note that Schlick would several times, in Rostock and later also in Vienna, hold classes on Schopenhauer and Nietzsche (see Schlick 2012 and Ferrari 2021). Iven (2008, 52–56) has documented that Schlick was at least acquainted with Nietzsche even in his early days, that is to say around 1897–1898.

[2] Nor would this crucial aspect undergo any change in Schlick's late conception of ethics, for it remains in his eyes undisputable that "Kant's categorical imperative, which demands that one act wholly independent of one's inclinations, demands what is impossible" (Schlick 1939, 62).

[3] Matthias Iven, the editor of the critical edition of Schlick's *Lebensweisheit*, quotes in this regard a passage of an unpublished draft where Schlick claims that the worst religion is that requiring prayers and promises, whereas the most sublime idea of religion can be found only when the name of God is not expressed (Schlick 2006, 185 note 25).

at once the spirit of youth, the spirit of an optimistic attitude towards the world (Schlick 2006, 190).

Interestingly enough, a rather similar point of view would be affirmed again in Schlick's late *Problems of Ethics* (published in 1930). Dealing anew with the question of pleasure as the basis for ethics, Schlick points out that religions show the tendency to consider pleasure as something belonging to "the realm of prohibited things". This means that a basic opposition arises for religion between "earthly and heavenly joys": whereas the former are worthless, only the latter are good, as for instance in the case of ascetism providing the highest feeling of religious pleasure. And Schlick significantly concludes: "It would be the task of a sensible education to return the heavenly joys to earth again" (Schlick 1939, 126). Schlick's endorsement of eudemonism is thus conceived in strong opposition to religious morality, which condemns the ethic of pleasure by charging it even of immorality. Nonetheless, Schlick maintains that if the religious promise of future happiness as compensation for human evil and sufferance could even be regarded as a form of eudemonism, in no way does this signify that this kind of future pleasure is at odds with our goals and feelings in this mundane world. It should be affirmed, by contrast, that just the hope for the future enhances our ethical rules here on the earth (Schlick 1939, 133–134). Put in other words, and in agreement with his early views, the future is for Schlick an essential aspect of ethical life, and thus religion is welcomed when it enlarges the horizon of the search for happiness.

But going back now to the early Schlick, it must be noticed that he repeatedly focused on the questions of ethics and life in his classes held in Rostock from 1911 onwards. In the winter semester of 1915/1916, Schlick held a lecture on *The Philosophy of present Time* (*Die Philosophie der Gegenwart*), which he had already delivered in the summer 1912 semester and would repeat in Vienna in the summer 1920 semester (albeit with slightly modified titles).[4] Opening his course, Schlick significantly states: "The problems of life, namely the questions concerning the sense, of goals and reasons of human action, of the essence of culture and so on are all questions of value" (Inv 006/A.8, 4). Concerning the contribution of "philosophers of value" to contemporary philosophy and partly to current debates about religion, Schlick refers mainly to James, Guyau, and Eucken. It is thus of some interest to take into account what Schlick maintains in this regard, bearing in mind his conviction that the task of contemporary philosophy consists of not only dealing with science, but also addressing the *Weltanschauungsfragen*.[5]

That Rudolf Eucken could to some extent have interested Schlick as a philosopher of culture is more than a supposition. Schlick had met Eucken in Jena in 1907, although he was quite disappointed by him (Iven 2008, 157). Eucken was indeed a very respected figure, also well-known abroad; some of his works were translated

---

[4]We refer to the manuscript of 1915/1916 preserved in Schlick's estate (signature Inv 006/A.8, 1–34)

[5]This is just the title of a course held in Rostock in winter 1917/1918 semester (preserved in Schlick's estate with the signature Inv 006/A. 9a typed and 9.b, 1–25 for the transcription).

in English and Italian and – last but not least – he was awarded the Nobel Prize for Literature in 1908. At the time of Schlick's visit in Jena, Eucken had just published one of his most read books, entitled *The Sense and the Value of Life* (*Der Sinn und der Wert des Lebens*), which offered a specimen of Eucken's philosophical insights. The core issue, "life", was investigated in its various articulations and tendencies, focusing in particular on the task of building a new form of life beyond both the domain of technology and the standardization of the human spirit (Eucken 1918; see also Graf 1997). Eucken was, in this sense, a critic of contemporary society engaged in a diagnosis of the evils and dangers for modernity from the point of view of a future "reign of spirit (*Reich des Geistes*)". This reign, Eucken stressed *ad infinitum*, had to be understood as the realization of the "monism of spirit" in opposition to the naturalistic monism professed by Ernst Haeckel, who was by a strange twist of fate also his colleague in Jena (Dathe 2000).

That said, one can explain why Schlick deals at length with Eucken in his class. Nonetheless, his account of Eucken's work is in part ironical and skeptical. Schlick appreciates not so much his philosophy of culture, developed in many books repeating again and again the core theses of his idealism. This kind of idealism, Schlick comments, goes back to Fichte and in part to Hegel, but unlike the German idealism Eucken maintains a more positive relationship with the sciences. The point is, however, that the "realm of spirit" eventually seems to fall into the old Platonic dualism between the human world and the eternal realm of ideas. Accordingly, Eucken's philosophy of religion too pays the price for such a dualism, despite his main idea of religion as *Weltanschauung* independent from its historical forms. Eucken embodies therefore – Schlick says in concluding his lecture in Rostock – another symptom of the "chaos" characterizing contemporary philosophy (Inv 006/A.8, 32–34).

A similar critical stance emerges from the illustration of James' philosophy of religion. The premise of Schlick's criticism can be found in his *Habilitationsschrift* of 1910 devoted to *The Nature of Truth in Modern Logic*. Schlick believes that James's account of truth is unacceptable for two main reasons. On the one hand, we can surely maintain that all true judgements have to be verified, but this does not mean, conversely, that all the judgements that are (or can be) verified are true. On the other hand, James's definition attributes to truth in general "a property uncharacteristic of it either in everyday or scientific language, namely that of mutability" (Schlick 1978, 67). From this point of view, the pragmatist conception of truth is false (or even "unscientific"), because it confuses the essence of truth with both the criteria for its verification and its practical consequences (Schlick 1978, 88). In his class in Rostock Schlick again endorses this argument, stressing in particular that judgements are useful because they are true, but they are not true being useful (Inv 006/A.8, 29). This critical stance is also valid in that which regards James' conception of religion.[6] According to Schlick James' account of religion is highly

[6]Schlick refers to James' mostly discussed book on *Pragmatism*, which appeared in 1907 and was translated into German in 1908. By contrast, it seems that Schlick has no acquaintance with

disputable for two main reasons. First, James' attempt to conciliate philosophy with religion assumes that religion – or rather any form of religion – can be true as a form of human life useful to make it better. In other words: even in the case of religion, usefulness is the criterion of truth. Second, given this false premise, Schlick discusses how James intends to justify the truth of religious experience taking into account a variety of feelings and emotional states (pathologies of the mind, mystical visions, unconscious *Erlebnisse*). The point is that similar fields of experience are accessible only through intuition, endorsing thereby a kind of radical empiricism that considers immediate experience as a source of knowledge. Thus, both James and Bergson represent that the philosophical trend of contemporary thought labelled as intuitionism is at odds with the very nature of knowledge and truth (Inv 006/A.8, 30).[7]

Lastly, but interestingly enough, it is rather such an original thinker as Jean-Marie Guyau that Schlick considers, and, unlike the case of Eucken and James, with warm sympathy. Guyau is surely, Schlick claims, a kind of French partner of Nietzsche, because of the main idea of life as an ongoing expansion of life itself, both of them endorsing being motivated by an "affirmative world-conception" (Inv 006/A.8, 30). The struggle Guyau's against intellectualism (quite similar to Nietzsche's) implies not only the view of morality "without obligation and sanction", but also the hope in a "irreligion of the future" as an expression of the overall solidarity of the individual with both humankind and the cosmos.[8] This particular sort of religion as irreligion, clearly in opposition to any cult, dogma or myth, seems to be congenial to Schlick, who speaks of Guyua with undeniable conviction. And it is no less important that Schlick agrees with Guyau's faith in the future, a core issue – as we have seen above – in Schlick's conception of both life and youth. Not accidentally, at the end of his course Schlick quotes a passage by Guyau, then reproduced on his grave: "Non, pas un de mes rêves peut-être ne sera perdu: d'autres les reprendront, les rêverons après moi, jusqu'à ce qu'ils s'achèvent un jour" (Inv 006/A.8, 32).[9]

3.2 Monism as Worldview: A Replacement of Religion?

On February 25, 1921, Schlick delivered a lecture at the *Werner-Siemens-Realgymasium* in Berlin. The lecture was devoted to the *Sense of life* (*Vom Sinn*

The Varieties of Religious Experience. A study in Human Nature (published in 1902; a German translation would only appear in 1914).

[7]Schlick's criticism of intuitionism is widely carried out in his paper of 1913 *Gibt es intuitive Erkenntnis?* in which he considers James along with Husserl and Bergson as the most influential spokesmen of the totally false idea that intuition can be regarded as a source of knowledge (see Schlick 1978, 141–152).

[8]Schlick takes into account here respectively Guyau (1885) and Guyau (1906).

[9]Schlick quotes and translates into German from *L'irréligion de l'avenir*, but we quote in turn from the original French text (Guyau 1906, 458).

des Lebens) and Schlick repeated it the following year in Kiel; however, the text would only be published in 1927, in the sixth issue of the journal *Symposion*.[10] The invitation to lecturing came from the Berlin Section of the German Monists' Association *(Deutscher Monistenbund)*, founded in 1906 under the guidance of Ernst Haeckel and Wilhelm Ostwald. Monism today appears to historians of science as a mostly forgotten player on the scene of German philosophy in the first two decades of 20th century, but its influence on the scientific philosophy was at that time really quite relevant, not least for its role in popularizing science.[11] Furthermore, in 1913 the Austrian Academic Monists‘ Association was founded in Vienna, and its importance as a backdrop for the Vienna Circle should be emphasized because of the "family resemblances between (natural-scientific) monism and the scientific world conception" (Stadler 2001, 187). Moreover, materialism too appeared to the authors of the Manifesto of the scientific world-conception as a sort of ancestor of modern empiricism (Carnap-Hahn-Neurath 2012, 90).[12]

It was typical of the followers of monism to regard ethics too as a substantive component of their worldview. In his *The Riddle of the Universe* (*Die Walträtsel*), first published in 1899 and thereafter reissued more than ten times (also as an out-and-out bestseller), Ernst Haeckel had repeatedly spoken of the ethical implications of monism. A new ethics would arise, Heackel suggests, only when some widely diffused prejudices were overcome. On the one hand, it should be said that the pretension of free will has to be replaced by a wholly deterministic stance (Haeckel 1919, 138). On the other hand, modern scientific thought must fight against the "strange idol" of Kant's categorical imperative (Haeckel 1919, 358). As a consequence, the rejection of Kantian ethics poses into question his "dogmatic" view of two reigns in opposition with each other: the empirical reality and the intelligible domain of ideas relegated into an intelligible world (Haeckel 1919, 402). In modern terms: no gap can subsist between the factual and the normative.

We can add, furthermore, a very insightful witness to such a monistic ethical attitude. On the occasion of the first Congress of the German Monists' Association held in Hamburg in 1911, Friedrich Jodl delivered a speech on monism and the present problems of culture (Jodl 1911). Jodl was Professor of Philosophy at the University of Vienna and died in 1914; it should be remarked that, after an interlude of 6 years, no less than Moritz Schlick took over his vacant chair.[13] According to Jodl, the proper task of monism as a cultural movement consists of addressing the problems of both inner life and the ethical questions of mankind. The social ethics

[10]We quote Schlick's text from the critical edition available in (Schlick 2008, 99–125).

[11]About this topic see Hübinger (1997) and Ziche (2000). The ethical and political implications of the monist movement are discussed by Lübbe (1974, 124–170).

[12]For his part, Rudolf Carnap would later remember both his readings of Haeckel and Ostwald and his "sympathy" for the Monists' Association in the early days of his philosophical apprenticeship (Carnap 1963, 7).

[13]On Jodl's role within the Viennese milieu see Uebel (2000, 292–294).

devoted to the progress of modern society is thus "the coronation" of the monistic view (Jodl 1911, 20). Jodl urged therefore his listeners in Hamburg to hold firmly the link between science and ethics as well as that between nature and culture: we defend science *and* ethics, Jodl suggested, but not science *without* ethics. Crucial was thus to emphasize the need for a new *Weltanschauung* able to pose as its own center the moral consciousness of men. In this regard Jodl also stressed the place of religion, which can be considered in two senses. On the one hand, the idea of God coincides with "the holy idea of men" according to his own essence (Jodl 1911, 34); on the other hand, the modern monism should not be understood simply as the enemy of religion: it is surely very far from religion, but in the sense of being irreligious, rather than anti-religious as such (Jodl 1911, 29).

Similar statements are in perfect harmony with Haeckel's *Weltanschauung* (as he called it), which implies a strong opposition to Christian religion by reclaiming rather to monism the proper status of religion: as Haeckel says, "our monistic religion" based on what is essential between science and religion (Haeckel 1919, 341–357). The human commandments of monistic religion are love, perseverance, compassion and help; they all permit one to conceive Monism as a "replacement (*Ersatz*)" of Christian religion, because "the Goodness of truth lives in the temple of nature" (Haeckel 1919, 347). It seems however still more striking that Haeckel emphasizes, on this basis, that the ground message of the Gospel ("love your neighbor as yourself") is "completely" in agreement with the monistic ethics (Haeckel 1919, 361). In the end, once traditional religion is demolished, the followers of Monism will be able to build "a palace of reason" (Haeckel 1919, 346).

It is precisely into this wide context that we have to place Schlick's lecture in Berlin. At first sight, one can wonder that Schlick was so close to the monistic movement; but it is enough to read the § 35 of the *General Theory of Knowledge* for a better understanding of the reasons why Schlick was committed to the German Monists' Association. In a passage worth quoting at length Schlick claims:

> We are thoroughly convinced that all the qualities of the universe – all being whatsoever – are of one kind in so far as they can be made accessible to knowledge by means of quantitative concepts. In this sense we embrace a monism. There is only one kind of reality, that is, we need in principle only one system of concepts to know all the things of the universe. And there do not exist in addition classes of things that this system does not fit.
>
> Such a monism seems to me to be as comprehensive and far-reaching as reason's need for unity might desire. At the same time, it is the only kind of monism that can be arrived at by epistemologically refined thinking. It has all the useful features that made nineteenth century materialism so successful with a public which, unburdened by epistemological scruples, found satisfaction in materialism's strong drive toward a unified, closed world picture (Schlick 1974, 326).

Therefore, the kind of monism Schlick appreciates is, properly speaking, a scientific monism, whereas he strongly rejects a metaphysical monism aiming to grasp, as the materialists of the 19th century did, the ultimate essence of matter. This pretention was, Schlick complains, "naïve, inadequate and philosophically erroneous" (Schlick 1974, 327). Although materialism shows "a healthy tendency", it needs exactly the revision that Friedrich Albert Lange had carried out in his *History of Materialism*, whose merits Schlick emphasizes just for having purged dogmatic materialism from its "pathology". To the Neo-Kantian Lange must be thus acknowledged the merit of an epistemological correction of materialistic viz. monistic stances (Schlick 1974, 327).

Interestingly enough, in the framework of these considerations Schlick briefly recalls the question of *Ignorabimus* that Émile du Bois-Reymond had posed in a famous speech delivered in 1872. For Du Bois-Reymond, an influential physicist whose research in physiology is epoch-making, claimed that the knowledge of nature cannot go beyond some limits (*Grenzen*), in particular in what concerns the essence of force and matter. For this reason scientists have to acknowledge that these two "riddles" are unsolvable, remaining thus resigned not only to a temporary *Ignoramus*, but to a perennial *Ignorabimus* (Du Bois-Reymond 1974, 76–77). Later on, in another lecture delivered in 1880, Du Bois-Reymond would spell out that the riddles he had spoken of were properly seven "transcendent" problems. At least four similar riddles would have never found a solution (that is to say, the essence of both matter and force, the origin of movement, the origin of sensation and the freedom of the will), whereas the other three could find a solution in the long run (the origin of life, the finality of nature, the origin of thought and language). All these questions or riddles can be regarded, Du Bois-Reymond claims concluding his lecture, as the articulations of only one great problem: the problem of the universe (*Weltproblem*) (Du Bois-Reymond 1974, 186–187).

Because of his bitter polemics against contemporary materialism, Du Bois-Reymond gained a central role in the wide discussion developing in Germany about both the limits of scientific knowledge and the domain of the Absolute that natural sciences can never explain. The religious implications of Du Bois-Reymond's *Ignorabimus* are quite evident and rapidly became the other side of an endless dispute.[14] Haeckel's vehement response to his colleague, ironically defined as "dictator" and "rhetorician" (Haeckel 1919, 186), should be taken into account here. Haeckel's response was quite dogmatic. The supposed seven riddles are reducible to one only: the riddle of substance. But, in turn, this problem is merely apparent: the "cosmic ground law", postulating the entanglement between matter and force (or energy) will reduce the whole reality according to mechanical causality governing both nature and human spirit (Haeckel 1919, 217–238, 376). The riddles of the universe are not riddles at all.

Schlick was well aware of the questions connected to the disputation about the *Ignorabimus*. In his *General Theory of Knowledge* (first edition 1918, second edition 1925) he offers a brief comment to this question, taking a polemical position towards Du Bois-Reymond's famous motto. By using quantitative concepts – Schlick says – science can proceed in knowing reality until an error arises compelling us to dismiss

[14] It should be remembered that a similar debate arose in England about Agnosticism (the term was coined by Thomas Huxley in 1869), which represents, to some extent, the other face of the German *querelle*. Nonetheless, in England the questions regarding science, faith and religion were discussed still more vivaciously than in Germany: see (Lightman 1987) and (Pyle 1995).

the method we are using. But science has not yet renounced such a method, which would as a consequence open the door to the idea of something that is in principle unknowable. "Nothing harms inquiry – Schlick argues – so much as the pronouncing of an *ignorabimus*, and we must be on our guard against uttering one prematurely" (Schlick 1974, 326). From this point of view Schlick was quite close to the monistic attack against Du Bois-Reymond, as was, after all, the case of some of the members of the Vienna Circle. The Manifesto of Vienna Circle tell us, for instance, that "the scientific world-conception knows no unsolvable riddles", because they are partly pseudoproblems and partly empirical problems to be verified through empirical science (Carnap–Hahn–Neurath 2012, 82). No differently, Carnap too proclaims that "when we say that scientific knowledge is not limited, we mean: there is no question whose answer is in principle unattainable by science" (Carnap 2003, 290).[15] There are questions unsolvable only from a practical point of view, but in principle all the problems correctly posed, Carnap explains, are answerable. Finally, Ludwig Wittgenstein should also be recalled here, his aphorism 6.5 in the *Tractatus logico-philosophicus* lapidary affirming: "When the answer cannot be put into words, neither can the question be put into words. The riddle does not exist. If a question can be framed at all, it is also possible to answer it". In a strict sense, Wittgenstein was surely not close to the Vienna Circle, but no doubt can subsist that in his statement the heritage of the controversy about Du Bois-Reymond's stance emerges (Bayertz 2007).

However, and besides this fascinating history, the crucial problem Schlick addressed starting from his "critical monism" was, roughly speaking, the following: if there is a very substitute of both the absolute and religion, this is surely *ethics* or, in other terms, the attempt to answer the question about "the sense of life" beyond any kind of riddle. It is, then, no surprise that in his lecture in Berlin Schlick appears as a genuine supporter of monism, albeit in his moderate version. Significantly, he goes back to his early views, stressing in particular that the "sense of life" in no way consists of the sacrifice of human inclinations imposed by moral duty, but rather of the search for happiness strictly connected to vital impulses belonging typically to youth. Schlick claims accordingly – and we have already underlined this theme above – that his own "naturalistic ethics" is conceived in total opposition to Kant (Schlick 2008, 120). Not accidentally, the major point of reference for Schlick is here the great German poet Schiller, who contemplated "play" as the very revelation of human nature. As a consequence, youth is not only a stage of life destined to be overtaken through the wisdom of the mature man; rather, it constitutes the sense itself of the whole life, a kind of perennial status (Schlick 2008, 116). As Schlick maintains, the proper goal of education is that childhood does not get lost

[15] Schlick too would share this position, in particular in the late essay "A new Philosophy of Experience", where he states that the classical body-mind problem, often considered as a typical example of *Ignorabimus*, is not resolvable inasmuch it has been posed in a false way; that is to say it is not, properly speaking, a very problem (Schlick 2008, 400–404).

in maturity, so that matured men too must respect the rule: "Let preserve the spirit of youth, because it represents the sense of life" (Schlick 2008, 125).

Following these suggestions Schlick refers to both his beloved authors Nietzsche and Guyau, but also to Emerson, recalling at once the Movement of Youth (*Jugendbewegung*) as well as Gustav Wyneken, a leading figure in the German cultural landscape at the turn of the 20th century.[16] Against this backdrop, one can comprehend that Schlick stresses how "the very core of life" is ethics, whose mission also consists of revealing the deep sense of life (Schlick 2008 119). According to Schlick, ethical life rests upon feeling and, in turn, ethical feeling is, in particular, a typical feature of youth (Schlick 2008, 122). Moreover, ethics is a part of that "play of spirit" (as Schiller said) to which knowledge too belongs.[17] The struggle for truth is the final goal (*Endzweck*) for the scientist, who employs all his forces in solving the riddles of reality independently from its usefulness (Schlick 2008, 106). Ultimately, the sense of life reveals itself both in the energy of young man and in the enthusiasm for knowledge.

It is not difficult to read Schlick's speech in terms of an immanent vision of life. Life needs no transcendent realm of values in providing its authentic sense: both the pursuit of happiness and the view to the future are sufficient to highlight the human quest for certainty in *this* world. This is the only way to give a sense to life, bridging thereby the gap between nature and culture, between "is" and "ought". This last point is a crucial one for Schlick, who again and again emphasizes how the separation between normative and factual sciences is simply "fundamentally false" (Schlick 1939, 17). Schlick therefore claims:

> Even if ethics were a normative science it would not cease because of this to be a science of facts. Ethics has to do entirely with the actual; this seems to me to be the most important of the propositions which determine its task. Foreign to us is the pride of those philosophers who hold the questions of ethics to be the most noble and elevated of questions just because they do not refer to the common is but concern the pure ought (Schlick 1939, 21).

To be sure, by emphasizing this way to conceive ethics, Schlick was compelled in general to a long tradition, namely to the "science of ethics" that French and English Positivism of the 19th century had put forward as the goal of a new approach to human morality. Nonetheless, one can ask whether something of the monistic ethics endorsed, for instance, by Haeckel and Jodl still resonates in Schlick's late suggestions.

---

[16]Note that Wineken and the *Jugenbewegung* were also influential for the young Carnap (Carus 2007, 52–56). See also Tuboly (2017).

[17]See also Schlick (1962, 14): "The scientist remains young because he searches again and again for something *new*".

3.3 Late Thoughts on Ethics and Religion: Schlick and Wittgenstein

Vienna, December 30, 1929. Wittgenstein meets Schlick at his home. Among other issues, Wittgenstein refers to Heidegger admitting that he "can imagine what Heidegger means by being and anxiety" (Waismann 1979, 68). This statement has to do with Wittgenstein's persuasion that there are questions destined a priori to be unanswerable, being mere nonsense. Of similar kind is, for instance, the "astonishment that anything at all exists"; and yet, Wittgenstein says, when we try to express such an emotional status through meaningful language "we do run up against the limits of language". Kierkegaard – the other figure evoked here by Wittgenstein, along with Heidegger – baptized this state of affairs as "paradox". But a similar paradox characterizes *ethics* too, because ethics again consists of "running up against the limits of language" (Waismann 1979, 68).

Wittgenstein's reflections during his conversation with Schlick are notoriously puzzling. At any rate, they are in agreement with the lecture on ethics Wittgenstein delivered in Cambridge in September 1929, where he had already pointed out that "ethics, if it is anything, is supernatural and our words will only express facts" (Wittgenstein 1965, 7). This at first glance strange affirmation recalls an aphorism occurring in *Tractatus* 6.421: "It is clear that ethics cannot be put into words. Ethics is transcendental. (Ethics and aesthetics are one and the same)". The conference in Cambridge would try to explain what Wittgenstein really meant. At stake here is the question of "absolute value"; and yet no linguistic expression can be found in order to explicate what the absoluteness of a value could signify. Such an expression is merely nonsense, for it alludes to a domain beyond the world. But for Wittgenstein, this is just the point: the nonsensicality of this proposition represents its very "essence". And Wittgenstein goes on:

> For all I wanted to do with [this nonsensical proposition] was just *to go beyond* the world and that is to say beyond significant language. My whole tendency and I believe the tendency of all men who ever tried to write or talk Ethics or Religion was to run against the boundaries of language. This running against the walls of our cage is perfectly, absolutely hopeless. Ethics so far as it springs from the desire to say something about the ultimate meaning of life, the absolute good, the absolute valuable, can be no science (Wittgenstein 1965, 11–12).

Wittgenstein's very hard question signalizes a wholly different intention from Schlick's idea of factual ethics. To begin with, the *incipit* of Schlick's late book on ethics leaves no room for doubt: "If there are ethical questions that have meaning, and are therefore able to be answered, then ethics is a science" (Schlick 1939, 1). No wonder, thus, that Wittgenstein immediately expresses his doubts about Schlick's ethics as science. In a letter to Schlick written in Cambridge on November 27,1930, he affirms: "I believe that from many points of view I will not agree with you" (Inv. 123 Wittg – 10). Some weeks later Wittgenstein meets Schlick and the talk is too about Schlick's book. Wittgenstein appreciates some remarks Schlick has devoted to theological ethics: what is good is good because God wills it? Or does He will it because good is good? (Schlick 1939, 11–12). For Schlick the former interpretation is the "more superficial", whereas the latter is "deeper". But Wittgenstein's contrasts with Schlick's opinion: good is what God commands, and this the right answer (Waismann 1979, 115).[18]

Moreover, a little later Wittgenstein emphasizes again, and in opposition to Schlick, that ethics has nothing to do with a conceptual discourse. In his own words:

> If I were told anything that was a *theory*, I would say, No, no! That does not interest me. Even if this theory were true, it would not interest me—it would not be the exact thing I was looking for.
>
> What is ethical cannot be taught. If I could explain the essence of the ethical only by means of a theory, then what is ethical would be of no value whatsoever [. . .]
>
> *For me* a theory is without value. A theory gives me nothing (Waismann 1979, 116–117).

The clash with Schlick's account of epistemological and psychological foundation of ethics is patent. One has only to recall some passages of Schlick's we have already quoted above to understand how deep the difference of *Problems of Ethics* is from Wittgenstein's main idea. What Wittgenstein refuses is just what Schlick endorses: the possibility to explain the nature of ethics.[19] By contrast, Wittgenstein maintains that such an explication can be false or not but is wrong in itself because it is precisely an explication (Waismann 1979, 117). In order to reconstruct Schlick's relationship with Wittgenstein this topic would surely require more attention than scholarship has devoted to it.

However, this is not the only point to deserve consideration. In the same conversation with Schlick held on 17 December Wittgenstein makes some remarks on religion. Wittgenstein argues that we are dealing with the question of language also in the case of religion, in terms quite similar to the puzzle of ethics. For the essence of religion, Wittgenstein suggests, is not bound to language, or, better said, religion is by no means a theory: it is also quite irrelevant that religious words could be true, false or meaningless. Talking about religion is not necessary, but religion in any case remains a form of life worthy of fully respect. As Wittgenstein says: "I do not scoff at this tendency in man; I hold it in reverence. And here it is essential that this is not a description of sociology but that I am speaking *about myself*" (Waismann 1979, 118).

---

[18]To notice is that Wittgenstein refers obviously to the German edition of Schlick's book, from which we too have drawn both the quotations (Schlick 2006, 365). Later on, in the English translation in part revised by Schlick himself, one finds a slightly, but significantly modified version of the passage quoted by Wittgenstein: "more superficial" is substituted by "one interpretation", and "deeper" by "perhaps profounder" (Schlick 1939, 11). However, Schlick's whole argumentation intends to compare Kant's "absolute ought" with the first way of theological ethics (good is good because God wills it). For Schlick, Kant's formal ethics is in agreement with this theological stance and is wrong just because of this convergence (Schlick 1939, 11). At any rate, Wittgenstein does not consider Schlick's Anti-Kantian argumentation.

[19]See Schlick (1939, 22–25): "Ethics seeks causal explanations" and, above all, Schlick (1939, 28–30): "The method of ethics is psychological".

It should be noticed that Wittgenstein anticipates here some arguments he will highlight later in his lectures on religious belief delivered in Cambridge. A very important issue is, for instance, Wittgenstein's examination of the belief in a Judgement Day. The religious person is not interested in what I describe about this belief, just because describing is totally different from grasping the object of religious faith (Wittgenstein 1999, 55). In this field, Wittgenstein explains, "controversies look quite different from any normal controversies. Reasons look entirely different from normal reasons" (Wittgenstein 1999, 56). Religion, no differently from ethics, cannot be tested as is usual for meaningful propositions, namely by recurring to hypotheses, high probability or knowing. By contrast, one speaks only of faith and dogma when religion is asked (Wittgenstein 1999, 57). However, similar suggestions represent only a part of Wittgenstein's approach to religious beliefs. By reading random thoughts Wittgenstein wrote from 1929 onwards, one finds a lot of observations regarding religion that illustrate the dramatic awareness Wittgenstein had of religious faith as "passion". The main reference is, significantly, Kierkegaard: Wisdom is passionless. By contrast Kierkegaard calls faith a "*passion*" (Wittgenstein 1998, 61).[20] It is useful to add, finally, that Wittgenstein shows just a dramatic way (*à la* Kierkegaard) of understanding religion as purely arising from the heart, from the soul, but in no way from "speculative intellect" (Wittgenstein 1999, 38). The Christian faith is only for one human being "who needs infinite help" and "suffers infinite distress" (Wittgenstein 1999, 52).

We have briefly sketched some of Wittgenstein's annotations on religion because they offer a framework for understanding the (possible) place of religion in Schlick's philosophical investigations. The most valuable document is perhaps one of Schlick's late papers, in which his indebtment to Wittgenstein is clearly stated. Indeed, in his very influential article "Meaning and Verification" (published 1936, but already finished in 1934), Schlick declares himself to be an enthusiastic follower of Wittgenstein's verificationism, according to which the meaning of a proposition consists of "the method of its verification" (Schlick 2008, 712). Schlick's core argument is here that verification means properly "possibility of verification", to say "verifiability", but not "verifiable here now" or "being verified now." In Schlick's own words: "Verifiability means possibility of verification" (Schlick 2008, 717–718, 720). Furthermore, it is quite crucial to bear in mind that for Schlick the possibility of verification means *logical* possibility. The empirical verification concerns the truth of a proposition, for instance its accordance with the laws of nature, or lack thereof. By contrast, the meaning of a proposition, i.e. the question about its sense, may be posed only in the framework of the *logical* possibility of verification.

Schlick employs, among other things, a very interesting example to illustrate his verification principle. Once it is stated anew that there is "no unfathomable mystery in the world", the alleged hopeless *Ignorabimus* being therefore ungrounded (Schlick 2008, 726–727), Schlick poses a question at first sight wondering: is

[20] See also Wittgenstein (1998, 71): "'Wisdom is grey.' Life on the other hand & religion are full of color".

there life after death? The problem had been suggested to Schlick by Clarence I. Lewis, who regarded it as a classical metaphysical problem escaping any empirical verification.[21] Contrary to Lewis, Schlick claims that the hypothesis that nobody can attend his own burial is not sufficient to reject the belief in survival after death as purely metaphysical. In the case of the existence of the other side of the moon, at stake is a physical question that cannot be verified by immediate empirical means, but is answerable in principle when we indicate one or many logical possibilities of verification (for instance by inventing a powerful rocket, able to reach the other face of the moon). The same can be said in the case of survival after death. We can imagine a "supercelestial space" allowing us to formulate the hypothesis that such a very peculiar space can exist. Like all hypotheses, this space is not absolutely demonstrable; however, the impossibility of reaching or perceiving it is solely an empirical problem. Schlick's conclusion is a little surprising: "The hypothesis of immortality is an empirical statement which owes its meaning to its verifiability and it has no meaning beyond the possibility of verification" (Schlick 2008, 733).

Schlick likewise believed that such a conclusion was wholly in the spirit of Wittgenstein's (alleged) verificationism. One could have some doubts about Schlick's conviction. At any rate he was at least far from Wittgenstein's remarks on religious beliefs, which are nothing but beliefs arising *within* the limits of language, as Schlick suggests. For Wittgenstein, by contrast, they live, as we have stressed above, *beyond* the limits (or the cage) of language. Hence, how could they be verified?

But at this point one has also to ask whether Schlick was in full agreement with his Viennese partners in denouncing the nonsense of metaphysics (and religious beliefs). The Manifesto of scientific world-conception had proposed an unequivocal rule:

> If a metaphysician or theologian wants to retain the usual medium of language, then he must himself realize and bring out clearly that he is giving not description but expression, not theory but communication of knowledge, but poetry and myth. If a mystic asserts that he has experiences that live above and beyond all concepts, one cannot deny this. But he cannot talk about them, for talking implies capture by concepts and reduction to scientifically classifiable states of affairs (Carnap-Hahn-Neurath 2012, 82).

To be sure, this hard polemic against religion and theology was shared by Carnap. In that same year, 1929, he delivered a lecture devoted to God and the soul as pseudo-problems characterizing metaphysics and theology. Release from similar questions requires a strong criterion of verification which prescribes that every statement has to be reduced to perceptions (Carnap 2004, 59). Once this is done, it results that both metaphysics and theology are senseless. A year earlier Carnap had presented the same thesis in his masterwork, the *Aufbau*, though articulating it more in detail. It is worthy noticing that Carnap again refers to the celebrated *Ignorabimus*, but stresses yet another formulation of the question in what concerns the critique of

[21]Concerning the debate between Lewis and Schlick on verifcationism I allow myself to refer to (Ferrari 2019).

metaphysics. There are surely unsolvable problems, Carnap argues, but they do not belong to science; by contrast the very riddles are the riddles of life (*Lebensrätsel*), but they are not veritable problems, representing rather "situations of practical life" (Carnap 2003, 297: trans. lightly modified). This way of reasoning about life as a source of riddles finds its most clear expression in the Preface to the *Aufbau*. That recently metaphysical and religious trends have again gained an influential position within contemporary culture is undeniable, according to Carnap. The strategy which scientific philosophy must employ against this reactionary movement consists of recognizing that "we too have 'needs in our soul' (*Bedürfnisse des Gemütes*) in philosophy" (Carnap 2003, XVII, trans. modified),[22] but similar needs usually lying at the basis of metaphysic and religion must be converted into the scientific maxim of "clarity of concepts, precision of methods, responsible theses" (Carnap 2003, XVII).

It is questionable whether Schlick was wholly in agreement with Carnap, although he surely endorsed the style of scientific philosophizing (Ramharter 2020, 378, 392). But the criterion of verification as Carnap intended it as well as the consequences he thereafter drew by reducing any statement to perceptions had as result the pitiless refusal of religious beliefs (as Neurath, for his part, did). Unlike Carnap, it seems that Schlick cultivated a more moderate view of religious beliefs, admitting their significance for human life similar, at least in general terms, to Wittgenstein's stance. To put it in other words, Schlick could have at least subscribe to Wittgenstein's opinion: "I do not scoff at this tendency in man; I hold it in reverence" (Waismann 1979, 118).

3.4 Concluding Remarks

Throughout his life Schlick remained a worldly-minded philosopher. In his mind, the sense of life lies both in our knowledge of the world and in our moral action. The pleasure brought by both these cornerstones of culture can wholly satisfy our need for happiness, once it is recognized that modern society has in any case to be protected from evil as well as from the loss of freedom. By contrast, faith and religious feeling belong to a realm Schlick supposed worth addressing from another perspective, though not in the sense of the largely disputed *Ignorabimus*. At stake is rather, in Schlick's late philosophy of culture, the guiding role of ethics for contemporary society in an age of crisis.

Schlick was very sensitive, indeed, to the dramatic events of his time. He was witness to the risc of Nazism in Germany and visited Italy many times in the era of the Fascist regime. The "spiritual situation of the time", as Karl Jaspers called it at the beginning of the 1930s, was cause for increasing worry to Schlick

[22]The expression "Bedürfnisse des Gemütes" is drown from Lotze's *Mikrosmos*, a standard work in German philosophy around the 1860s.

too (Jaspers 1979). But unlike Jaspers, whose book Schlick quotes elsewhere, his diagnosis was not deeply pessimistic. To be sure, he was fully convinced that great dangers menaced Western civilization, stretching from the ideology of *Blut und Boden* to the power of technology. Nonetheless, as paradigmatic endorser of liberalism, democracy and pacifism, Schlick had great faith in the ethical rescue of European mankind. "Only goodwill – he says in his draft on nature and culture – can be the ultimate principle of union; a State which has been formed according to this principle is the true *Civitas Dei*, and all the States resting on other principles are the *Civitas Diaboli*" (Schlick 1952, 107). A similar stance is assumed by Schlick in the face of technology too, because evil is caused not by technology itself, but by its abuse (Schlick 1952, 123). Like Cassirer, Schlick too believed that ethics has to be the guideline of technology, according to the moral principles lying at the core of modern civilization (Cassirer 2012). But Schlick looked at a similar task without any reference to a possible function of religion or to religious consciousness as a basis for a renewed idea of humanity. If Schlick was well aware of "the shadow of tomorrow" Johan Huizinga spoke of in 1936, his view of the destiny of the West was oriented towards a *Civitas Hominis* founded on ethical principles (and much less in the sense of Spengler's decline of the West).[23]

But, in the end, what remains of religion within Schlick's whole work? Schlick seems to reveal two different attitudes towards religion (or, one could be tempted to say, "irreligion"). On the one hand, he endorses a verificationism in the sense, broadly speaking, of Wittgenstein or, to some extent, of Carnap at least until the end of the 1920s. The question about death and immortality seems to justify such an interpretation, although Schlick in his late paper we have considered is dealing rather with logical vz. possible verifiability. On the other hand, however, Schlick is open to another perspective, according to which religious feelings really are experienced by man, albeit without any commitment to whatsoever historical form of religion.

This twofold upshot of Schlick's remarks on religion certainly deserves more detailed discussion, having to do not only with Schlick, but regarding rather the cultural implications of scientific philosophy or, more specifically, of Viennese empiricism. May be, the concluding (and provisional) remark of this story could be assigned to Schlick himself. In his estate there are, among very many other things, two aphorisms illustrating clearly enough Schlick's double face concerning religion. In the former aphorism Schlick says:

> Once philosophy asked about the primal ground of being, about God's existence, about immortality and liberty of the soul, about the sense of the world and the guiding principles of action – but now we ask about nothing at all other than "what do you mean properly?" (Inv. A. 175. 067).

By contrast, in the latter aphorism Schlick tell us:

---

[23]Schlick is indeed very critical of Spengler, who was by contrast one of Wittgenstein's favorite authors (Schlick 1962, 17; see Wittgenstein 1998, 16).

> The very core of every religion is reverence (*Ehrfurcht*); everything else is secondary and irrelevant. But do not talk about the replacement (*Ersatz*) of religion or churches – religion in itself, that is to say reverence, must be conserved and cannot be replaced by anything; the same is valid regarding churches, if these have the meaning of meeting places for both contemplation and reverence, which nobody can get into without leaving out any feeling of either hate or humiliation, everybody having rather decided to think of this place only with peace, good and light, pure thoughts. How I missed the sound of the Sunday bells as I was in the United States! (Inv. A. 175. 068).

Is perhaps this longing for "Sunday bells" that is still worthy of religious reverence for Schlick?

References

Bayertz, Kurt. 2007. "Das Rätsel gib es nicht". Von Emil Du Bois-Reymond über Wittgenstein zum Wiener Kreis. In *Weltanschauung, Philosophie und Naturwissenschaft. Vol. 3. Der Materialismus-Streit*, ed. Kurt Bayertz, Myriam Gerhard, and Walter Jaeschke, 183–203. Hamburg: Meiner.

Carnap, Rudolf. 1963. Intellectual Autobiography. In *The Philosophy of Rudolf Carnap*, ed. Paul A. Schilpp, 3–84. La Salle: Open Court.

———. 2003. *The Logical Structure of the World and Pseudoproblems in Philosophy*, translated by Rolf A. George. Chicago/La Salle: Open Court.

———. 2004. *Scheinprobleme in der Philosophie und andere metaphysikkritische Schriften*, ed. Thomas Mormann. Hamburg: Meiner.

Carnap, Rudolf, Hans Hahn, Otta Neurath. 2012. *The Scientific World-Conception. The Vienna Circle*. Translated by Thomas Uebel. In *Wissenschaftliche Weltauffassung. Der Wiener Kreis*. Hrsg. von Verein Ernst Mach (1929). Reprint der Erstausgabe. Mit Übersetzungen ins Englische, Französische, Spanische und Italienische. Herausgegeben mit Einleitungen und Beiträgen von Friedrich Stadler und Thomas Uebel. Wien-New York: Springer.

Carus, A.W. 2007. *Carnap and Twentieth-Century Thought. Explication as Enlightenment*. Cambridge: Cambridge University Press.

Cassirer, Ernst. 2012. Form and Technology. In *Ernst Cassirer on Form and Technology. Contemporary Readings*, eds. Aud Sissel Hoel and Ingvild Folkvordpp, translated by Wilson McClelland Dunlavey and John Michael Krois, 15-53. New York: Palgrave Macmillan.

Dathe, Uwe. 2000. Rudolf Eucken – ein Gegner des Monismus und Freund der Monisten. In *Monismus um 1900. Wissenschaftskultur und Weltanschauung*, ed. Paul Ziche, 41-59. Berlin: Verlag für Wissenschaft und Bildung

Du Bois-Reymond, Emil. 1974. *Vorträge über Philosophie und Gesellschaft*, ed. Siegfried Wollgast. Hamburg: Meiner.

Eucken, Rudolf. 1918. *Der Sinn und der Wert des Lebens*. sechste Auflage ed. Leipzig: Quelle & Mayer.

Feigl, Herbert. 1937/1938. Moritz Schlick. *Erkenntnis* 7: 393–419.

Ferrari, Massimo. 2019. Lewis and Schlick. Verificationism between Pragmatism and Logical Empiricism. *European Journal of Pragmatism and American Philosophy* XI (1). https://journals.openedition.org/ejpap/1516.

———. 2021. Moritz Schlick zwischen Schopenhauer und Nietzsche (forthcoming).

Graf, Friedrich W. 1997. Die Positivität des Geistigen. Rudolf Euckens Programm neoidealistischer Universalintegration. In *Kultur und Kulturwissenschaften um 1900. II. Idealismus und Positivismus*, ed. Gangolf Hübinger, Rüdiger vom Bruch, and Friedrich W. Graf, 53–85. Stuttgart: Steiner Verlag.

Guyau, Jean-Marie. 1885. *Esquisse d'une morale sans obligation ni sanction*. Paris: Alcan.

———. 1906. *L'irréligion de l'avenir. Étude sociologique*. Paris: Alcan.

Hübinger, Gangolf. 1997. Die monistische Bewegung. Sozialingenieure und Kulturprediger. In *Kultur und Kulturwissenschaften um 1900. II. Idealismus und Positivismus*, ed. Gangolf Hübinger, Rüdiger vom Bruch, and Friedrich W. Graf, 246–259. Stuttgart: Steiner Verlag.
Haeckel, Ernst. 1919. *Die Welträtsel. Gemeinverständliche Studien über monistische Philosophie*. 11. Auflage. Leipzig: Kröner Verlag.
Iven, Matthias. 2008. *Moritz Schlick. Die frühen Jahre (1882–1907)*. Berlin: Parerga.
Jaspers, Karl. 1979. *Die geistige Situation der Zeit*. Berlin/New York: de Gruyter.
Jodl, Friedrich. 1911. *Der Monismus und die Kulturprobleme der Gegenwart*. Vortrag gehalten auf dem Ersten Monisten-Kongress am 11. September 1911 zu Hamburg. Leipzig: Kröner.
Lightman, Bernard. 1987. *The Origins of Agnosticism. Victorian Unbelief and the Limits of Knowledge*. Baltimore: The John Hopkins University Press.
Lübbe, Hermann. 1974. *Politische Philosophie in Deutschland. Studien zu ihrer Geschichte*. München: Deutscher Taschenbuch Verlag.
Mormann, Thomas. 2010. Zwischen Weisheit und Wissenschaft. Schlicks weites philosophisches Spektrum. *Grazer Philosophische Studien* 80: 263–285.
Pyle, Andrew, ed. 1995. *Agnosticism. Contemporary Respones to Spencer and Huxley*. Bristol: Thoemmes Press.
Ramharter, Esther. 2020. Die strenge Trennung. Religion im Umfeld des Wiener Kreises. In *Die Geburt der Moderne aus dem Geist der Religion, Weltanschauung und Moderne in Wien um 1900*. Ed. Rudolf Leeb and Astrid Schweighofer, 377–398. Frankfurt am Main: Vandenhoeck & Ruprecht.
Schlick, Moritz. 1939. *Problems of Ethics*. Trans. David Rynin, New York: Prentice-Hall.
———. 1952. *Natur und Kultur*. Aus dem Nachlaß hg. von Josef Rausche. Wien: Humboldt Verlag.
———. 1962. *Aphorismen*, ed. von B. Hardy Schlick. Wien: Im Selbstverlag des Herausgebers.
———. 1974. *General Theory of Knowledge*. Trans. by Albert E. Blumberg, with an introduction by Albert E. Blumberg and Herbert Feigl, Wien: Springer.
———. 1978. *Philosophical Papers*, vol. I: *(1909–1922)*, eds. Henk L. Mulder and Barbara F.B van de Velde-Schlick, trans. P. Haeth. Dordrecht/Boston/London: Reidel.
———. 2006. *Kritische Gesamtausgabe*. Abteilung I, vol. 3. *Lebensweisheit. Fragen der Ethik*, ed. Mathias Iven, Wien/New York: Springer.
———. 2008. *Kritische Gesamtausgabe*. Abteilung I, vol. 6, *Die Wiener Zeit. Aufsätze, Beiträge, Rezensionen 1926–1936*, eds. Johannes Friedl and Heiner Rutte. Wien/New York: Springer.
———. 2012. *Kritische Gesamtausgabe*. Abteilung II, vol. 5.1, *Nietzsche und Schopenhauer (Vorlesungen)*, ed. Matthias Iven. Dordrecht/Heidelberg/London/New York: Springer.
Stadler, Friedrich. 2001. *The Vienna Circle. Studies in the Origins, Development, and Influence of Logical Empiricism*. Wien/New York: Springer.
Tuboly, Adam Tamas. 2017. Carnap's *Weltanschauung* and the *Jugendbewegung*. The Story of an Omitted Chapter. In *Integrated History and Philosophy of Science. Problems, Perspectives and Case Studies*, ed. Friedrich Stadler, 129–144. Cham: Springer.
Uebel, Thomas. 2000. *Vernunftkritik und Wissenschaft: Otto Neurath und der erste Wiener Kreis*. Wien/New York: Springer.
Waismann, Friedrich. 1938. Vorwort. In Moritz Schlick. *Gesammelte Aufsätze*, VII–XXXI. Wien: Gerold & Co.
———. 1979. *Wittgenstein and the Vienna Circle. Conversations recorded by Friedrich Waismann*, ed. Brian McGuinness, translated by Brian McGuinness and Joachim Schulte. Oxford: Blackwell.
Wittgenstein, Ludwig. 1965. A Lecture on Ethics. *The Philosophical Review* 74 (1): 3–12.
———. 1998. In *Culture and Value. A Selection from the Posthumous Remains*, ed. Georg Heinrik von Wright. Oxford: Blackwell.
———. 1999. In *Conversations on Aesthetics, Psychology and Religious Belief*, ed. Cyril Barrett. Oxford: Blackwell.
Ziche, Paul, ed. 2000. *Monismus um 1900. Wissenschaftskultur und Weltanschauung*. Berlin: Verlag für Wissenschaft und Bildung.
Zilsel, Edgar. 1937. Moritz Schlick. *Die Naturwissenschaften* 25: 161–167.

Chapter 4
The "Continuous Line from the Formulations of the Magicians to the Formulations of the Sociologists". Otto Neurath on the Anthropology of Magic and Religion

Marco Brusotti

In 1921, Neurath ascribed to Hermann Keyserling a rare "cosmopolitanism that does not want to erase differences": The *Reisetagebuch eines Philosophen* (Keyserling 1919) "strives to present mankind as a whole, not as a whole that can be recognized as such from the outside, but as a whole that each of us can experience to a certain degree within ourselves. Like Proteus, the traveling westerner becomes an Indian, the Indian a Chinese, the Chinese a Japanese, the Japanese a westerner again. The journey around the world is a journey around your own soul and the diversity of your being" (Neurath [1921a], in GS 1, 199).[1] This diversity-friendly cosmopolitanism may seem recklessly overconfident about the Protean nature of human beings in general and westerners in particular. Are we really capable of experiencing mankind – as a whole and in its various cultural specifications – "to a certain degree within ourselves" (Neurath 1921a, in GS 1, 199)?

This optimistic attitude may come as a surprise because Neurath is a notoriously fierce critic of *Einfühlung* and *Verstehen*. Rather than relying on empathy, the social-behaviourist, though not a reductionist,[2] is keen on finding empirically testable correlations that allow for predictions. Thus, Neurath's point of view at the beginning of the thirties seems far removed from that of *Humankind* [*Menschheit*]

[1] I have used English translations to the extent that they are available, modifying them if necessary, especially when the existing English edition is abridged (e.g., *Empirical Sociology*). I myself have translated the passages from those among Neurath's writings of which there is no English translation. The abbreviation GS stands for Neurath's *Gesammelte Schriften* (= Neurath 1981, Neurath 1991, Neurath 1998).

[2] On Neurath's non-reductionist conception of 'physical social sciences', see Uebel 2007. On Neurath's sociology, cf. also Uebel 1997 and Uebel 1991, esp. Fleck 1991, [Paul] Neurath 1991.

M. Brusotti (✉)
Dipartimento di Studi Umanistici, Università del Salento, Lecce, Italy
e-mail: marco.brusotti@unisalento.it

E. Ramharter (ed.), *The Vienna Circle and Religion*, Vienna Circle Institute Yearbook 25, https://doi.org/10.1007/978-3-030-76151-6_4

(1921), written almost a decade earlier. Behind his striking programmatic statements against *Einfühlung*, however, lies a more nuanced and sophisticated position that becomes more sharply contoured if we take account of his concrete engagements with anthropological and sociological literature. His little-known critical confrontation with Bronislaw Malinowski shows that he appreciates the anthropologist's practical virtue of empathy, even if social-behaviourists do not seem to be participant observers like the eminent field researcher.

The Malinowski review (Neurath 1931c) shows that the new developments in anthropology over the last decade had not entirely escaped his attention. In 1931, however, he was, in certain respects, still more in tune with the declining evolutionary tradition than with the anthropologists of his generation.[3] The task of understanding foreign cultures, which later was to become so urgent, had not yet been the focus of attention in the anthropology with which he was best acquainted. Rather than understanding foreign cultures in the plural, the central issue in the genetic framework of nineteenth-century anthropology had been reconstructing and explaining an alleged 'primitive' culture in the singular. Living non-European peoples (especially the so-called *Naturvölker* or, still more disparagingly, the 'savages', 'primitives', etc.) were thought to be less distant from this purely conjectural 'primitive culture' than (urbanized, 'civilised') Europeans were. That 'we' (the West, the civilized, etc.) have religion (faith) and science (knowledge) was taken for granted, but what about 'them'? Did 'they' have anything other than superstition? Did the presumed ancient 'primitive culture' know anything that deserved to be called 'religion'? There was no generally shared position in the deeply ethnocentric but controversial discussion of this range of issues, but, rather, a common background for heated debates.

They came to revolve around the concept of 'magic'. Throughout the present paper, the word 'magic' is intended *de dicto* and not *de re*: not as referring to a definite sociological reality, to an object of anthropological research to be found 'out there', but as a (now outdated) category of western thought. With its long history as a polemical label in various interreligious or interconfessional controversies and/or as a term for marginal phenomena on the fringe of 'modern society', the word 'magic' involved a host of misunderstandings that persisted in the anthropological debate well after the end of the nineteenth century.[4] The word 'magic' quite often stood for what westerners found most difficult to understand in 'primitive culture'. As the epitome of exotic and shocking customs and beliefs, 'magic', even more than religion, became the hermeneutic sticking point.

---

[3]In the present paper, the terms 'evolutionist' / 'evolutionary' stand for the anthropological tradition from E. B. Tylor to J. G. Frazer, not for Darwin's theory of evolution. The difference is important because nineteenth- and early twentieth-century 'evolutionary' anthropology was permeated with pre- and extra-Darwinian views (think of Herbert Spencer). On its complex relationship to Darwin's theory, cf. Stocking 1987.

[4]For a critical discussion of the concept of magic and the cultural roots of this analytical category in the European tradition, see Tambiah 1990.

Neurath firmly rejects theories that make this hermeneutic task seem too demanding. He insists that the difference between westerners and the peoples then studied by anthropologists should not be overstated. "Many primitive life forms" would cease to seem strange to us if we became "used to seeing our own way of life 'objectively'" (Neurath 1931c, 100). "One often deliberately presents magical action as being stranger than it is" (Neurath 1930a, in Neurath 1983, 34; GS 1, 373). Theorists should refrain from misleading explanations that exaggerate the strangeness of the actions and procedures called 'magical'. But are they *correctly* called so? Does Neurath intend to give a theory of 'magic'? Or does he, rather, venture to suggest that the use of this very word 'magic' belongs to the means by which anthropologists present the agents and their actions as stranger than they actually are? There is a certain oscillation in his texts. The tendency to give a positive evolutionary theory of magic may prevail, even if he only apparently takes the concept of 'magic' for granted; although his main aim is not to do altogether without it, he does put forward arguments that call the anthropological concept of 'magic' itself into question.

Lucien Lévy-Bruhl's theory of the two mentalities provides a peculiar justification for the more general disposition to present the so-called 'primitives' as being stranger than they actually are: The French philosopher sees a deep divide separating their 'mystical' or 'prelogical' mentality from our 'logical' mentality. As Quine afterwards, Neurath rejects Lévy-Bruhl's theory. Against it, he sets two quite different views of anthropology. One is Malinowski's functionalism, the cutting edge at the time, though Neurath finds Frazer's quite old-fashioned anthropology even more congenial.[5] It may seem strange that, in comparison to Lévy-Bruhl, the author of *The Golden Bough* should play the role of the more inclusive anthropologist. Frazer is certainly not less Eurocentric, and his paternalistic standpoint does not represent a viable alternative to Lévy-Bruhl's sociological approach, which is in many ways more modern. Neurath, however, does not count the Scottish anthropologist among the scholars who overstate the difference between modern westerners and the so-called 'primitives'.

Rather, to show that 'magic' only apparently goes back to a completely different 'mentality', he builds on Frazer's analogy between magic and technology. This analogy, a cornerstone of the well-known developmental scheme (magic, religion, and science), involves a principled distinction between magic and religion.

Not least because Neurath embraces the analogy between magic and technology and the principled distinction between magic and religion, his reflections on this range of subjects are in many respects antipodal to Wittgenstein's much better-known sustained criticism of *The Golden Bough*. Neurath and Wittgenstein share the assumption that 'magic' does not go back to anything like a completely different

---

[5]James G. Frazer (1854–1931) and Lucien Lévy-Bruhl (1857–1939) are more than a couple dozen years older than Neurath (1882–1945), whereas Alfred R. Radcliffe-Brown (1881–1955) and Bronislaw Malinowski (1884–1942) belong to his generation. On Frazer, cf. Ackerman 1987, Fraser 1990a, Fraser 1990b. On Neurath and Lévy-Bruhl, see below p. 77 ff., on Neurath's late acquaintance with social functionalism, p. 103, n. 100.

'mentality'. However, Wittgenstein rejects Frazer's analogy between magic and technology and deconstructs his sharp distinction between magic and religion (cf. Brusotti 2011, 351 f.).

The two Viennese philosophers deal with these issues almost simultaneously. Neurath gives a lecture on "Magic and Technology" in January 1931, and Wittgenstein begins writing down the first of his remarks on *The Golden Bough* in June of the same year. This shows the Scottish anthropologist's continuing popularity not only in broad circles of readers but also among philosophers. Indeed, the comparison with Neurath indicates that it is not so strange that Wittgenstein picked up *The Golden Bough* as late as 1931. (He will keep reading it well after the early thirties.) The *Abridged Edition* had been translated into German just a few years earlier, in 1928.[6] And in 1931, the aging Frazer, who suddenly went blind in May of that year, was not yet an 'embarrassment' for his colleagues; on the contrary, he was a publicly revered and celebrated icon, even if his approach was largely outdated – at least for field workers and the new generation of social anthropologists.

In a 2011 paper (Brusotti 2011), I compared and contrasted Neurath's and Wittgenstein's engagement with Frazer (and Spengler).[7] The more general conflict between their philosophical approaches is evident in the widely divergent stance they take to Frazer's developmental scheme and to other main tenets of *The Golden Bough*. The present text, which does not go into Wittgenstein, deals with Neurath's quest for a non-metaphysical sociology of magic and religion and takes into account a broader spectrum of anthropological positions known to him in the late twenties and early thirties.

4.1 "Marxism of a Jesuit"

Even if it is mandatory for a sociology of religion to move away from metaphysical preconceptions, it is not, in principle, impossible to conduct non-metaphysical empirical research on religion from a religious point of view. Neurath advances this methodological thesis in a better-known review essay from 1931. Under the

[6]Which edition of *The Golden Bough* might Neurath have read: the monumental twelve-volume third edition (1911–1915) or the popular *Abridged Edition* (1922)? The German translation of the latter (Frazer 1928) was recently available. This could be why Frazer's thought figures are prominent in Neurath's writings from 1928 onwards. By the way, even if there is no indication that Wittgenstein ever used the German translation, the appearance of the latter may have contributed to his decision to read the original English text in 1931.

[7]I don't think that Neurath's debt to Frazer had been noticed before Brusotti 2011. Neurath's reflections on "Magic and Technology" were practically forgotten, apart from a few specialists. A limit of my earlier contribution, however, is that it focuses almost exclusively on the author of *The Golden Bough*. This is also true of a later study that sees a connection between Neurath and Frazer. Cf. Josephson-Storm 2017, chapter 9, 245–258. – I have treated Wittgenstein's remarks on Frazer's *The Golden Bough* in Brusotti 2014 as well as in a number of essays; see, for example, Brusotti 2016a, 2016b.

well-chosen heading "Marxismus eines Jesuiten [Marxism of a Jesuit]" (Neurath 1931d), he discusses *Scholastik, Puritanismus und Kapitalismus* [Scholasticism, Puritanism and Capitalism] by Johann Baptist Kraus. Despite the glamourous title, Neurath does not really mean that the Jesuit historian is a Marxist, but only that he "methodologically <seeks> the middle line between Max Weber and Karl Marx" (Neurath 1931d, in GS 1, 404) and *"can successfully represent certain sides of the materialistic conception of history against Weber*" (Neurath 1931d, in GS 1, 405).

Following up on O. Spann, incidentally, one of the sociologists opposed by Neurath, Kraus claims that Weber does not make a clear enough distinction between "interpretive understanding" and "causal explanation", and, by equating them to one another, misunderstands his own "meaningful interpretation" (Kraus 1930, 12) as a causal explanation.

Neurath agrees with Kraus that the method employed by Weber is not able to provide causal explanations. In "Marxism of a Jesuit", the sociologist is the main target, and Neurath does not exactly apply the principle of charity when he understates the difference between Weber and, say, Werner Sombart.[8] In Neurath's eyes, a Jesuit like Kraus, who openly professes all Catholic dogmas, rather succeeds in keeping away from metaphysical preconceptions than 'bourgeois' sociologists like Weber and Sombart, "who can express their sympathy for the divine and the spiritual, which is a refined form of the divine, only by penetrating everything and anything with the doctrine of the 'power of the spirit' [...]" (Neurath 1931d, in GS 1, 401). Maybe we see the Jesuitism of a Marxist here. For Neurath follows Kraus. In *Sociology in the Framework of Physicalism*, the argument that Weber "endowed the spirit with a kind of 'magical' power" (Neurath 1931e, in Neurath 1983, 83; GS 2, 556) is ascribed explicitly to the Jesuit scholar. The objection, which, as Neurath himself concedes, forces Weber's view "into the materialistic garb", is that the latter attributes to the 'spirit of Protestantism' the 'magical' power to explain capitalism, i.e., to "the words and writings of the theologians" the potential to "change the habits of an age".[9] While, according to Weber, "[t]he ideas shape the economy"

[8]According to Turk, Neurath tries "to distinguish his own 'scientific sociology' from Weber's *verstehende* Sociology by placing Weber squarely in the tradition of the idealist neo-Kantians, drawing a line from Dilthey through Windelband and Rickert, then on to Weber himself." "Neurath misstated Weber's position": Even if Neurath acknowledges Weber's scientific level, he interprets him as a mere continuator who simply adopts "the neo-Kantian division between the nomothetic and the idiographic" rather than trying "to circumvent and transcend" it (Turk 2018, 79; cf. the whole chapter "Out of the German Historical School: The divergent paths of Otto Neurath and Max Weber", in Turk 2018, 61–82).

[9]"This is approximately the view that is quite opposed to that of Max Weber. The 'effect' of the teachings of individual priests is thereby estimated to be very low, whereas with Weber, if one forces his view into the materialistic garb, it would be claimed that the words and writings of the theologians change the habits of an age – without explaining why the doctrines and writings themselves grew up then and there [...]" (Neurath 1931a, in Neurath 1973, 384; in GS 1, 494) – On Kraus cf. Neurath 1931e, in Neurath 1984, 83–84; GS 2, 556–557. With reference to Weber, a late text still mentions the "interesting criticism, e. g., by Roman Catholic writers" (Neurath 1944, 48, n. 24) like Kraus (and Fanfani).

(Neurath 1931d, in GS 1, 401), for Kraus it is the other way round: He traces puritan religiosity back to the economic habits of traders and lenders. Neurath agrees with him that "in the economic field, practice must be given priority over theories" (Neurath 1931d, in GS 1, 404; quotation from Kraus 1930, 247, n. 292) and that Weber fails to do this.

Is Neurath's appreciation of the Jesuit historian due only to the specific polemical context and to the contingent circumstance that Kraus provides him with insights and arguments against Weber's approach? I do not think so. For the same judgment is also found, more generally formulated, in *Empirical Sociology*: A non-metaphysical sociology does not necessarily have to be a monopoly of "untheological research" (Neurath 1931a, in Neurath 1973, 391; GS 1, 502).

> Reports by missionaries about primitive peoples are often difficult to use, because in the description of what they observed the concepts they used colour a great deal, and because they failed to record much that would be important for a theory uninfluenced by theology. The report of such a missionary is in itself already a theoretical result. One must, so to speak, disintegrate the products of alien theories in order to reassemble the parts afresh and to complete them as best one may.
>
> Still, it can occasionally happen that exactly the theological attitude will register with special precision what untheological research will not regard as important until a later stage. Christian missionaries in particular have in recent years achieved important ethnological results quite free from metaphysics. (Neurath 1931a, in Neurath 1973, 391; GS 1, 502)

In the last decades of the nineteenth century, and the first of the twentieth century, missionary ethnographies were, indeed, far from negligible. In the German-speaking area, the so-called Vienna School of Ethnology deserves mention. The cultural-historical direction of the Steyler missionaries (Societas Verbi Divini, S.V.D.) Wilhelm Schmidt and Wilhelm Koppers dominated German-language ethnology until the early fifties.[10] Today's anthropologists are quite dismissive of their 'extreme diffusionism'.[11] Current scientific standards are stricter and the bias of missionary ethnographies is even more evident than at Neurath's time. Nevertheless, his appreciative words are no less justified than his criticism. What deserves to be emphasized is that he strives for balance on this point: Religious studies and religion are not categorically mutually exclusive. Against metaphysical assumptions, Neurath is very strict. However, with regard to the difficult and tense

[10]Schmidt (1868–1954) founded the *Institut für Völkerkunde* at the University of Vienna in 1929. The Institute was headed by Koppers (1886–1961), who was professor of ethnology at the same university between 1928 and 1938. On their 'extreme diffusionism', see Harris 1968, 379–392; Rössler 2007, 10; Madey 1992, 509–511. Among the many missionaries, only three should be mentioned here: Carl Friedrich Theodor Strehlow, the head of the Finke River Mission in Hermannsburg (Australia), and Diedrich Westermann, the author of the grammar of the Ewe language published in 1907, both Lutherans, as well as the Anglican priest Robert Henry Codrington.

[11]To the diffusionists belongs also the museum ethnologist Fritz Graebner, discussed below, p. 67 ff., who, however, was neither a clergyman nor active in Vienna. While Wilhelm Schmidt acknowledged him (cf. Harris 1968, 383), Malinowski led a "crusade against Graebnerian 'museum moles'" (Harris 1968, 520).

relationship between religion and *Religionswissenschaft*,[12] his position is much more moderate than sweeping preconceptions about the Vienna Circle in general and him in particular would lead one to expect.

The late twenties and early thirties are the years of Neurath's "closest association with Marxism" (Turk 2018, 79). He suggests that a Marxist attempt to contextualize Marxism is as legitimate as Kraus' historical project; "a Marxist can show how Marxism depends on the basis of production without having to doubt the content of Marxism" (Neurath 1931d, in GS 1, 405).[13] To bring this line of argument into play in favour of Marxism is a main concern of Neurath. However, the analogy to Kraus applies only up to a point. The Jesuit is not a 'Marxist' in studying his own confession: His book is the critical study of a confession by a scholar who belongs to another. Neurath himself remarks that it would be more difficult (although not impossible) to study one's own confession or religion in the same detached way.

4.2 The "Strictly Scientific Anti-metaphysical Anthropologists"

Needless to say, both *Einfühlung* and *Verstehen* belong to the 'dangerous', 'dubious', misleading words Neurath would include in his famigerate *Index verborum prohibitorum*.[14] "With Weber, time and again, the empathic immersion appears in place of science. Over this poetic activity there is no test, no control, nor does it belong in a scientific account" (Neurath 1931a, in Neurath 1973, 357; GS 1, 463). Neurath denies that the humanities have a peculiar organ, namely, '*Einfühlung*', '*Verstehen*'. "Empathy is not a scientific tool, nor is a special 'understanding' a special aspect of 'humanistic' science" (Neurath 1931a, in Neurath 1973, 361; GS 1,

[12]The term *Religionswissenschaft* had already been in use for a long time, but Neurath does not employ it.

[13]The 'bourgeois Marxist' Karl Mannheim denies just this. Marxists, he maintains, are not able to apply Marxist sociology to Marxism itself (cf. Neurath 1929b, in GS 1, 350). What matters to Neurath is that this "application of its own method to itself" (Neurath 1929b, in GS 1, 355) is possible and not self-defeating. He takes account of the difference between context of discovery and context of justification, even if he does not explicitly endorse Reichenbach's distinction. (According to Nemeth (2007, 292), Neurath respects this distinction as one between descriptive and normative inquiries.) An historical contextualization does not cancel Marxism: "Dependence on the historical situation does not mean deviation from the truth in any direction" (Neurath 1931d, in GS 1, 405). Like Mannheim, Neurath makes sociologists and their theories the subject of sociology. What sets him apart from the 'bourgeois Marxist', however, is that the sociology of knowledge that Neurath advocates investigates knowledge's conditions of origin, without reducing it to them.

[14]On the "elimination of *Einfühlung*", see below p. 76 f. Reisch 1997 includes only 'empathy' in his reconstruction of Neurath's *index*. However, in both passages given by Reisch, *Verstehen* is mentioned, too. Cf. Neurath 1931e, 70–71. On Neurath's critique of the concept of *Einfühlung*, cf. among others Neurath 1931a, in GS 1, 463–464, 468, 472; Neurath 1932, in GS 2, 567.

468). The concept of *Einfühlung* must not be invoked to mark a great divide between natural and social sciences. The strict dualism of *Geisteswissenschaften* and natural sciences "is, in the last resort, *a residue of theology*" (Neurath 1931e, in Neurath 1983, 69; GS 2, 544). Even if ungenerous towards Weber, this reproach can be rightfully directed against Spann or Sombart. In any case, Neurath's thesis is radical: There is no distinguishing characteristic of the humanities, and nothing justifies the strict separation between the 'mental' and the natural sciences (cf. Neurath 1931e, in Neurath 1983, 68–69; GS 2, 543). "Sociology is also a natural science, there are only natural sciences at all!" (Neurath 1930b, 718).[15]

The word '*Einfühlung*' may be a top candidate for the *index*, and Neurath's prevailing policy may consist of avoiding a term rather than acting as "a terminological ropedancer" (Neurath 1941; on 'ropedancing', cf. Reisch 1997), but he concedes that '*Einfühlung*' could still be used without leaving "the physicalist sphere" (Neurath 1931e, in Neurath 1983, 74; GS 2, 548) as long as what it stands for is not expected to have "some peculiar magical power beyond ordinary induction" (Neurath 1931e, in Neurath 1983, 73; GS 2, 548). This means that empathy should be clearly subordinated to the concern to find empirically testable lawlike correlations that enable scholars to make forecasts. Inspired by anthropology rather than laboratory psychology,[16] Neurath's social behaviourism is laxer than Watson's classic behaviourism, but still has the ambitious task of finding "lawful connections, *correlations*" (Neurath 1932, in GS 2, 566). The "*strictly scientific anti-metaphysical anthropologists*" should be able to find correlations "between mode of production, social order, cult, literature of a people" (Neurath 1932, in GS 2, 566)[17]; sociologists investigate, "for example, under which condition there is matriarchal right, ancestor worship, agriculture and other phenomena" (Neurath 1931e, in Neurath 1983, 77; GS 2, 551). Which contemporary trend did fit in with the idea that sociology and even anthropology should look for this sort of complex correlation? Is this what anthropologists were actually doing? Was Neurath still referring to the old comparative method? Or was he aware of the then-new functionalist anthropology? And to what extent?

---

[15]"That is, *sociology is not a 'mental science'* nor a 'science of mind' (Sombart), [...] but *as social behaviourism* sociology *is part of unified science*." ["Das heißt: *Soziologie ist nicht eine 'Geisteswissenschaft'* oder 'Geistwissenschaft' (Sombart), [...] sondern sie *ist als Sozialbehaviorismus ein Teil der Einheitswissenschaft*."] (Neurath 1931e, in Neurath 1983, 71; GS 2, 545). The late Neurath himself (1944, 8) translates *Geisteswissenschaften* as "mental sciences" (in quotation marks).

[16]"The external perspective of Neurath's behavioristics has more in common with that of the ethnologist than the classical behaviourists" (Nemeth 2007, 292). Here, Nemeth comments on the following passage: "We have fine questionnaires as far as preliterate tribes are concerned but hardly any when we try to ask sociologists how they themselves behave in arguing and writing" (Neurath 1944, 43). In the forties, these questionnaires, as Rivers had edited them some time earlier, were still used in area studies. See below p. 95.

[17]Neurath could find correlations "which are accessible to a clean mathematical formulation" in the book by Alfredo Niceforo he reviewed in 1930.

4.3 "Mother Right and Magic". Neurath (and Russell) Reading Malinowski's *Sexual Life of Savages*

A short text, probably known only to specialists, can help us answer this question. Like "Marxism of a Jesuit", "Mother Right and Magic" is not intended primarily for an academic audience.[18] The review discusses a book by one of the key figures of twentieth-century anthropology. Bronislaw Malinowski's groundbreaking fieldwork on the Trobriand Islands resulted in numerous books and essays. Among them, three monographs are particularly outstanding. The first and best known is *Argonauts of the Western Pacific* (Malinowski 1922). Together with Alfred Radcliffe-Brown's *The Andaman Islanders*, it marked a (not merely) symbolic turning point in the history of anthropology. *Coral Gardens and Their Magic* (Malinowski 1935) is the last of Malinowski's most relevant Trobriand monographs. Neurath reviews the German translation of the middle piece of the triad, *The Sexual Life of Savages in North-Western Melanesia* (Malinowski 1929a; Malinowski 1929b).

Why should a review of *The Sexual Life of Savages* be called "Mother Right and Magic"? The heading may mislead the reader to think that the correlation between mother right and magic is a thesis advanced by Malinowski. However, this is not the case. Strangely, this correlation is not explained in any detail anywhere; the review simply hints that 'magic' is highly valued in the matrilinear society that Malinowski describes. The passage that may come closest to an explanation merely lists three topics important for Neurath (mother right, production, magic) and claims that they play a role in Malinowski's book.[19] Actually, the review title corresponds to a developmental hypothesis formulated by Neurath himself in *Empirical Sociology*: (1) here a strange just-so story "explains" "the origin of agriculture"; (2) from this "production order [Produktionsordnung]", the author derives first "mother right" as a social form and then "magic" as a fixed habit.[20] Neurath's genealogy of soil

[18]Neither text was published in a scientific or philosophical journal. "Mother Right and Magic [Mutterrecht und Magie]" appeared in *Arbeit und Wirtschaft* [Labour and economy], a still existing journal, whose general purpose, then as now, is to deal with economic issues from the perspective of workers and employees. "Marxism of a Jesuit" was published in *Der Kampf: sozialistische Revue* [The Struggle: Socialist Revue], the monthly of the Social Democratic party, a politically engaged magazine with high intellectual standards. - "Mutterrecht und Magie" is not published in the *Gesammelte Schriften*; it is listed in Neurath's bibliography in Stadler 1997, 762.

[19]"Not only do you get to know the lives of people who are bound together by maternal right, but you also become familiar with their production, you see how the spouses and children are supplied with food, especially by the woman's family, and you constantly learn what 'magic' means among these primitive peoples, the spell, the ceremony, which is used by the primitive people in much the same way as our technical skills. What we know as primitive art are creations that serve certain purposes in life. To certain forms a charm is attached, certain objects are used for protection and other purposes" (Neurath 1931c, 99–100).

[20] "How, for example, can one explain the growth of agriculture in this framework? The collectors and hunters bring food home. Tubers, corn, and waste are thrown away. New life forms on the waste-heaps, and one gets used to treating wasteheaps like 'fields',

tilling from the dumping of waste is a weird version of a speculation in one of the "theoretical books on ethnographic questions" (Neurath 1931c, 102) to which the review refers at the end: *Das Weltbild der Primitiven* [The Worldview of the Primitives], by the extreme diffusionist Fritz Graebner, the prototype of the 'museum moles' Malinowski disdained.[21]

The author of *Empirical Sociology* agrees with Graebner not only on the genealogy of soil cultivation but also with regard to the emergence of mother right in agrarian societies as well as in terms of the complementary conjecture that "father right" occurs in cultures "with agriculture less predominant" (Neurath 1931a, in GS 1, 474–475; cf. Graebner 1924, 48–49).[22] The review confirms that, in general, even if not without exception, a "social order based on mother right [mutterrechtliche

---

> systematically. The habit of throwing away tubers and, later, corns becomes agriculture. The man hunts, the woman, spatially restricted by pregnancy and rearing, tills the field. The daughter takes over the mother's field. To belong to a good field becomes important. Belonging to a certain group of fields, i.e. a group of women, characterizes the individual. "Mother right" emerges.
>
> This extrapolation makes one tribe superior to another that is still living on collecting and hunting. Victory consists in the spreading of the acquired habit: this becomes stabilised, in a certain sense 'magic'. The adoption of the 'magic' and of the method of production are identical. The magicians are conscious representatives of the new habits and spread fear and fright as well as hope and confidence. Clearly, certain customs give superiority to groups in which they prevail; the custom of loathing dead animals and dead men often helps to avoid infection. To differentiate loathing and fear makes modern men superior to primitive peoples. We dose the caution before machines and adapt it to our success." (Neurath 1931a, in Neurath 1973, 377; GS 1, 486–487)

The editors of Neurath 1973 do not translate the passage on the emergence of mother right and add in square brackets the distancing remark "and this hypothetical explanation would explain the genesis of 'matriarchy', in Neurath's view" (Neurath 1973, 377).

[21]Neurath refers to "Graebner's small presentation of the world view [Weltanschauung] of the primitives, intended also for non-experts" (Neurath 1931c, 102). The "small presentation" is Graebner 1924, the first volume of the first *Abteilung* (*Das Weltbild der Primitiven und die Philosophie des Morgenlandes* [The World View of the Primitives and the Philosophy of the Orient]) of the *Geschichte der Philosophie in Einzeldarstellungen* [History of Philosophy in Individual Presentations] (1924–1930), edited by Gustav Kafka.

> At the beginning of their formation there is the invention of soil cultivation, the discovery, apparently made by women, that food waste - initially tubers - placed in favourable soil, sprouts again, and that the new plants, if properly cared for, produce a richer yield than their wild ancestors. The new invention allows the man to place by far the greater half of the economic burden on the woman; [...]. Since, however, the kindred women naturally carry out their planting work together, it becomes of increasing economic importance which group of women one is related to. And this is apparently the basis of the peculiar custom of emphasizing maternal kinship, the so-called maternal right. It is only gradually that from this point on women also gain increasing social importance. (Graebner 1924, 33)

[22] There are areas with 'matriarchy' and a certain form of agriculture, and others with agriculture less predominant and 'patriarchy'. One can distinguish areas of the shield, of the bow, one can highlight bow peculiarities, as Ratzel began to do in a very systematic way.

Gesellschaftsordnung]" is repeatedly found "among soil tilling peoples" (Neurath 1931c, 99). In the further course of the text, however, just this explanation of matrilinear social systems by soil tilling is questioned. The Trobrianders would be an exception, for their 'form of production' does not consist just of soil tilling; within limits, they also know cattle breeding (they keep pigs). Thus, Neurath himself remarks that the society described in *The Sexual Life of Savages* does not corroborate the hypothesis.

Needless to say, Malinowski maintains neither the connection between agriculture and mother right nor any such general thesis. Rather than with him, the reviewer agrees with Graebner. As already said, both (1) the aforementioned genealogy of soil cultivation and (2) the general thesis on the connection between soil tilling and mother right correspond approximately to views exposed in *The Worldview of the Primitives*. Neither thesis goes back to Malinowski.

Graebner is certainly no Marxist. Nevertheless, Neurath sees a possible Marxist explanation of a certain form of society (matrilinear social systems) in the connection to a certain form of production (agriculture). With such developmental conjectures, the idea that the "*strictly scientific anti-metaphysical anthropologists*" (Neurath 1932, in GS 2, 566) look for complex correlations takes on a Marxist twist. However, the thesis of a correlation between soil tilling and mother right is not in itself typically Marxist. It does not go back to Friedrich Engels (or to his main anthropological source, Lewis Henry Morgan).[23] Rather, Neurath is quite close to Heinrich Cunow, a minor Marxist ethnologist, mentioned at the end of the review. Cunow's thesis is a bit more sophisticated: He points out that what determines the form of family and society is the entire economic system rather than agriculture alone.[24] More than in the details, he and Neurath agree on the approach.

> It is obvious that there is a close connection between Christianity and the machine, and equally, that fetishist cults are absent or weak where coloured people have adopted the machine. We see that 'matriarchy' occurs mostly with farming tribes, but also sometimes with non-farming tribes. Even a superficial knowledge of human social life shows us that cults, customs, and ceremonies can be transferred. When we meet one of the Australian aborigines who lives like his neighbours it can happen that he recites Our Father and makes the sign of the Cross. [...] (Neurath 1931a, in Neurath 1973, 367; GS 1, 474–475)

The assumption that there is a historical relationship between "the bow and arrow of Indonesia and of West Africa" goes back to Friedrich Ratzel. (The passage in which Neurath mentions Ratzel has been left out of the English translation in Neurath 1973. On Ratzel cf., for instance, Harris 1968, 383.) Ratzel's student, Leo Frobenius, continued these morphological studies. Such 'historical relations' also inspired Graebner (cf. Harris 1968, 383–384). On the "human and cultural cover" and the relation between diffusionism and stage theories, see below p. 91, n. 80.

[23]Engels draws just the opposite conclusion. See below p. 71 f., n. 28.

[24]Cunow sees the social position of women gaining far greater importance "with the gradual transition of the hunter peoples to the cultivation of land" (Beginning Chap. IV). "*Despite the so-called mother succession, there < is > no question of an actual 'mother right', let alone of a maternal rule, on the lowest stages of development.* Maternal rule, where it is to be found, is based on the position of women in the economic system, primarily on their indispensable activity as farmers of plantations and fields" (Cunow 1929, 114). But only "primarily": "It is not only

But what about Neurath's further thesis, the one that gives the review its title, the correlation of mother right and magic?[25] Neither Graebner nor Malinowski advances this conjecture. The latter's book begins just with a paragraph on "The Principles of Mother-Right" (Malinowski 1929a, 3; Malinowski 1929b, 3).[26] But what does Malinowski mean with this term? Bachofen's primeval gynaicocratic rule is not at issue (not even for Neurath). Malinowski describes a matrilineal and virilocal society on the Trobriand Islands, a description that goes back to his field research there. His book is about a society in the then-present (more precisely, a decade earlier: He had stayed on the Trobriand Islands for the last time in 1919).

Neurath, of course, does not overlook the fact that Malinowski describes a specific (then) contemporary society of which he is a participating observer. The reviewer sees that the anthropologist is not interested in genetic investigations: "It should be an advantage of this book that the author does not make any hypotheses about the origin of these life forms" (Neurath 1931c, 102). Despite the admission that this abstention from hypotheses is an "advantage", Neurath himself is definitely interested in genetic investigations.

Characteristic of this interest are the beginning and the end of his review: They set a sort of Marxist framework for the discussion of Malinowski's book. Already, the very first sentence mentions Engels' *Urfamilie*,[27] and the short text closes with a reference to Cunow's "Marxist analysis" (Neurath 1931c, 102), which is also cast in genetic terms. Like Engels' *Origin of Family*,[28] Cunow's *On the*

agriculture, but the whole economy that determines the form of society and family" (Cunow 1929, 70). Cunow rejects the idea of a primeval gynaicocratic rule; he is more in agreement with Engels on the development of patriarchy. The equilibrium between the sexes is actually already shifting with the "expansion of agriculture and horticulture" (Cunow, 126), and Cunow sees a "development of male dominance among the pastoral peoples" (Cunow 1929, 146).

[25]Graebner considers animism, which represents a later stage than pre-animistic magic, to be the "main substantive phenomenon in the worldview of the peoples with mother right" (Graebner 1924, 50). According to him, animism has replaced the magic of analogy, which is characteristic of more primitive societies than the agricultural ones (i.e., of hunter-gatherers, according to the views of the time).

[26]In the German translation that Neurath reviewed, this paragraph, the first of the first chapter, is called "Die Grundlagen des Mutterrechts" (Malinowski 1929b, 2). In general, Eva Schumann translates Malinowski's "matrilineal" into "mutterrechtlich". Cf. Fritz Kramer's comment in the revised German edition (Malinowski 1979, 434).

[27]There is probably a strategy behind this. "Kulturbeschreibung auf materialistischer Basis [Culture Description on a Materialistic Basis]", another review of a non-Marxist author for *Arbeit und Wirtschaft*, begins in the same way. Neurath's short text on the German translation of a book by Alfredo Niceforo (*Kultur und Fortschritt im Spiegel der Zahlen* [Culture and Progress Reflected in Numbers]) opens with a quote from Engels' *The Condition of the Working Class in England* (cf. Neurath 1930b, 717).

[28]*The Origin of Family, Private Property and the State* (Engels 1884), which is one of Engels' most widely read books, is based on Lewis H. Morgan's *Ancient Society* (Morgan 1877). Engels draws on the *Morgan-Konspekt*, the notes Marx probably made from the summer of 1880 onwards (IISG, Marx-Engels-Nachlaß, Signatur B 162/B 146; cf. MEGA I/29, Apparat, 586 ff.). While Neurath

Primeval History of Marriage and Family (Cunow 1912) searched for prehistoric beginnings and origins; his then more recent book, *Love and Marriage in the Life of Peoples* (Cunow 1929), still shared this main concern of a whole ethnological tradition. The various competing theories emphasized that only a protracted gradual development had led from a (speculatively deduced or postulated) primeval state to the monogamous family that had come into being relatively late in the history of humanity and that was still in the process of development. Among Victorian and Edwardian ethnologists, the original "group marriage" or "communal marriage" had been a favourite obsession and the subject of a heated and controversial debate.[29] Even for Graebner, who considers the primeval "promiscuity" (Graebner 1924, 11) to be a pure fantasy product, origins were still an issue.

On the contrary, what the author of *The Sexual Life of Savages* is definitely *not* concerned with is just this sort of unwarranted musings about primeval times. In his "Special Foreword to the Third Edition", Malinowski categorically rejects such "speculations about the 'origins of marriage'" and "imaginary modes of human mating" (Malinowski 1932, xxxii). The anthropologist on the field is no "reconstructive theorist, whether evolutionary or diffusionist" (Malinowski 1932, xxxi), and must evade "the obsession of any reconstructive diffusionist or evolutionary doctrines" (Malinowski 1932, xxxi).

True, this third edition does not appear until 1932, a year after Neurath's review, but already the first edition was extremely cautious on this point. However, this did not prevent Bertrand Russell from understanding Malinowski's main thesis as speculation about a general original state. The philosopher writes in *Marriage and Morals* (1929):

> Malinowski maintains – and it is difficult to see how his opinion can be controverted – that all mankind must have passed through the stage where the Trobrianders are now, since there must have been a period when paternity was nowhere recognized. Animal families, where

---

traces mother right back to agriculture, Engels draws just the opposite conclusion. In the context of his speculative reconstruction of the emergence of private property, he claims that precisely the development of agriculture and cattle breeding, with division of labour and accumulation of wealth (herds, etc.) as consequences, brings about the "overthrow of mother right" and the "introduction of father right" in Morgan's middle stage of barbarism (MEGA I/29, 257).

[29]Morgan postulated a "primeval condition, where unrestricted sexual intercourse within a tribe prevailed, so that every woman belonged to every man, and every man to every woman equally" (Engels [1884], 152), According to Morgan, five successive forms of family can be distinguished on the long road to monogamy, each one a limitation of the previous form. – Engels' preface to the fourth edition of his *The Origin of the Family* (1891) reconstructs the history of research on the 'original family' from Bachofen's *Mutterrecht* (1861) via J. F. MacLennan, whom Engels sharply criticizes, to Morgan (cf. MEGA I/29, p. 135). Cunow begins both of his books with the history of the question about the original family. The first chapter of his later monograph is called "How One Used to Imagine the Original Forms of Marriage and Family" (Cunow 1929, 7–27). Cunow gives Engels (and Morgan) an important place in his reconstruction but is far from simply agreeing with him (see note 24 above). – In 1914, the young Malinowski had also written an overview of the "Sociology of the Family" (cf. Malinowski 1914).

> they include a father, must have a like basis, since they cannot have any other. [. . .] (Russell 1929, 24.)[30]

Russell (1929, 18–19) mentions, in particular, three books by Malinowski: *Sex and Repression in Savage Society* (1927), *The Father in Primitive Psychology* (1927), and (the first edition of) *The Sexual Life of Savages in Northwestern Melanesia* (1929). Russell is not completely on the wrong track when he reads the first two books in the sense of the evolutionary tradition. For a while, Malinowski had been struggling to free himself from the spell of origins. Certainly, already in the *Argonauts* (1922), he had actually left behind the evolutionary approach; however, even later books like *Sex and Repression* and *The Father in Primitive Psychology* still had not wholly emancipated themselves from evolutionary views. Notwithstanding, *The Sexual Life of Savages* is a different case. In the "Special Foreword to the Third Edition", Malinowski, whose tendency towards diplomacy is well known, mentions Russell's reception without directly criticizing it.[31] Not even a page later, however, the anthropologist firmly denies any evolutionary intention. A main thesis of his is that the Trobrianders are ignorant of physiological paternity, or, in Neurath's words, that they "know nothing of the fact that procreation takes place by the male" (Neurath 1931c, 100). Now, in the "Special Foreword", Malinowski is adamant that in his book he is "not pronouncing any opinion as to whether there was any 'original ignorance of paternity' throughout primitive mankind and whether the Trobrianders still suffer from it" (Malinowski 1932, xxi-xxii). One cannot generalize their case to all 'primitives' including 'our' remote ancestors.[32]

[30]This passage can be found close to the end of the second chapter "Where fatherhood is unknown" (Russell 1929, 14–24). The title of chapter three is "The Dominion of the Father" (Russell 1929, 25–32). In later editions, the headings are more technical: "Matrilineal Societies" (title of chap. 2) are set against "Patriarchal Systems" (title of chap. 2).

[31]"Bertrand Russell fully appreciated the functional significance of the Trobriand facts as regards paternity and made use of them in his pioneering work on *Marriage and Morals*" (Malinowski 1932, xxi).

[32]Malinowski's theory of ignorance contradicts Morgan's basic assumptions, for it denies to the Trobrianders what the latter presupposes as a matter of course: the seemingly obvious knowledge about the role of the male in procreation. Neurath agrees with Malinowski's explanation of the Trobrianders' ignorance. However, unlike Russell, the Viennese does not even mention the controversial theoretical consequence that was the central issue in the then-widespread debate surrounding the anthropologist's books on 'primitive' sexual life. From the Trobrianders' ignorance of the father's role, Malinowski followed that Freud's Oedipus complex is not universal. Rather than in *The Sexual Life of Savages*, however, which hardly touches upon Freud's Oedipus complex, this main theoretical concern was in the foreground in earlier publications, among them the two other books mentioned by Russell (Malinowski 1927a, 1927b; Malinowski 1924 also deals with psychoanalysis). (On the very last page of *The Sexual Life of Savages*, Malinowski refers the reader to *Sex and Repression in Savage Society* (Malinowski 1927b); cf. Malinowski 1929a, 572; Malinowski 1929b, 395). Without even touching on this issue, Neurath leaves it at the vague remark that Malinowski "apparently has doubts about much of Freudian psychoanalysis, which he describes as very stimulating" (Neurath 1931c, 102). Here, the reviewer refers to Malinowski's statement that "in the course of <his> inquiry <he> had to reject far more of psychoanalytic doctrine than <he> could accept" and that "even a theory which has, in the light of investigation,

This vehement rejection of evolutionary implications, however, takes at least in part "the form of a recantation" (Malinowski 1932, xxii). Even if already in 1929 Malinowski had left behind this previous position, his later "Special foreword to the third edition" suggests that the break with the ethnological tradition due to his new functionalist approach should have been stated more explicitly in the first edition of *The Sexual Life*.

Already, the long title of the book reveals Malinowski's ongoing struggle. General terms like "savages" and "natives" coexist with more precise geographical indications: "Savages in North-Western Melanesia", "Natives of the Trobriand Islands". Sweeping terms like "savage", "primitive", and even "native" had the function of giving the book a more general meaning, thus meeting the expectations of the readers. Rather than in the depiction of a specific small people living in a peripheral place, they were interested in what they assumed to be the life of "savages" and "natives" in general as well as of our own remote forebears.

Nevertheless, especially Malinowski and Radcliffe-Brown were bringing about a paradigm shift. At that time, however, even in Britain and the United States, the then-current state of anthropological research was by no means as clear as it is now in retrospect. Furthermore, since the early twenties, an increasing gap had been gradually forming between the new Anglo-Saxon anthropology and German-language scholarship: The extreme diffusionism of the Vienna school of ethnology was a specific 'dead end'. Therefore, misunderstandings are characteristic of this transitional period, rather than being simply Neurath's personal shortcomings.

His affirmative review testifies that he had not entirely overlooked recent developments in contemporary anthropology. However, it should be clear by now that "Mother Right and Magic" is marked by a range of compromises between the state of research known to Neurath and Malinowski's new anthropological approach. The Viennese does not seem to be intimately familiar with the anthropologist's production beyond the book he is discussing; and of the other new releases to which he refers, only one was on the cutting edge of research: Robert and Helen Lynd's *Middletown: A Study in Contemporary American Culture* (Lynd 1929).[33] The remaining literature cited, even if then-recent, consists of old-style books like

---

to be partly rejected can stimulate and inspire" (Malinowski 1929a, 386). There is no clue that Neurath, who here almost literally quotes the reviewed book, has taken note of any of Malinowski's texts dealing with psychoanalysis. He probably was not especially knowledgeable about the anthropologist's role in contemporary debate. Neurath, who knows *Totem and Tabu*, begins his review mentioning Freud's attempt to show "that certain primitive institutions, 'taboo', 'totem' and others, are related to phenomena we know of in neurotics" (Neurath 1931c, 99). Referring to the same psychoanalytic assumption, the author of *Empirical Sociology* more carefully leaves open "[h]ow far the magic attitude is related to the attitude of our neurotics (Freud)" (Neurath 1931a, in GS 1, 428). On Malinowski's increasingly critical examination of psychoanalysis, cf. Stocking 1995, 277–283.

[33]A few years after Neurath's review, Robert and Helen Lynd published a second book (*Middletown in Transition: A Study in Cultural Conflicts* = Lynd, Lynd 1937). Both monographs are mentioned in *Modern Man in the Making* (Neurath 1939, 144). *Middletown* is also quoted in Neurath 1939, 157, 158.

Graebner's and Cunow's. Hence, the background against which Neurath reads Malinowski's new monograph does not fit it, not least because the anthropologist's production had overtaken this state of research. That Neurath was not entirely aware of this path-breaking newness confirms Malinowski's general disappointment with the reception of *The Sexual Life of Savages*.[34]

4.4 Where are the Theories?

Neurath is aware that Malinowski is not a Marxist, and the conclusion of the review discreetly points to this 'limit':

> Reading such a book can certainly not replace Marxist analysis, as Cunow gives it, but it can provide a stimulating insight into the motley and fragmentary nature of even a detailed description that forms the basis of a theory. (Neurath 1931c, 102)

The claim that after reading Malinowski's book even detailed anthropological descriptions look motley and fragmentary is somehow ambiguous. Does it concern only the then-usual accounts, on which Marxist theories were also based? Or does it also apply to Malinowski's description, though the review acknowledges that he sets new standards of anthropological representation? In any case, Neurath's main point is that describing is not enough and that a Marxist analysis is indispensable.

More in general, he evidently misses theory in Malinowski's book. There may be several reasons for this. The anthropologist stands for a major breakthrough in field research, and his strengths lie in method rather than in his general theories about basic human needs. In the first edition of *The Sexual Life of Savages*, these theories, which swiftly leap from the Trobriand Islands to humanity in general, are hardly to be found anyway. It contains only a few theoretical explanations.

More decisive than the content of Malinowski's book are Neurath's expectations. He calls not only for ethnographic descriptions but also for theories that disclose general, although not necessarily exceptionless, laws of historical development (such as the alleged correlation between agriculture, mother right, and magic). In many respects, he shares the traditional – evolutionist and diffusionist – view that the main task of anthropology is to provide us with world-historical insights into the early stages of human development and into the distant primeval origins of, say, family, state, and religion.[35] Thus, Neurath subscribes to a central ethnocentric stereotype of evolutionary and diffusionist ethnology: The contemporaries he calls 'primitives' come somehow close to being representatives of our own prehistoric ancestors. "The life of the primitives is also important because it shows us what our

[34]Cf. his "Special Foreword to the Third Edition" in Malinowski 1932, xix.

[35]With similar background assumptions, an old-fashioned anthropologist like Frazer, who was benevolent to Malinowski and had written the preface to the *Argonauts*, argued for the received division of labour between the ethnographer on the spot and the theorist at home, even if he surely did not strive for a Marxist analysis.

own past may have looked like, which is still our present in so many ways" (Neurath 1931c, 99).[36] Neurath's entire background, his cultural-historical approach, his Marxist views, etc. lead him to pursue questions of developmental history, while just this sort of genetic investigation was becoming problematic for the new generation of anthropologists.

In 1931, he could find theory, and especially Marxist theory, only in old-style books like Cunow's. In later years, 'basis' and 'superstructure' will end up in Neurath's *index* of dubious concepts better to be avoided, and he cannot be said to be dogmatically committed to economic determinism.[37] At the time, however, he does not reject the Marxist teaching "on the dependence of the 'ideological superstructure' on the 'substructure of the mode of production' [über die Abhängigkeit des 'ideologischen Überbaues' vom 'Unterbau der Produktionsweise']" (Neurath 1930b, 720). In "Mother Right and Magic", already the first paragraph, which begins with a reference to Engels, suggests "the connection between mode of production and fashioning of life [Lebensgestaltung]" (Neurath 1931c, 99). As shown above, the review sees in a specific mode of production – soil tilling – the common basis of both "mother right" and "magic"; elsewhere, not only in the West, it is the capitalist mode of production that conditions the way life is fashioned.[38] No matter how strict this "connection" between mode of production and form of life is thought to be, with Neurath, it gives us a key to ethnological understanding in as far as it establishes a closeness between 'us' and the so-called 'primitives'. Peoples like the one Malinowski describes "are no longer as distant from us as before; we are increasingly learning to see through the connection between mode of production and fashioning of life" (Neurath 1931c, 99). Some of their customs and beliefs, which at first might seem astonishing or even ludicrous, become understandable if we take into account that their form of production and social structure are completely different. Thus, 'magic' ceases to be "mysterious" if it is explained with reference to a specific form of production: as "*the poor, hardwon techniques of social existence* of primitive peoples, especially agricultural ones" (Neurath 1931a, in Neurath 1973, 320; GS 1, 427). What may resist interpretation goes back to the difference in economic and social structure, not to another 'mentality'. This historical materialistic approach is a way of getting closer to a people like the

[36]The restriction that, in many ways, this distant prehistory is still our present does not call the stereotype into question, but belongs to it. Any evolutionary ethnologist, including Frazer, would have agreed.

[37]On basis and superstructure, as well as on Neurath's criticism of economic determinism, cf. Sandner 2007, 149–150. Reisch 1997 includes 'basis' and 'superstructure' in his reconstruction of the *index* with reference to Neurath 1941, 225.

[38]"*The living standards [Lebenslage] of the proletariat, its happiness or misery* are conditioned by the capitalist order of life." (Neurath 1928a, in Neurath 1973, 289; GS 1, 286). Already in 1917, Neurath promises "a *general doctrine of the orders of life* [*Lebensordnungen*]" (Neurath 1917, 129). About the terms *'Lebenslage'* and 'Lebensstimmung', especially in "Economic Plan and Accounting in Kind [Wirtschaftsplan und Naturalrechnung]", see Hegselmann 1979, 25–26.

Trobrianders. However, this is not the only hermeneutic key to distant worlds in Neurath's review.

4.5 "Elimination of Empathy"?[39]

Neurath's theoretical mistrust of the concept of '*Einfühlung*' does not prevent him from appreciating Malinowski's practical virtue of empathy. As we have seen, Neurath concedes that the term '*Einfühlung*' could still be used as long as what it stands for is not "expected to have some peculiar magical power beyond ordinary induction" (Neurath 1931e, in Neurath 1983, 73; GS 2, 548). Of such magical expectations, he suspects Weber, but not Malinowski.

Neurath will have counted the latter among the anthropologists who meet social behaviourist standards, even if they do not explicitly commit themselves to this method.[40] Can Malinowski's approach really be passed off as 'social behaviourist' in disguise? Neurath understands 'social behaviourism' rather loosely:

> Social behaviourism is getting his due [...] wherever everything that is said can be replaced by protocol statements, in which, for example, researcher's travel observations, remarks backed by photographs, films, etc. are mentioned. (Neurath 1932, GS 2, 567)

"Mother Right and Magic" praises Malinowski because his factual report can be related to the 'protocols' of his field research, even if this rather popular text avoids the overly technical term 'protocol'.

> Since he [Malinowski; MB] not only reports facts, but usually even describes how he arrived at his knowledge, the reader can sometimes form his own judgement on important points. (Neurath 1931c, 100)

Neurath must have noticed that Malinowski's methodical ideal requires the field-worker to take the native's point of view. Nevertheless, the sharp critic of '*Einfühlung*' is surprisingly affirmative about the anthropologist's approach. Neurath himself points out how important it is to "get to know foreign peoples as much as possible from their point of view, to feel at one with them" (Neurath 1931c, 99). The Viennese praises Malinowski's effort "to bring the life of the Trobriander closer to us through parallels from our life" (Neurath 1931c, 102). "The pleasing addition to the text are the numerous photographs [...] everything passes us by, a piece of distant life that in this way becomes part of our life, of the life of mankind" (Neurath 1931c, 102).[41] This idea – humanity's common life is also our life – recalls

[39]"Elimination der Einfühlung" is the header of page 57 of the original edition of *Empirical Sociology*.

[40]"The successful anthropologists, national economists and sociologists constantly use behaviorist methods without always knowing it themselves" (Neurath 1932, in GS 2, 568).

[41]On "protocol statements, in which [...] remarks backed by photographs [...] are mentioned", cf. the cited passage from *Sozialbehaviorismus* (Neurath 1932, in GS 2, 567). Among the many

Neurath's apology of Keyserling's difference-friendly cosmopolitanism a decade earlier.

4.6 Mysticism and Empathy. Lévy-Bruhl on Alien Mentalities

In 1931, Neurath points out that 'we' – he and his time – no longer assume that peoples like the Trobriander are extremely distant from us. 'We' see that their thinking deviates from ours "more in details than in the broad lines" (Neurath 1931c, 99); 'we' are better able to recognize our close relationship to the so-called 'primitives' than earlier centuries. Even in the twentieth century, however, not every scholar agrees that their 'mentality' is not altogether different from ours. The spread of the word 'mentality' goes back to Lévy-Bruhl, to whom the school of the *Annales* owes this central term of its mental history. The French thinker, who had been made known in German-speaking countries by a Viennese philosopher, Wilhelm Jerusalem, belongs to the opponents mentioned by name in Neurath's review. Here, Lévy-Bruhl epitomises the tendency to make the so-called primitives seem more alien than they actually are: "[. . .] Some ethnologists, such as Lévy-Bruhl, do their part to contrast the thinking of primitives with our thinking as sharply as possible" (Neurath 1931c, 99).[42] Neurath hints at Lévy-Bruhl's dichotomy between the 'mystical' or 'prelogical' mentality of the primitives and our 'logical' mentality.[43] Malinowski himself rejects any approach that places the natives "in the 'prelogical stage of mental development'" (Malinowski 1932, 454). Neurath will have recognized this transparent allusion to Lévy-Bruhl.

To the French philosopher, he opposes not only Malinowski (in "Mother Right and Magic"), but also Frazer. The following statement in *Empirical Sociology* matches up with the Scottish anthropologist's own arguments against Lévy-Bruhl, although Neurath will probably not have taken notice of them.[44] "The human of the magical form of life has no special mode of thinking (Lévy-Bruhl), we are of his flesh and blood (Frazer)" (Neurath 1931a, in Neurath 1973, 321; GS 1, 427). Neurath clearly takes Frazer's side but leaves the anthropologist's position strangely blurred. Lévy-Bruhl himself, even if he ascribes a pre-logical mentality

---

studies dedicated to Malinowski's photographs, only one should be mentioned here: Young 1998 contains about 200 previously unpublished photographs.

[42]Actually, Lévy-Bruhl is not an ethnologist, but a then-prominent philosopher. Jerusalem had translated his *Les fonctions mentales dans les sociétés inférieures* (1910). This German edition with the title *Das Denken der Naturvölker* (Lévy-Bruhl 1921) was also intended as an action against Husserl's phenomenology. On this translation and in general on the reception of Lévy-Bruhl's concept of 'mentality' in German-speaking countries, cf. Brusotti 2017.

[43]The late Lévy-Bruhl will revise this earlier position: Both 'mentalities' - the 'mystical' and the 'logical' - can actually be found everywhere, among 'us' as well as among 'them'.

[44]On Frazer's objections against Lévy-Bruhl, cf. Brusotti 2014: 156.

to the so-called 'primitive', is far from denying that we are "flesh of his flesh and blood of his blood"; for all his Eurocentrism, the committed *Dreyfusard* is not a racial theorist. Neurath's statement, though rather vague, corresponds roughly to Frazer's self-image. The author of *The Golden Bough* proceeds from the unity of the human mind: In this evolutionary view, the law of similarity and the law of contact (or contagion), the same association laws that among the primitives lead to homoeopathic (or imitative) and to contagious magic are principles of our own thought – hence, the Scottish anthropologist's criticism of Lévy-Bruhl's idea of a 'prelogical' mentality.

To not make magic look stranger than it is, Neurath builds on Frazer's analogy between magic and technology and refers to his developmental scheme, 'magic, religion and science'. The Malinowski review relies only to a certain extent on this analogy and does not employ Frazer's well-known triadic theory. However, both play an important role in a few other texts from the same years.

4.7 Neurath and Frazer on Magic and Religion[45]

Carnap ascribes to Schlick the "view that religion belonged to the childhood phase of humanity and would slowly disappear in the course of cultural development" (Carnap 1978, 35). Does this idea belong to the common religion-critical background of the Vienna Circle? Neurath only partially shares it: He believes that humanity will no longer be religious in adulthood, but not that it was so in its "childhood". In the manifesto of the Circle, the hope for a thoroughly secular culture of the future corresponds with the far more problematic speculative idea, borrowed from Frazer, that the early beginnings of civilisation already were in a 'secular', i.e., non-religious or pre-religious condition; with a twofold historical caesura, humankind first entered a religious state and now – thousands of years later – leaves it again:

> The representatives of the scientific world-conception [...] confidently approach the task of removing the metaphysical and theological debris of millennia. Or, as some have it: returning, after a metaphysical interlude, to a unified picture of this world which had, in a sense, been at the basis of the belief in magic, free from theology, in the earliest times. (Neurath 1929a, in Neurath 1973, 317; GS 1, 314)

This three-stage theory – belief in magic, theology-metaphysics, and scientific world conception – must not be confused with Comte's well-known scheme, which distinguishes a religious-theological, a metaphysical-philosophical, and a scientific-positivist stage. According to Comte, metaphysics is basically just a kind of theology; thus, there is a certain *continuity* between the first and second stages. *The Scientific Conception of the World*, on the contrary, emphasises here just the *break* between "the belief in magic, free from theology, in the earliest times", and the

[45]The now following pages (chapters 7–11) are derived from Brusotti 2011, 333–344.

"metaphysical interlude". The pamphlet is based on another evolutionary scheme, namely, on that presented by Frazer since the second edition of *The Golden Bough*.

The formulation with which *The Scientific Conception of the World* introduces this historico-philosophical thesis suggests that not all authors support it.[46] However, Neurath, the principal drafter, attaches great importance to this issue. The annotated bibliography of the manifesto ascribes the thesis to him, informing the reader that already the book he had published a year before, *Lebensgestaltung und Klassenkampf* [*Personal Life and Class Struggle*] (1928), had dealt with the "return to ways of thinking of the early times after a theological-metaphysical interim age" (Neurath 1929a, GS 1, 323).[47] During the following years, Neurath often articulates this thesis, among others in *Empirical Sociology*, whose first section is called "From Magic to Unified Science [Von der Magie bis zur Einheitswissenschaft]", in "Ways of the Scientific World-Conception", his first signed essay for the journal *Erkenntnis* (Neurath 1930a), as well as in lectures for a broader public like "Magic and Technology" (Neurath 1931b).[48] "We distinguish" – he explains in this lecture – "three ages, one magical, one technical and in between a theological one" (Neurath 1931b; GS 2, 529).

> Seen in the perspective of humankind, the pre-magician is defeated by the magician, the magician by the theologian, and the theologian by the scholar. (Neurath 1931a, in Neurath 1973, 378; GS 1, 488)

Rather than inspiration, Comte is a target here. Neurath criticizes those who, like the French philosopher, "regard the comprehensive scientific conception as a young movement, which would have to sweep out a tradition of millennia and therefore requires a totally new attitude" (Neurath 1930a, in Neurath 1983, 33; GS 1, 372).[49] On the contrary, "basic elements of the scientific world-conception were already present in the early age of humankind" (Neurath 1930a, in Neurath 1983, 33; GS 1, 372). Proving this early presence is "not indifferent from the educational and psychological point of view" (Neurath 1930a, in Neurath 1983, 33; GS 1, 372); for against this background it seems more plausible that one could be able to "revitalise

[46]The phrase is: "Oder, wie einige meinen: [...]" ["Or, as some have it"] (Neurath 1929a, in Neurath 1973, 317; GS 1, 314). This means that the thesis is not uncontroversial and that not all authors of *Wissenschaftliche Weltauffassung* support it. "It was presumably Neurath who composed its first version, and while it was revised by Carnap with help from Feigl, Neurath and Waismann, in the end it did receive the name suggested by him: *The Scientific World-Conception. The Vienna Circle.*" (Haller 1991b, 41). On this teamwork, cf. the note by Marie Neurath (Neurath 1973, 318, note 2).

[47]See below n. 52.

[48]Neurath gave this popular talk on January 21, 1931 in the *Verein Ernst Mach*. The unsigned report on the lecture was published in *Erkenntnis* in the same year under the title "Magie und Technik" (Neurath 1931b). In the same journal, the historico-philosophical thesis is also formulated in the likewise unsigned report on the lecture "Unified Science and Marxism" (Neurath 1930c). On magic in a later programmatic paper, in which Neurath presents the Vienna Circle to the French public, see below n. 63.

[49]On Neurath and Comte, cf. Brusotti 2011, 333–335; Josephson-Storm 2017, 255.

[wiederbeleben]" (Neurath 1930a, in Neurath 1983, 33; GS 1, 372) the scientific world view.[50]

4.8 Frazer: Magic, Religion, and Science

From the second edition (1900), *The Golden Bough* distinguishes magic, religion, and science as the three successive evolutionary stages of the human mind.[51] What is specific to Frazer, however, is far less this famous threefold distinction than a certain conception of the historical relationship between the members of the triad. Frazer postulates a primordial "Age of Magic" (Frazer 1922, 56) – corresponding to the stone age – "when man trusted to magic alone for the satisfaction of such wants as transcended his immediate animal cravings" (Frazer 1922, 54): "an Age of Religion has thus everywhere [...] been preceded by an Age of Magic" (Frazer 1922, 56). Thus, this early phase of human development is a pre-religious one. In *The Golden Bough*, magical thought and action are not simply a sphere of life among others, but stand for the totality of primitive beliefs and customs; magic is a general, early phase of development of the human spirit: a 'wild science' and 'wild technology' that is still free of theological elements. "In short, magic is a spurious system of natural law as well as a fallacious guide of conduct; it is a false science as well as an abortive art" (Frazer 1922, 11). Theoretical and practical magic – that is, beliefs and rites – relate to each other like science and technology: The former seeks to explain nature, the latter to control it. (Practical) magic is an ineffective technique based on superstitious explanations.

Religion develops not from magic, but from its *failure*. Humans first attribute unlimited power to themselves and try to control nature by their own strength. Finally, however, they must realize that their magical arts remain ineffective.

[50]Cf. also the conclusion of "Magic and Technology" (Neurath 1931b; GS 2, 531) mentioned below, p. 91.

[51]On the origin of Frazer's threefold distinction, see Fraser 1990a, 117 ff. - The subtitle of the first edition of *The Golden Bough* (1890) is *A Study in Comparative Religion*. Only from the second edition on, it becomes *A Study in Magic and Religion*. Frazer did not aim at a three-stage doctrine from the beginning: In the first edition, he had considered magic as a sub-area of religion. The impulse to turn magic into an independent earlier stage came from field research: Spencer's and Gillen's report on the Australian Arunta/Aranda (today, rather, Arrernte) suggested to Frazer the thesis of the evolutionary precedence of magic. Thus, even if his three-stage doctrine looks like an anthropological adaptation of pre-Darwinian evolutionary schemata, he arrived at this most famous of his theses in a different way. Apparently, he was only later informed of similarities to Hegel, whom he did not hold in high esteem. As Frazer puts it, he agrees with the German philosopher "that in the mental evolution of humanity an age of magic preceded an age of religion, and that the characteristic difference between magic and religion is that, whereas magic aims at controlling nature directly, religion aims at controlling it indirectly through the mediation of a powerful natural being or beings to whom man appeals for help and protection" ("Hegel on Magic and Religion" [Appendix to the first volume of the third edition], in Frazer 1911, vol. 1, 423–426, here 423; cf. Hegel 1965, 279 ff.).

Magicians had tried to control nature directly and, if they believed in gods at all, to subjugate them by resorting to impersonal powers. The priest tries to reconcile the deities and invokes the longed-for effects from them.[52] Magic is, thus, replaced by religion in its leading function, and though it does not wholly disappear, it remains only a secondary phenomenon.[53]

Finally, science provides insight into the real causal connections. Humankind gradually comes to understand the errors of religion and returns to a simpler empirical application of the principle of causality, making a new attempt to conquer nature on his own, through science and technology. By no means the only Eurocentric consequence of Frazer's global evolutionary scheme is that this third and, for the time being, last developmental stage of the human mind has been reached only by western culture.

With the second edition of *The Golden Bough*, Frazer's prestige reached its zenith. However, the thesis of a pre-religious, exclusively magical phase of cultural development never found broad acceptance.[54] The idea of pure unadulterated magic was an artificial construction – and this is true of Frazer's concept of magic in general. Besides the deep Eurocentric prejudices with which his representation of the Other is riddled, the thesis of the evolutionary precedence of magic has other valences. Above all, Frazer has a critical agenda: From the very beginning, he intends to question religion (though he does this more or less openly depending on the edition of *The Golden Bough*). Like Tylor, he wants to show that, except for a relatively small cultural and scientific elite, primitive magic superstitions are preserved practically everywhere in Christianity, even if mostly only as unconscious *survivals*, as incomprehensible remnants of outdated cultural stages. However, the author of *The Golden Bough* goes further than Tylor. From the assumption that magic is the primitive stage and antithetical to religion, Frazer follows that man has no innate religious disposition. Although Tylor too separates magic and religion, the historical precedence of magic is not compatible with his theory of animism; he sharply rejects magic as "one of the most pernicious delusions" and does not feel the slightest inclination to give it priority.[55] Frazer does just this. Even if, in his developmental schema, religion is supposed to represent progress in relation to the purely magically determined primitive worldview, (not only the subtext of) *The Golden Bough* tends, in a decisive respect, to prefer the earlier magical stage

[52]As Neurath puts it in *Personal Life and Class Struggle*: Comparable to physical forces, the demons of magic were instruments of humans (magicians), whereas in religion, humans (priests) are instruments of the gods. Neurath begins by arguing that the relationship between religion and science changes over time: Early religion was much closer to science than later religious developments. Then he goes on to claim that pure magic had been even more affine to science than early religion. Cf. Neurath 1928a, in GS 1, 273–274; not translated in Neurath 1973.

[53]Cf. Frazer's letter to Baldwin Spencer of 28 November 1898 (Frazer 2005, 135 ff.; cf. especially 137 f.); see Stocking 1995, 143.

[54]Among the critics were Lang, Marett, Hubert and Mauss, as well as Durkheim.

[55]Cf. Tambiah 1990, 45.

of development to the later religious one. Neurath will follow Frazer just in this respect.

4.9 "From Magic to Unified Science"

Because of the "view that basic elements of the scientific world-conception are already contained in magical practices", Neurath has been compared to Feyerabend.[56] However, Neurath does not deny the possibility of a principled demarcation between magic and science[57] and only exceptionally employs the analogy to magic in a science-critical sense.[58] Rather than seeing in him a 'precursor' of Feyerabend, it seems more appropriate to highlight Neurath's historical relationship with Frazer.

In the texts from 1928–1931, figures of thought from *The Golden Bough* play an important role, even if explicit references to the Scottish anthropologist are sparse.[59] Neurath embraces Frazer's close analogy between magic and technology, distinguishes with him contagious and imitative magic,[60] elaborates on the opposition magic versus religion, and adopts Frazer's scheme of development, including the view that primeval times did not yet know religion and that the latter is only an interlude. Neurath attaches great importance to Frazer's thesis that the early belief in magic is free of religious ideas and shares the anthropologist's assumption that magic "once encompassed the whole of life" (Neurath 1931a, in Neurath 1973, 329; in GS 1, 435), even if he adds a question mark to this original autocracy of preanimistic magic.[61]

---

[56]"Neurath shares with Paul Feyerabend the view that basic elements of the scientific world-conception are already contained in magical practices; and that these could be revived under certain external conditions; but Neurath already published these views nearly 50 years ago. [...]" (Haller 1991a, 35). That is, in 1928.

[57]Haller (1991a, 35) knows this.

[58]"Not infrequently certain scientists behave as if they could provide people with many more correlations than is scientifically possible. Sometimes they resemble religious leaders or magicians of the past, pretending to have some superhuman knowledge and cunning" (Neurath 1944, 27).

[59]Frazer is explicitly referred to in *Empirical Sociology* (cf. GS 1, 427; see above p. 77) and in *Ways of the Scientific World-Conception*, even if just here Neurath relativizes the anthropologist's actual thesis (Neurath 1930a, in Neurath 1983, 33; GS 1, 372; see below n. 61). On a later mention of Frazer, see below p. 91.

[60]Neurath's "Ähnlichkeits- und Berührungsmagie" (Neurath 1930a, in GS 1, 374) or "Berührung[s-] und Ähnlichkeitszauber" (Neurath 1930a, in GS 1, 373) is the German translation of Frazer's two types of sympathetic magic: *contagious magic* (law of contact) and *homoeopathic* or *imitative magic* (law of similarity). I have modified the English translation (in Neurath 1983, 35) accordingly.

[61]"In the early period, religion is accompanied by a powerful *magic* (see *Frazer* and others), which was once perhaps the sole ruler" (Neurath 1930a, in Neurath 1983, 33; GS 1, 372). Frazer's actual thesis is the one Neurath weakens with a "perhaps", the assumption being that magic "once encompassed the whole of life" (Neurath 1931a, in Neurath 1973, 329; in GS 1, 435). Just in relativizing this thesis, however, Neurath displays an approach similar to Frazer's. The Scottish

Must this idea of a "belief in magic, free from theology, in the earliest times [von Theologie freien Zauberglauben der Frühzeit]" (Neurath 1929a, in Neurath 1973, 317; in GS 1, 314) necessarily go back to Frazer? Graebner uses almost the same words when he contradicts Wilhelm Schmidt, who holds just the opposite thesis, namely, that "primitive monotheism is older than the preanimistic belief in magic [der präanimistische Zauberglaube]" (Graebner 1924, 144, n. 4). Graebner defends Preuß and Marett against Schmidt's criticism and gives historical priority to the "primordial belief in magic [uralten Zauberglauben]" (Graebner 1924, 10), which is "a non-animistic belief in magic [ein nicht animistischer Zauberglaube]" (Graebner 1924, 10), "a preanimistic world view, especially expressed in magic beliefs and customs [Zauberglauben und Zauberbräuchen]" (Graebner 1924, 16).

'Preanimistic' magic was so called because it was supposed to precede religion, which Tylor had defined as animism (belief in spirits). *Against Frazer*, Marett, to whom the term 'preanimistic' goes back, had assumed that the magical and the religious were not yet separated in the original stage. According to preanimists like Marett and Preuß, animistic religions were not preceded by magic in a pure state (Frazer), but by a still undifferentiated 'magic-religious' stage.[62] With Neurath, however, 'pre-animism' does not stand for Marett's (emotionalistic) 'magic-religious', but only for Frazer's (intellectualistic) 'magic', a stage in which the religious is not yet there at all.[63]

According to Father Schmidt, on the contrary, a "primordial monotheism" precedes even pre-animism. In rejecting this "primordial monotheism", Neurath agrees with Graebner. Actually, however, Graebner seems closer to Marett than to Frazer (whom he does not even mention in this context). Thus, it is *The Golden Bough* that offers Neurath an alternative to the position of the main exponent of the Vienna school of ethnology. In the capital city of *Urmonotheismus*, Neurath maintains the evolutionary priority of magic.

Like Frazer, he concludes from these magical "preanimistic periods" (Neurath 1930a, in Neurath 1983, 34; GS 1, 373) that religion and metaphysics are not the fulfilment of innate human needs, but only an interlude, in which "besides the earthly-finite, observable world, a *transcendent*, supernatural, incomprehensible one" (Neurath 1931b; GS 2, 529) is assumed.

Neurath goes even beyond Frazer because although he speaks of "three ages" (Neurath 1931b; GS 2, 529), he distinguishes three "revolutions", which actually

---

anthropologist has, indeed, a rather loose attitude towards his own theories: He takes them far less seriously than the abundant ethnographic material he has assembled. On the affinity between the priority Frazer accords to the material and Wittgenstein's arguments against *The Golden Bough*, cf. Brusotti 2014, 205 ff.

[62]On this hypothesis and its early criticism, cf. Stocking 1995, 168.

[63]In a later programmatic text, Neurath explains that the scientific language he is looking for "actually goes back to forms that have a very ancient history, the forms of preanimist magic" in which - he believes - "nothing metaphysical" could be found (Neurath 1935, 11). Even in the Renaissance, when one encounters "some thesis formulated without metaphysics", this thesis often is "derived from magic" (Neurath 1935, 18).

implies *four* ages. With Neurath, not in *The Golden Bough*, magic itself already represents “a revolution” compared to an even earlier stage of “non-magic [Nicht-Magie]” (Neurath 1931b; GS 2, 530).[64] In this early analogy of the scientific revolution, “the pre-magical human [der vormagische Mensch]” (Neurath 1931b; GS 2, 530) had to succumb; “the magician” triumphed “over the not-yet-magician [Noch-nicht-Magier]” (Neurath 1931a, in Neurath 1973, 378; GS 1, 488).

The revision of Frazer’s theory, which Neurath implicitly proposes, goes in the opposite direction from Marett’s afore-mentioned assumption of the ‘magic-religious’. According to Marett, the magical and the religious were both present from the beginning, although still undifferentiated. With Neurath, however, both were still missing in the remotest age.[65] Very early humankind ignored animistic religion (contrary to what Tylor thought), but also a ‘magic-religious’ *mana* (Marett) and even (Frazer’s) ‘pure’ preanimistic magic.[66]

This conjecture of a pre-magical stage is not the only point where Neurath diverges from Frazer’s main hypothesis. At times, actually, the author of *The Golden Bough* himself seems ready to fall back on the more cautious theory of the mere primacy (and not full autocracy) of magic in the magical age. Likewise, Neurath occasionally assumes that religion already existed in the magical age, though it was subordinate to magic. What changes in the later religious age is that religion gains the priority and magic assumes the ancillary function religion had in the previous

[64]The other two revolutions are “*the revolution of the unmagical groups against the magical groups*” (Neurath 1931a, in Neurath 1973, 378; GS 1, 487) and “the *revolution of the social technicians* against the *theologically supported groups of the past*” (Neurath 1931a, in Neurath 1973, 378; GS 1, 488). In *Magic and Technology*, Neurath also asks, “But who will defeat the scholar?” (Neurath 1931a, in Neurath 1973, 378; GS 1, 488).

[65]“Originally humans confronted the world without judgement; gradually they came to seek for causes and effects and to emphasize certain actions which are later called magic when one no longer regards them as really effective. Originally there is no difference between the shooting of an arrow and the artistic drawing of an animal to be hunted, in order thereby to attract it. But today the shooting of arrows is classified with technology and the painting of the image with magic” (Neurath 1928a, in Neurath 1973, 285; GS 1, 283). With a kindred example, Wittgenstein (1931) intends to suggest just the opposite, namely, that the humans Frazer calls ‘savages’ are well aware of the difference: “The same savage, who stabs the picture of the enemy apparently in order to kill him, really builds his hut out of wood and carves his arrow skillfully and not in effigy” (Wittgenstein 1993, 125). On Wittgenstein’s argument, cf. Brusotti 2014, 157 ff.; for a comparison with Neurath, cf. Brusotti 2011, 348–352.

[66]Karl Beth (1914) had assumed an indifferent “pre-magical and pre-religious period” [“prä-magische und präreligiöse Periode”], but unlike Neurath, he rejected Frazer’s priority of magic: According to Beth, magic and religion could be of the same age. (On the hypothesis of a pre-magical stage, cf. also: Kaplan 1927.) In any case, Neurath is not alone when he assumes a ‘pre-magical phase’. Thus, his explanations are not based only on a direct reading of Frazer. Even if Neurath does not take note of or ignores the scathing criticism of Frazer’s hypotheses, which was already voiced at that time, variations on his theories (the hypothesis of a ‘pre-magic phase’) or terminological details (e.g., the explained use of ‘preanimistic’) reveal that he also knew about the controversial debate about *The Golden Bough*, at least to some extent.

stage.[67] Thus, Neurath distinguishes "preanimistic magic, probably the oldest," and "animistic magic" (Neurath 1931a, in Neurath 1973, 321; GS 1, 427–428).

As we shall see, Neurath is more appreciative of magic than the author of *The Golden Bough*. However, even if they do not merely differ in tone, their deeper agreement should not be overlooked. Already, *The Golden Bough* tends to value the archaic magical stage more than the later religious worldview (cf. Frazer 1922, 48–49; Connor 1990, 67; Fraser 1990a, 131). To the Scottish anthropologist, science seems more akin to magic than to religion. Science ties in again with the mundane framework of magic; they share the correct basic attitude, for both assume that inescapable causal laws govern the world: "in nature one event follows another necessarily and invariably" (Frazer 1922, 49), "the processes of nature are rigid and invariable in their operation" (Frazer 1922, 51). In the end, even if the evolutionary stages are (at least) three, there are only *two* parties struggling: Magic (in former times) and science (today) on the one hand, and religion on the other hand, are "the two conflicting views of the universe" (Frazer 1922, 51).

In this point, Neurath follows Frazer. Neurath, too, sees the new scientific world-conception picking up where its distant predecessor left off and taking up that magical phase again after the religious intermezzo.

4.10 Magic and Technology

"Thus the analogy between the magical and the scientific conceptions of the world is close" (Frazer 1922, 49). Likewise, Neurath claims that "the magical way of thinking is related to that of the modern scientific world-conception" (Neurath 1930a, in Neurath 1983, 34; GS 1, 373), above all: more closely related than religion and metaphysics are. "Magic has to do with the finite, the empirical" (Neurath 1930a, in Neurath 1983, 35; GS 1, 374). Thus, both magic and science proceed "from earthly, finite statements"; "the kind of connection is basically the same" even if the "contents" of science are new "in detail" (Neurath 1931b; GS 2, 529).[68] Unlike the "uncontrollable formulations" of later theologians about God, magical and other hypotheses of the early times "could be controlled by experience" (Neurath 1930c, in GS 1, 369). In contrast to the physicist, "the magician" sticks to tradition, but like the physicist, "the magician generally works finite changes [...] which can be perceived and therefore checked by everybody" (Neurath 1930a, in Neurath 1983, 34; GS 1, 373).

---

[67] "[...] Now there are technicians at the side of priests who replace the old magicians. Magic turns [...] partly into theology that used to be as much a side phenomenon as magic becomes now" (Neurath 1931a, partly in Neurath 1973, 378; GS 1, 488). In the modern technical world, "praying and using amulets" (Neurath 1931a, in GS 1, 488) are among such side phenomena.

[68] This is not specifically logical empiricism. According to Frazer's old-fashioned associationism of Humean provenance, the laws of association are the same everywhere, even in magic and science. Only the premises differ.

Frazer takes up an intellectualistic figure of thought from Tylor: the 'primitive philosopher' who speculates about the causes of natural events like thunder and lightning; in *The Golden Bough*, however, this 'philosopher' assumes a rather weird, lunatic aspect (cf. Stocking 1995, 145 f.). With his more favourable view of the 'primitives', Neurath is again closer to Tylor. Indeed, he goes much further. Neurath's primitive magician is first and foremost a rational empiricist, "just as the modern engineer or physician" (Neurath 1930a, in Neurath 1983, 35; GS 1, 374). The "magician" "is to be compared to a technician, a psychoanalyst or a surgeon" (Neurath 1930a, in Neurath 1983, 33; GS 1, 373).

The comparison sometimes sounds weird, for example, when Neurath calls the "buffalo dance" a "technical measure" (Neurath 1930a, in Neurath 1983, 34; GS 1, 373). For him, however, just this is the key to understanding, thus allowing the anthropologist to dismiss 'mystical' interpretations. "Magic may seem alien to us at first glance; still, in a certain respect, it is closer to modern physics or biology than theological thinking" (Neurath 1930a, in Neurath 1983, 33; GS 1, 373).

More explicitly and unambiguously than Frazer, Neurath prefers the early belief in magic to religion. By dubbing magic as "the bastard sister of science" (Frazer 1922, 50), Frazer maintains that they are related, but also emphasizes that they are *opposites*: magic stands for false, science for true causality. "It is therefore a truism, almost a tautology, to say that all magic is necessarily false and barren; for were it ever to become true and fruitful, it would no longer be magic but science" (Frazer 1922, 50).

Neurath apparently sticks to this definition, but turns the connection between magic and false causality from the *material* to the *formal mode* of speech: "We call all the supposedly effective habits of the primitives, which we reject, 'magic', whereas we call those we accept 'technology' or 'modern custom'" (Neurath 1931a, in GS 1, 484).[69] Similarly, we call certain "measures [Maßnahmen]" "ceremonies" only "if we think them 'ineffective'" (Neurath 1931a, in Neurath 1973, 319; GS 1, 426). Whereas Frazer takes for granted that we are right in doing so, Neurath intends to question this current use of the term 'magic'. What he thus criticizes as arbitrary is just the Scottish anthropologist's use of the word! Although evidently inspired by *The Golden Bough*, Neurath's consideration in the formal mode ends up turning against Frazer.

Leaving Neurath for a moment, Frazer's definition has its well-known shortcomings. It counts all false or misapplied empirical knowledge as magic. The author of *The Golden Bough* overlooks the difference between *false* empiricism and magic: Not all false theories and ineffective techniques belong *ipso facto* to magic; empirical notions may turn out to be false or misapplied, but this does not allow us to classify them as 'magic'. It depends on how the agents themselves view

[69]This passage is not translated in Neurath 1973, 376. "Those arrangements of ancient times which today we recognize as 'effective' we usually count as 'primitive technology'; [. . .] What we regard as useless we mostly call 'magic', and what we approve we call 'science'" (Neurath 1931a, in Neurath 1973, 322; GS 1, 428).

them. The anthropologist should look at the whole practice including 'the words they use'. Malinowski implicitly follows this criterion: the Trobrianders distinguish magic (*megwa*) from both effective and ineffective techniques.[70]

Neurath goes a different way. Whereas Frazer does not distinguish between magic and *false* empiricism, Neurath questions that the early times distinguished between magic and empiricism *tout court*. With him, the equation between magic and technology is legitimate; only the identification of magic with *ineffective* technology is arbitrary. He warns against distinguishing on the basis of real or alleged effectivity:

> [...] in order to assess the conduct and viewpoint of that time, it would be wrong to establish an antagonism between magical ceremonies and technology and perhaps to separate early technology from early magic by counting what we accept today as technology and what we reject, from the standpoint of modern knowledge, as magic. With respect to behaviour everything at that time belonged to the same level. (Neurath 1930a, in Neurath 1983, 35; GS 1, 374)

Strangely enough, the concept of 'magic act' [Zauberhandlung] is alien to the magical age: Neurath's 'primitives' do not yet distinguish between magic and technology. *We* draw this distinction. *We* speak of 'magic acts'. For *we* distinguish between the effective and the ineffective among their procedures, calling the first 'technology' and the latter 'magic'.[71] 'Magic' is a concept *of ours* and depends on how we as external observers judge the 'effectiveness' of the customs. Such judgements do not coincide with those of the age or culture object of inquiry.

By ascribing a proto-scientific worldview and a technical rationality to magic, the social reformer Neurath goes far beyond Frazer's condescending attitude. Neurath never equates 'magic' with 'superstition', whereas in *The Golden Bough*, magic is "the ancient system of superstition" (Frazer 1922, 11) and the primitive worldview is as a whole superstitious.[72] The Scottish anthropologist, even if he is anything but consistent, insists that humankind's first attempt to explain and master nature *fails*: the paradox is that his 'savages' seem to know only false theories and ineffective techniques. Neurath reverses the claim: Most primitive techniques are, after all,

---

[70]S. F. Nadel explains that "any study of primitive magic" must solve "the problem of distinguishing between magical procedures proper and others which represent merely empirical knowledge misapplied": "Whatever other facts we should consider, the crucial evidence must clearly be the people's own conceptions, and hence the words they use" (Nadel 1957, 196). According to Nadel, Malinowski uses this criterion, albeit without elaborating on it explicitly. Cf. Brusotti 2014, 181.

[71]Cf. Neurath 1928a, in Neurath 1973, 285; GS 1, 283.

[72]The *Abridged Edition* never gives something like a definition of the ubiquitous term 'superstition' (which is not even a lemma of the index). With Frazer, magic is superstition because, as we have seen, it is false and ineffective. In this, 'magic'/'superstition', that in an older tradition is the opposite of religion, is, rather, the opposite of science. *The Golden Bough* intends to show that (Christian) religion is haunted by magical survivals and cannot fully dispel superstition. Interestingly, however, the *Abridged Edition* never directly addresses religion or Christianity itself as superstition. Frazer endorses "the distinction between religion and superstition" (1922, s. 92) i.e., between religion and magic, claiming that this distinction "has emerged" (1922, s. 92) only "at a late period" (1922, s. 92).

effective. Only a few – as, by the way, some of our own – are ineffective, or so they seem *to us*, and *we* call them 'magic'.

Neurath emphasizes that judgements on effectivity are fallible and may need revision. The *historical* difference between modern and ancient does not coincide with the *substantial* distinction between 'effective' and 'ineffective'. The latter distinction depends on the state of research; a more advanced age will probably conclude that a few of our 'modern' techniques are 'ineffective'.[73] The example of "suggestion therapy", shows "how quickly our assessment of what 'true' technology is often changes" (Neurath 1930a, in Neurath 1983, 35; GS 1, 374).[74]

Even if Neurath insists on the variability of customs and traditions, he is far from making a relativism out of Frazer's theory. The difference between magic and science/technology does not lie only in the eye of the modern beholder. A purely analytical distinction that boiled down to a mere *façon de parler* would not be able to support genetic hypotheses. For this to work, the distinction must be *de re* and not just *de dicto*. Otherwise, he would end up simply with two scientific/technological ages separated by a religious interlude. However, Neurath still has the conceptual means to draw a line of demarcation between magic and science. Even if he does without Frazer's opposition of scientific (true) and magic (false) causality, he substitutes for it a whole range of differences between the magical and the technical age. The diagram attached to the report on his lecture "Magic and Technology" shows, as the only common ground between the two ages, the "earthly-finite" "integration [Eingliederung]" of tool *and* social technology. Incidentally, the tool technique is "earthly-finitely" integrated into society during all three ages, the middle theological era included, in which, unlike in the other two stages, the social technique is transcendently integrated. Magical and technical ages differ in turn (1) in the justification (by tradition in the former and by experiment in the latter), (2) in the degree of "logisticalization [Durchlogisierung]" (weak in the magical age and high in the other two), and (3) in the (respectively weak and high) "control by experience" (Neurath 1931b, in GS 2, 531). Thus, Neurath does not deny that it is possible to make a principled distinction between magic and science.

[73]Neurath relativizes the distinction between magic and science, but not the divide between magic and religion. He is far from admitting that this latter distinction too is applied to the phenomena by the western observer. Marett, despite his still evolutionistically oriented thought style, had clearly stated this point against Frazer. For a current analysis of the concept of religion, in which the question of Eurocentrism is addressed, see Figl 2003, 71 ff.

[74]On "suggestion therapy", cf. also Neurath 1931a, in GS 1, 484 (not translated in Neurath 1973, 376). Josephson-Storm (2017, 258–265) sees a connection between Neurath's views on magic and the interest in parapsychology by *other* members of the Vienna Circle and claims that Hahn, Schlick, von Mises, Feigl, and Carnap were not simply involved in uncovering deceptions. Josephson-Storm does not really make clear of what the connection with Neurath is supposed to consist. Not unlike Frazer, Neurath is adamant in rejecting any 'mystical' interpretation of magic. This also holds true for his judgements on suggestion therapy.

In comparison to Frazer, however, he definitely shifts the emphasis to the *affinity* between science and magic – and even more between 'us' and 'the primitives',[75] though he keeps using this latter concept as a matter of course.[76]

4.11 A Social Conception of Magic

The Durkheimians had rejected the developmental scheme outlined in *The Golden Bough* and clearly showed the sociological inadequacy of Frazer's individualistic and intellectualistic approach. Therefore, the stage theory that the sociologist Neurath adopts was already considered unsociological well before the beginning of the thirties. However, he does not take up Frazer's stage theory without substantially modifying it. Within the framework of his empirical sociology, which is also inspired by Marx, thought-figures from *The Golden Bough* receive a socio-theoretical interpretation far removed from Frazer's 'mental anthropology'. Neurath gives a Marxian twist to the analogy between magic and technology by explaining magic with reference to a specific form of production: as the technology especially of primitive agricultural peoples.

In his sociology, technology embraces "tool technology [Werkzeugtechnik]" and "social technology [Gesellschaftstechnik]" (Neurath 1931b; GS 2, 530). Magic had been both. Hence, whereas Frazer's magicians resemble weird natural scientists (when they are not simply impostors), Neurath's magicians are also proto-sociologists and social planners. Correspondingly, the Viennese provides a different explanation of the transition from magic to religion (and then from religion to science). In *The Golden Bough*, the transition to religion goes back to 'the birth of humility', as Marett dubs it: Finally, humankind was forced to admit the evident failure of magic as a *tool* technology. In Neurath's view, the transition was, rather, due to magic's inadequacy as a *social* technology (cf. Neurath 1931b; GS 2, 530). In the early days, magic had been the motor of rationalization, especially as innovative social technology. Like social engineers, magicians coordinated social processes by means of ritual control systems; for this reason, magical societies were superior to pre-magical ones. Hence, Neurath's claim: "There is a continuous line from the formulations of the magicians to the formulations of the sociologists" (Neurath 1931a, in Neurath 1973, 382; in GS 1, 492).

Later, however, magic was subsided by religion just in its function as a social technology. The coordinating power of single magicians was not comparable to that of religious institutions; "the 'old civilisations' of Egypt, Mesopotamia, India,

[75]Neurath reads Frazer himself in this more open sense when he compares him favourably with Lévy-Bruhl (see above p. 77 f.).

[76]Neurath (with Lévy-Bruhl, Graebner, etc.) rather uses 'primitives', whereas Frazer clearly opts for 'savages' (on Frazer's criticism of Lévy Bruhl in this regard, see Brusotti 2014, 69, n. 22). Both are obviously arbitrary (and discriminatory) categorizations by western observers. Cf. Fabian 1983, 18.

China, America" (Neurath 1931a, in Neurath 1973, 379; in GS 1, 489) display the superiority of religion in coordinating social processes through ritual systems of rules.

Thus, religion replaced magic "[w]ithin large areas with comprehensive organisation" (Neurath 1931a, in Neurath 1973, 379; in GS 1, 489). In these "old civilisations", "magical extrapolations" "were reduced [zurückgedrängt]" (Neurath 1931a, in Neurath 1973, 379; in GS 1, 489). "With a structure like that, peoples are superior to little tribes with their special cults and magical constraint in all technical matters" (Neurath 1931a, in Neurath 1973, 379; GS 1, 489).

With Neurath, who strives for a money- and market-free economy in kind [Naturalwirtschaft], ancient economies are to serve as models for future developments; "a central bureaucratic state" in which "a large-scale economy in kind [Großnaturalwirtschaft] played a significant role" can be found "in the Nile valley at the beginning of the third millennium B.C." (Neurath 1909, in Neurath 2004, 122; GS 4, 145).[77]

In the magic age, there had been no separation between magic and technology. In the religious age, however, the new distinction between *theology* and (tool) technology emerged.[78] The latter became a 'field' on its own, emancipated from the old magical constraints (cf. Neurath 1931b; GS 2, 530).[79] Where the magician used to be, there were now technicians on the one hand and priests on the other. The same religion that freed tool technology, in fact, constrained *social* technology (cf. Neurath 1931b; GS 2, 530). Hence, the task of modern science and especially of empirical sociology is the "liberation of social technology from theological ties" (Neurath 1931b; GS 2, 530).

Neurath gives a more sociological valence to the stage theory of *The Golden Bough* but is far from dismissing it. This may be all the more surprising as the author of *Anti-Spengler* (1921) was already very sceptical about stage theories, not just about *The Decline of the West*, but also about the hypotheses of Frobenius and even Lamprecht. Later on, however, Neurath, who continues to count Spengler to the scholars who use "the step theory" only "for ingenious parallels" (Neurath 1931a, in Neurath 1973, 331; GS 1, 437), takes quite seriously stage theories like the one put forward by Franz Müller-Lyer in *Phasen der Kultur* [1906] (cf. Neurath 1931a,

[77]On Egypt, cf. Neurath 1909, in Neurath 2004, 122–127; GS 4, 145–149. On the relationship between economy in kind and administrative economy, as well as on "comparative economics", which is to accompany a "doctrine of life orders", cf. the preface, in Neurath 2004, 120–121; GS 4, 137–138.

[78]"This unified science is the substitute for magic which also once encompassed the whole of life, although in the meantime the split into theology and technology has noticeably emerged" (Neurath 1931a, in Neurath 1973, 329; GS 1, 435).

[79]Neurath goes beyond Frazer with this idea that "theology had a rather liberating effect" "in the field of tool technology" (Neurath 1931b; GS 2, 530) that had been tied up by magic (Neurath 1930a (in Neurath 1983, 35; GS 1, 374) refers to Max Weber). Christianity was no longer "bound to countless daily customs"; it prescribed "only very specific ceremonies", but left "for the rest, however, free way for technical development" (Neurath 1931a, in GS 1, 488; not translated in Neurath 1973, 378).

in Neurath 1973, 331; GS 1, 437). The author of *Empirical Sociology* goes beyond the claim that "one cannot reject the strict step theory from the outset" (Neurath 1931a, in Neurath 1973, 332; GS 1, 437). Such a theory can be "a form of empiricist representation" and a fruitful one, even if only "as a first approximation"; the step theory stands for "the most vigorous turning away from metaphysics" (Neurath 1931a, in Neurath 1973, 332; GS 1, 437).

> One can elaborate the step theory of comparative history, one can do more with a comparative ethnology, whose object ultimately is mankind, one can bring out the territorial distribution of types of culture, their overlapping, etc. (cf., e.g., Frobenius, Schmidt, Graebner) [. . .]. (Neurath 1931a, in Neurath 1973, 333; GS 1, 438)[80]

If Neurath adopts Frazer's developmental scheme, which was already controversial at the time, this is due primarily to his enlightening concern and his interests in cultural policy. He is attracted to an image of cultural development in which religion appears as an interlude that has to be overtaken.[81] That is why *The Golden Bough* is interesting for him (and a provocation for Wittgenstein). The lecture on *Magic and Technology* concludes:

> But it is essential that technology does not have to clear away millennia-old rubble with the help of the standard science, but in very many ways ties in with primordial habits of magic, which are overlaid by the theology of only a few centuries. *Technology is a continuation on a higher level.* (Neurath 1931b; GS 2, 531)

A good dozen years later, Frazerian inspirations have markedly receded. As late as 1944, Neurath still acknowledges the Scottish anthropologist's relevance, but recognition is now mixed with reservation.

> The importance of Westermarck, Frazer and other scholars who tried to present rich comparative material, more or less arranged, may be based on their capacity for destroying fixed opinions on so-called 'human nature.' (Neurath 1944, 27)[82]

That Frazer's approach is outdated will have not escaped the author of *Foundations*, who does not seem utterly convinced by the way the rich ethnographic material is

---

[80]The anthropologists mentioned here – Frobenius, Schmidt, and Graebner – stand for diffusionism. Frazer abhors it. Hence, this passage betrays a certain difference between Neurath's and Frazer's positions. Neurath's stage theory – as well as his conception of the "human and cultural cover [Menschen- und Kulturdecke]" (Neurath 1931a, in Neurath 1973, 332; GS 1, 437) – has a strong diffusionist twist.

[81]On theology as "a stage of transition [Durchgangsstadium]", cf. Neurath 1931a, in Neurath 1973, 324; GS 1, 430. Frazer is clearly more conservative than Neurath. Since the first decade of the twentieth century, he feels threatened by the impending social upheavals and increasingly emphasizes that dangerous irrational forces are slumbering and continue to work under the thin surface of civilization. Just the upheavals Frazer is afraid of are a reason to hope for Neurath, who is more confident about the outcome of the perceived struggle between science and religion and has a more positive attitude not only towards the 'primitives' but also towards the working class.

[82]Edvard Westermarck (1862–1939), who belongs to the evolutionary tradition, had written *The History of Human Marriage* (Westermarck 1891), a classic monograph that, although outdated, was relevant to the theme of Neurath's Malinowski review. On Westermarck, cf., for instance, Stocking 1995, 151–163.

"arranged" in *The Golden Bough* or in *Totemism and Exogamy*. What does Neurath mean by Frazer's (and Westermarck's) "capacity for destroying fixed opinions"? *The Golden Bough* described 'savages' that, even if very different from 'us', were supposed to be close to our own ancient forebears. Allegedly, they stood for humankind's (and Europe's) remote past and, thus, challenged the self-image of modern western civilisation. A wide popular reception – and Freud himself – saw Frazer's (alleged) discovery of modernity's primeval background as comparable to and affine to the revealing of repressed feelings in psychoanalysis.

In the early thirties, Neurath sees, in Frazer's view of primeval magic, a proof that religion is not inscribed in 'human nature' from the very beginning. He follows the Scottish anthropologist in considering primeval 'magic' as religion-free, theology as merely an intermediate stage, and religion as a world-historical parenthesis. He tends to minimize the difference between the 'primitive' magician and the modern scientist and has a more favourable view of the peoples he still calls 'primitives' than the evolutionary British tradition leading up to Frazer. From Neurath's open-mindedness, however, 'magic' benefits rather than religion.

4.12 Weakening the Analogy Between Magic and Technological Measures

Whereas "Magic and Technology" is rather focussed on prehistoric origins, "Mother Right and Magic" addresses the actual forms of life of then-contemporary non-European peoples. Even in this Malinowski review, however, there is a noticeable tendency to interpret 'magical' actions essentially as technical measures. The following statement passes off Neurath's own theory as the content of the reviewed ethnography. "One learns continuously what 'magic' means for these primitive peoples, the spell, the ceremony, which are used by the primitives in much the same way as our technical actions [Handgriffe]" (Neurath 1931c, 100). An argument for the analogy between 'us' and the Trobrianders is that even a few measures currently used in contemporary western medicine are likely to prove ineffective in the future. In the Trobriands, "[t]here is a wealth of potions and rules to follow, just as we use all kinds of ointments, concoctions and medicines, visits to baths, suggestions, analyses etc., some of which will probably be declared ineffective by a later medicine" (Neurath 1931c, 101–102).

Neurath's view of magic as primitive technology corresponds rather to Frazer's conjecture than to Malinowski's theory, which ascribes different functions to magic and technology. Accordingly, magic, religion, and science do not stand for three separate stages of human development but coexist in all societies.[83] The

[83] *The Golden Bough* is a palimpsest: In addition to the 'official' theory that science is a late stage of development reserved for the West, Frazer advances the incompatible thesis that magic and science coexist in all societies. This thesis is usually associated with Malinowski, who argues it clearly and

Trobrianders do not employ 'magic' (*megwa*), say, for the rather trouble-free coastal voyage, but only for the far more dangerous and unpredictable deep-sea voyage. They use *megwa* only where their technological skills reach their limits. As often, Malinowski generalizes observations about his Argonauts of the Western Pacific to 'primitives' on the whole, even if, interestingly, the Trobrianders had a word he translated as 'magic' (*megwa*), but none he could translate as 'religion' (cf. Tambiah 1990, 68).

Neurath and Malinowski share the view that 'primitives' are somewhat rational empiricists. Malinowski, however, counts this rational empiricism as 'science' rather than 'magic', which has a different application context and a different function (reducing anxiety). Neurath instead denies that 'primitives' distinguish between 'science' and 'magic', which is why he holds rational empiricism for a general feature of their way of life.

In any event, in *The Sexual Life of Savages*, Neurath could hardly find general theoretical statements contradicting his Frazerian interpretation of magic. Malinowski does not present his theory of magic in this book. Furthermore, he frequently describes kinds of magic that do not necessarily resist the interpretation as technical measures, for instance, love magic.

Nevertheless, in Neurath's review, the analogy between magic procedures and technical measures, even if still prominent, is repeatedly weakened and restricted:

> And when magic ceremonies are reported, we often forget that our technical and extra-technical measures will not appear much different to a Martian who does ethnological studies in Europe. Think what ceremonies a skier has, from greasing the skin to using certain *preferred colors* of clothing, to using certain strangely colorful ribbons and other aids, whose importance for the ride is difficult to explain (Neurath 1931c, 99).

Here, the analogy between two technologies, primitive (magic) and modern, gives way to a comparison between rituals, between "magical ceremonies" and a modern skier's "ceremonies", both of which have technological and non-technological aspects; in both cases, the involved "measures" are not just technological ones. "Even for us, moments other than technology and hygiene are decisive. [...]" (Neurath 1931c, 101–102). Thus, the form a skier's equipment assumes does not have exclusively obvious technological reasons, but is, among others, aesthetically and socially motivated. Things like fashion play a role. Magical procedures, too, have reasons that are not easy to classify and that might roughly correspond to the ones that are involved when a skier chooses his equipment.

---

firmly in his *Magic, Science and Religion* (Malinowski 1925). In this point, however, he ties in with Frazer, even if not with his 'official' theory. Strangely enough, one can say at the same time that in this point, Malinowski (1) opposes Frazer (his 'official' theory) and (2) uncritically follows him (his 'inofficial' theory) (Cf. Brusotti 2014, 156–157), The many inadequacies Malinowski's theory is fraught with are not at issue here. Among others, his unhistorical universalising claim that all peoples know something like 'science' involves an unqualified use of this term counting, as 'science', any form of empiricism, how unsystematic it may be (cf. Nadel 1957:198).

Thus, it does not seem difficult to understand 'magic'. We have only to compare it to our own everyday practices. This is one of two – positive – answers that Neurath's review gives to the hermeneutic question:

(1) We understand other cultures if a (Marxist) theory allows us to see through the causal "connection between mode of production and fashioning of life" (Neurath 1931c, 99); as weird as 'magic' may seem to us, we can explain it through the peculiarities of agriculture plus mother right.
(2) 'Magic' is self-explanatory anyway. The general human qualities at work in what we call 'magic' can easily be highlighted through simple "parallels [Parallelen]" to our form of life, for instance, to a modern skier's doings in their technical and aesthetic aspects.

Do such parallels make 'magic' seem natural independently from any causal explanation? Do comparisons make 'magic' plausible even before the Marxist (or whoever) explains it by agriculture plus mother right? Wittgenstein argues just this: Descriptive elucidation is independent of causal explanation. According to him, these two models stand for two quite different conceptions of 'explanation'. Neurath does not see the two ways of making magic plausible as an alternative but, rather, as complementary, though he does not explicitly state this. He does employ comparisons with descriptions that are no less imaginary and fictional than Wittgenstein's. However, he clearly appreciates Malinowski's extensive field research and is far from rejecting an empirical approach in which descriptions are the result of objective observation and comparisons can be tested. Rather than seeing in anthropology a merely descriptive approach, he insists on the causally explanatory task of anthropological investigation. In this last regard, he goes far beyond Malinowski himself.

The "language of ethnology" (Neurath 1931c, 100), as Neurath conceives it, must be 'objective'. We may understand foreign customs by analogy to ours: 'magic' by analogy to skiing. However, Neurath wonders whether we understand practices with which we are well acquainted objectively enough. Current practices like skiing should be described in the same 'objective' language that anthropologists have devised to describe 'primitive magic'. Would our use of this objective language not involve us also understanding skiing by analogy to magic, familiar customs by analogy to foreign ones? At least, the new way of describing is supposed to highlight that far from being determined only by technology, our customs, too, have other, say, aesthetic and ceremonial aspects. In any case, one must switch to the social-behaviouristic language to understand a foreign culture or even one's own.

4.13 Mars and Marx. An Alien Anthropologist on Victorian Rule Following

Neurath's "ethnologist from Mars" (Neurath 1931c, 100) already appears in Malinowski's book, where an "ethnographer from Mars" must deal with a proverbially prudish informant, a "mid-Victorian middle-aged gentleman or spinster" (Malinowski 1929a, 506).

> Imagine an ethnographer from Mars inquiring of our respectable gentleman (or spinster) about matrimonial morals in England. He would be told that monogamy is the one and only form of marriage, that chastity is required of both parties before marriage and that adultery is strictly forbidden by law, morals, manners, and our code of honour. All this is, in a way, quite true: it embodies the authorised ideal of religion and morality. And if the Martian went on to inquire whether adultery occurs in practice, our gentleman (or spinster) would resent the question as an implied insult and would grow cold or hot over it. [...] If the Martian, versed in the modern method of field-work, as recommended by some schools of Anthropology, should proceed to the 'concrete manner of questioning,' he might really get into trouble. To the concrete inquiry: 'How many times have you slept with your friend's wife, and how often has your wife slept with another man?' – the answer would not be verbal but behaviouristic. And the Martian, if in a position to do so, would enter in his note-book: 'The natives of the terrestrial planet never commit adultery; [...].' This statement would be obviously one-sided, and yet the terrestrial informant was in no way trying to deceive the inquirer. [...] The anthropologist with his rapid and at times penetrating questions, with his insistence on fact and on concrete detail, arouses the same reactions as would the hypothetical inquirer from Mars among us. [...] (Malinowski 1929a, 506–508)

Malinowski's Martian stands for "the hasty field-worker, who relies completely upon the question-and-answer method" (Malinowski 1929a, 509); the reason for his misunderstanding is the inadequacy of the methods of field research in which he was trained. The one-sidedness of his report shows the failure of the then-common use of ready-made forms such as the Royal Anthropological Institute's *Notes and Queries on Anthropology*. The Martian's shortcomings show that in the "divergence between the hearsay method of collecting evidence and first-hand experience of savage life, we have a very important source of ethnographic error" (Malinowski 1929a, 506).

The "hearsay method" does not take into account the "discrepancy between theory and practice" (Malinowski 1929a, 472), common (not only) to Trobrianders and mid-Victorians.[84] The imaginary ethnographer from Mars gives a one-sided report because he does not even ask whether and to what extent the people really follow the 'official' rules. "For in actual life rules are never entirely conformed to, and it remains, as the most difficult but indispensable part of the ethnographer's work, to ascertain the extent and mechanism of the deviations" (Malinowski 1929a, 509).

[84]Malinowski's analogy between the Trobrianders and the proverbially prudish mid-Victorians is not without irony against the then-common prejudice against the allegedly unrestrained 'savages'.

An awareness of the "discrepancy between theory and practice" (Malinowski 1929a, 472) may also be necessary with regard to the Trobrianders' 'supreme taboo', the incest between brother and sister, for sometimes, though quite seldom, even this taboo is violated. However, in "A Savage Myth of Incest", the last chapter of *The Sexual Life of Savages*, the main point is not the *real* discrepancy between what humans *say* and what they *do*. Rather, Malinowski describes an *apparent* discrepancy *internal to what is said*, more precisely, "a singular rift in traditional doctrine, a dogmatic inconsistency, which makes love and the magic of love derive from brother and sister incest" (Malinowski 1929a, 536). Why should what is most valued (love and love magic) stem from what is most condemned (brother and sister incest)? Is the myth in contradiction to Trobriand morality? Malinowski resolves the dilemma by describing the function of the myth: It praises the power of love magic, which is able to overcome any obstacle, even the supreme taboo, the most abhorred form of incest. The origin of love magic testifies to its power. The contradiction between myth and morality is mere appearance.[85]

Neurath agrees with Malinowski that it is a major difficulty for the ethnologist to understand whether and to what extent people really follow the 'official' rules that they nominally acknowledge and profess. For the Trobriand myth, the reviewer devises an analogy from (then) modern Christianity:

> In doing so, the author [Malinowski; MB] distinguishes more carefully than many other ethnologists between the rules preached by the natives and the behaviour that is common practice. Think how incorrectly a Martian would report about us, for example if he only related that sexual intercourse before marriage is rejected, adultery even punished. Malinowski shows that a number of very strict 'taboos' [...] are circumvented in love life, for example by using a counter spell, [...] Many primitive life forms seem strange to us because we are not used to see our own way of life 'objectively'. An American sociologist has recently published a book 'Middletown' in which he talks about the people of his time roughly as an ethnologist would. Just think how the lives of many Central Europeans would look in the language of ethnology. Not to mention the shaving ceremonies, which are accompanied, for example, by strange, vacuous sayings about the weather and politics. The use of certain shapes of the hat, the movement of this piece of furniture when you meet people, provides a chapter of its own. And some people will be very upset when they read, without the religious phrases they are familiar with, that on certain days they 'eat their God', and that the Martian ethnologist claims that the Son of God is miraculously the son of an unmarried woman, although without miracles she would be blamed for having a child. There are parallels to all this among the Trobrianders. There, the most terrible thing a person can do is to have sexual intercourse with his sister, and yet the

[85]"In the social sciences positive results form the backbone of success. [...] Malinowski expected, in accordance with Freud's hypothesis, a Melanesian tale telling of a brother-sister marriage (just because the brother-sister taboo is particularly strong). Negative instances could hardly shake such a hypothesis, but the finding of such a tale corroborated it" (Neurath 1944, 26). Perhaps from memory, this late interpretation emphasizes the agreement between Malinowski and Freud: They share the hypothesis that the most strictly forbidden incest must be the subject of a myth. Malinowski sees this common hypothesis confirmed when he finds the relevant myth in the Trobriand Islands, with the sole difference being that here, the myth concerns the incest between brother and sister (not, as with Freud, between mother and son). Thus, the late Neurath interprets Malinowski differently than in the 1931 review.

legend tells that a particularly strong love spell originates from a love relationship between brother and sister. The peculiar and familiar mixture of life and rule, of truthfulness, of conscious and unconscious hypocrisy, through which traditional forms are dissolved, comes out excellently in Malinowski's work. [...] (Neurath 1931c, 100)

Neurath starts by picking up what Malinowski says about his Martian ethnographer. The alien begins by producing a report that is correct but unpalatable to the described people and, so, involuntarily offends them. Finally, however, he ends up agreeing with the mid-Victorians. However, just this agreement involves a misunderstanding: He gives a one-sided description that flatters them, for it reports only the official rules, acquiescing in their idealised self-image and mistaking their professed ideal for the reality of their life.

Neurath's Martian anthropologist begins where Malinowski's ends: He takes for granted what the Europeans say and reports "incorrectly" about the same issues. Neurath's Martian's first report and Malinowski's Martian's second report make the same mistake: They take the natives at face value, overlook the real application of the rules, and miss the role they actually play in European life. After this flawed description, which leads to an apparent agreement, Neurath's Martian goes on to report a baffling contradiction he finds in Christianity. This second report – about Christians who 'eat their God', etc. – has just the opposite outcome, namely, an evident disagreement with the natives. This report does not state the rules in the way they would put them. They do not recognize themselves therein. The "Martian point of view" (to use Wittgenstein's term) on contemporary Christians is just not their point of view.

The review leaves it at the mere statement of the (alleged) contradiction, without giving further explanations. Is the Martian's second report supposed to be correct? If so, it holds up a mirror to Europeans like, say, Montesquieu's *Persian Letters*. It gives the real rules unnoticed by the agents, and these really are inconsequent. If the new report really has this critical function, (a) Neurath's Martian does not employ a faulty method like Malinowski's, and (b) Neurath does not see the Trobriand myth in the same way as Malinowski, who points out that it *only apparently* contradicts Trobriand morality. For Neurath's Martian report on the Son of God is not supposed to stand for a merely apparent contradiction, arising just because "the ethnologist from Mars" employs an inadequate method. Neurath does not mean that the Martian's view from afar is a distorted representation of Christian ways of life – hence, the difference between Neurath and Malinowski. (1) In general, Malinowski points out that a *real* discrepancy between professed norms and actual behaviour exists everywhere (his Martian overlooks *this* discrepancy). (2) With reference to the Trobriand myth, Malinowski claims that, here, there seems to be an inconsistency between norms and that *this* inconsistency is *merely apparent*.[86] Neurath does not take care of this interpretation and mentions the Trobriand myth and his Christian analogy as cases of (1), claiming that here we find the same discrepancy between

[86]By the way, in this later chapter of *The Sexual Life*, the Martian does not show up.

theory and practice as anywhere. Neurath's Martian does not hold this inconsistency for merely apparent, and this does not mean that he misunderstands us.

While Malinowski's Martian simply stands for a false method leading to a deep misunderstanding, Neurath's Martian is a more ambivalent figure. In his first report, he, too, employs a similarly misleading method; his second report, however, stands for a methodological *ideal.*

How so? Neurath sees the core epistemological problem in the asymmetry between the anthropologists' 'emic' view of their own society and their 'etic' view of foreign societies. The reason why the then-usual 'European' anthropologists[87] exaggerate the gap between western and other societies is that they look at the former 'from within' and at the latter 'from outside'. They consider only other societies 'objectively', and not their own. "Many primitive forms of life seem strange to us because we are not used to seeing our own way of life 'objectively'." (Neurath 1931c, 100). If we managed to look at our own way of life from afar, the hermeneutic problems with 'primitive' societies and their seemingly alien ceremonies would disappear.

The "inhabitant of Mars who carries out ethnological studies in Europe" (Neurath 1931c, 99) would look at our (*not at his own*) society from an external point of view. On the Europeans whom he explores, he would adopt the same external point of view from which European ethnologists look at the Trobrianders. Such an external point of view is not necessarily objective and unbiased. European ethnologists do have a prejudice in favour of their own society. Would the "ethnologist from Mars" (Neurath 1931c, 100) not be biased in favour of Martian culture? Neurath does not even mention this possible prejudice. What matters in the review is only that the Martian would not be hampered by positive prejudice regarding European culture. Strangely enough, however, Neurath does *not* give him the task of comparing the Europeans and the Trobrianders. Given this task, the Martian would be in a privileged position, for he would not belong to one of the two parties he had to compare. He would not surmise any substantial difference between the Europeans and the Trobrianders: 'We' and 'they' would be on a par. For him, there would be the symmetry between Europeans and Trobrianders for which Neurath advocates. From his uninvolved standpoint, the "ethnologist from Mars" (Neurath 1931c, 100) would be able to see and adequately appreciate the "parallels from our life and from that of the Trobrianders" (Neurath 1931c, 102) that escape current 'European' anthropologists and are important to Neurath.

However, this freedom from bias would come at a price. Both cultures he would have to compare would be alien to him. If he wanted to know *both of them* "from their point of view" (Neurath 1931c, 99), he would get much more work to do than Malinowski. However, the ethnologist from Mars does not even try to be an insider of the foreign societies. While Malinowski's method is participant observation, Neurath insists on a rather impersonal observation from an external

[87] Neurath does not mean Malinowski.

standpoint.[88] The Viennese is concerned primarily with the anthropologist turning off his "native point of view" towards his own culture as much as possible. In the name of symmetry, Neurath opts for an etic perspective as a matter of course: Only the etic perspective is objective, while emic views are inescapably biased.

Taking an external standpoint requires switching to a social-behaviouristic language different from that used by the agents, whether these belong to a foreign culture or to one's own. Neurath's Martian does not use the everyday language of the people he investigates; he disturbs them because he does not employ "the religious phrases they are familiar with" (Neurath 1931c, 100). Instead, he describes their customs and beliefs in the objective "language of ethnology" (Neurath 1931c, 100). Neurath sees a similar detachment as a methodological peculiarity of Marxism: "in its description of interrelationships expressible as laws", it "makes as little use as possible of what people state about themselves, their 'processes of consciousness', their 'ideology'" (Neurath 1931e, in Neurath 1983, 80; GS 2, 553).

4.14 *Middletown* and the "Social Anthropology of Contemporary Life"

Neurath's empirical sociologist, who sees western society with the same distant gaze with which western ethnologists observe foreign societies, has among his models Robert and Helen Lynd's pathbreaking study. "An American sociologist recently published a book 'Middletown', in which he speaks about the people of his time roughly like an ethnologist" (Neurath 1931c, 100).[89] Indeed, *Middletown* is – as Clark Wissler qualifies it in his short foreword – "a pioneer attempt to deal with a sample American community after the manner of social anthropology" (Lynd 1929, VI).

---

[88]This reverberates in their approach to language. Malinowski's contextualist method of describing the "use of the word" (Malinowski 1929a, 461), that is important for Wittgenstein (see Brusotti 2014: 346–353), does not seem to appeal to Neurath. Sceptical about the use of queries, Malinowski insists that the use of, say, the word *bomala* (taboo) (Malinowski 1929a, 461) or of the "word pair" (Malinowski 1929a, 464) *bwoyna* and *gaga* (good and bad) must be described with all its "shades of meaning" (Malinowski 1929a, 465) as well as with their manifold applications in different contexts. Neurath's review does not take notice of this method.

[89]Neurath uses the singular masculine ('An American sociologist' [Ein amerikanischer Soziologe]), because the first edition of *Middletown* mentions only the husband as the proper author ('Robert S. Lynd *with* Helen M. Lynd'). Later editions list both spouses as authors, although still in reverse order ('Robert S. Lynd & Helen M. Lynd'). According to *Modern Man in the Making* (1939), Robert and Helen Lynd adopt "the approach of the cultural anthropologist" (Lynd 1929, 3) and their "social anthropology of contemporary life" provides "a vivid picture of a city moving towards modernity" (Neurath 1939, 144, note to p. 45), approaching "the life of the people [...] as a unit complex of interwoven trends of behavior" (Lynd 1929, 3, quoted (as p. 4) in Neurath 1939, 144, note to p. 45). This corresponds to Neurath's own social-behaviorism.

> To most people, anthropology is a mass of curious information about savages [...]. What is not realized is that anthropology deals with the communities of mankind, [...]; and whatever may be the deficiencies of anthropology, it achieves a large measure of objectivity, because anthropologists are by the nature of the case 'outsiders.' To study ourselves as through the eye of an outsider is the basic difficulty in social science, **and may be insurmountable**, but the authors of this volume have made a serious attempt, by approaching an American community as an anthropologist does a primitive tribe. (Lynd 1929, VI; emphasis mine)

Wissler shares this ideal of objectivity: The basic, even if probably unattainable, methodological objective of sociology is to study one's own society "as through the eye of an outsider" (Lynd 1929, VI). In the first chapter of their introduction, the Lynds themselves acknowledge "the danger, never wholly avoidable, of not being completely objective in viewing a culture in which one's life is imbedded [sic; MB]" (Lynd 1929, 3). Thus, what they are striving for is "to gain precisely that degree of objectivity and perspective with which we view 'savage' peoples" (Lynd 1929, 5).[90] In this respect, the greatest difficulties lie not in understanding *foreign* cultures, *but in understanding our own culture*. Neurath shares this assessment of the situation.

Malinowski too noticed *Middletown*. "'Anthropology begins at home'" (Malinowski [1938], vii), he writes a little over a decade after Neurath's review, welcoming the fact that the Lynds and other researchers "are directing the technique, method, and aims of anthropology on to our own civilization" (Malinowski [1938], vii).[91] Malinowski claims that, even if an insider may be biased, "an educated and trained member of a community is entitled to observe it with profit and competence" (Malinowski [1938], vii). To that extent, the anthropologist endorses the Lynds' efforts to legitimize research conducted by insiders; however, what Malinowski is keen to emphasize is that this applies globally, not only to research on western societies conducted by westerners but also to an ethnography of the Kikuyu written by a Kikuyu anthropologist.[92]

---

[90]"By viewing the institutional life of this city as simply the form which human behavior under this particular set of conditions has come to assume, it is hoped that the study has been lifted on to an impersonal plane that will save it from the otherwise inevitable charge at certain points of seeming to deal in personalities or to criticize the local life. For, after all, having one's accustomed ways scrutinized by an outsider may be disconcerting at best. [...] To many of us who might be quite willing to discuss dispassionately the quaintly patterned ways of behaving that make up the customs of uncivilized peoples, it is distinctly distasteful to turn with equal candor to the life of which we are a local ornament. Yet nothing can be more enlightening than to gain precisely that degree of objectivity and perspective with which we view 'savage' peoples. Even though such a venture in contemporary anthropology may be somewhat hazy and distorted, the very trial may yield a degree of detachment indispensable for clearer vision" (Lynd 1929, 3–4).

[91]"'Anthropology begins at home' has become the watchword of modern social science. Mass-observation and 'Northtown' in England; 'Middletown' in U.S.A. [...] all these are directing the technique, method, and aims of anthropology on to our own civilisation" (Malinowski [1938], vii).

[92]"'Anthropology begins at home' is as true of Africa as of Europe" (Malinowski [1938], viii). Malinowski's text is an introduction to a monography on the Kikuyu written by a native: Jomo Kenyatta, a trained anthropologist whom he had mentored at the London School of Economics and who would later become the first president of Kenya.

However, Malinowski himself, who conducts his fieldwork on the Trobriand Islands as an outsider, does not need such a legitimation. The Lynds, on the other hand, intend to objectify community life in the contemporary United States and to observe this familiar society in an "impersonal" way and as far as possible from an external point of view (among others by making use of quantitative methods). While they legitimate their investigation by arguing for the insider's ability to take the point of view of the outsider,[93] Malinowski, whose ethnographies deal with a foreign society, rather insists on the outsider's attempt to come somehow closer to taking the point of view of the insider. An ethnologist like Malinowski may start from a position far outside the foreign society but aims to reduce the gap and, *even if he never becomes an insider*, his method is *participant* observation. He wants to see things as far as possible "from the native point of view" (Malinowski 1929a, 453), i.e., to view the facts "in their true relation to tribal life" (Malinowski 1929a, 453). This is the ethnographer's "final goal": "to grasp the native's point of view, his relation to life, to realise *his* vision of *his* world" (Malinowski [1922], 25).

Both Malinowski and the Lynds want what is most difficult: The Lynds' main objective is an 'etic' view of their own society, Malinowski's an 'emic' view of a foreign society.[94] This does not mean that the Lynds' view of society is exclusively etic and Malinowski's exclusively emic. However, they set different accents.

In this respect, Neurath is closer to the Lynds. Like them, he is concerned with objectivity: What Malinowski achieves on the Trobriand Islands, European sociologists should accomplish in investigating their own societies. Equipped with the appropriate scientific methods, Neurath's ethnologist from an alien planet, who would effortlessly consider both societies from an external point of view, would easily meet the Lynds' expectations. Neurath does express his commitment to the task of "getting to know foreign peoples from their point of view as far as possible" (Neurath 1931c, 99). However, his ethnologist from Mars does not strive for a "native point of view" (Malinowski 1929a, 453). Neurath's main concern is that anthropologists largely eschew the emic bias. His anthropologists, whether they come from Europe or Mars, do not really seem to be participant observers à la

[93]*Middletown* does not mention Malinowski. The model most referred to belongs to an earlier generation of anthropologists: "Rivers begins his study of the Todas with an account of the ritual of the buffalo dairy, because 'the ideas borrowed from the ritual of the dairy so pervade the whole of Toda ceremonial'. A similar situation leads to the treatment of the activities of Middletown concerned with getting a living first among the six groups of activities to be described" (Lynd 1929, 21). These six groups correspond to W. H. R. Rivers' sixfold classification of social groupings (cf. Lynd 1929, 4 n.). For other references to him, cf. Lynd 1929, 5, n., 132 n., 249 n. Rivers (cf. Hviding and Berg 2014) was an important field worker and experimental psychologist who, among others, had taken part in the Cambridge Anthropological Expedition to the Torres Straits led by A. C. Haddon.

[94]Malinowski does not use the term 'emic'. The distinction between 'emic' (through the eyes of an insider) and 'etic' (from the point of view of an external observer), which goes back to Kenneth Pike, is terminologically based on the difference between 'phonemic' and 'phonetic'. One critic of "the emic bias" was Marvin Harris. Cf. the chapter "Emics, Etics, and the New Ethnography", in: Harris 1968, 568–604.

Malinowski. Neurath's appreciation for the anthropologist's practical empathy does not involve a real commitment to his participant method.

4.15 The "Gradual Convergence of All Forms of Life"

One reason why the social behaviourist is concerned primarily with setting aside his "native point of view" towards his own culture to the extent possible is that, according to Neurath, cultural specificity will soon belong to the past. "The modern economy destroys the primitive cultures, it eliminates the medieval ancient cultures of the Chinese and Indians and creates a unified world of the proletariat!"[95] Neurath is critical of "colonial politics", questions whether "Europeanization" is really "necessary" (Neurath 1931c, 99), and knows very well that it usually represents "a serious damage to primitives" (Neurath 1931c, 99). However, even if he sees in the "Unification of Mankind" a result of capitalism,[96] he finally welcomes "the gradual convergence of all forms of life" (Neurath 1931c, 99). For he hopes that capitalistic globalisation will lead to a proletarian reaction in the sense of the Socialist International: the rise of a global class consciousness.[97]

With this optimistic and modernist drive, Neurath sees the peoples he keeps calling 'primitives' more positively than Frazer and the evolutionary tradition but is less sensitive to cultural differences than the British and American anthropologists of his own generation. His forecast that the specificity of local cultures will soon belong to the past is linked to this tendency to underestimate the scope of cultural diversity. In 1921, he endorses Keyserling's "cosmopolitanism that does not want to erase differences" (Neurath [1921a], in GS 1, 199), but actually this very cosmopolitanism, which is certainly worthy as an ethical attitude, carries with it a certain propensity for underrating cultural differences. Neurath *rightly* criticizes Spengler for exaggerating and essentializing them.[98] Emphasizing such differences seems suspicious to the Viennese, reactionary, nationalistic or worse. Alas, this

[95]"Die moderne Wirtschaft zerstört die primitiven Kulturen, sie beseitigt die mittelalterlichen Altkulturen der Chinesen und Inder und schafft eine einheitliche Welt des Proletariats!" (Neurath 1928b, 676).

[96]"Unification of Mankind" is the heading of a chapter of *Modern Man in the Making* (1939, 19–42).

[97]"Wer ernstlich die *Internationale* ersehnt, muß sich heute schon wünschen, fremde Völker möglichst von ihrem Standpunkt aus kennenzulernen, sich mit ihnen eins zu fühlen, um auch zu begreifen, auf welchem Wege die allmähliche Annäherung aller Lebensformen wohl vor sich gehen werde. Gegenwärtig ist die Europäisierung meist eine schwere Schädigung der Primitiven. Ist sie notwendig? Die großen Probleme der Kolonialpolitik tauchen auf." (Neurath 1931c, 99).

[98]The author of *Anti-Spengler* seems a bit more sceptical than the admirer of Keyserling: Even if he does not straightforwardly deny one's ability "to let" alien feelings "arise" within oneself (Neurath 1921b, 80; in GS 1, 187), he points out that it may be hard to verify how successful the corresponding attempts really are. Cf. Brusotti 2011, 347–348.

scepticism towards cultural diversity, even if understandable in the historical context of the time, is inclined to overshoot the mark.

The author of *Empirical sociology* intended to pursue "'political economy' and 'history' as 'ethnology of our age'" (Neurath 1931a, in GS 1, 424; not translated in Neurath 1973). This ethnology of modernity was not without analogies to "the social anthropology of contemporary life" (Lynd 1929, VI, quoted in Neurath 1939) envisaged in *Middletown*. This comprehensive programme remains a central issue in the emigrant's late writings, in which the anthropological investigation of "money reckoning" (Neurath [1944], 39) figures as "a piece of modern ethnology":

> We should suggest looking at markets and finance and at the whole reckoning in money as an institution like any other, such as funeral rites, golf, rowing, and hunting. [...] We should treat this whole field of money reckoning anthropologically as a piece of modern ethnology. (Neurath [1944], 39)

In *Foundations of the Social Sciences*, written some thirteen years later than the texts with which my paper was mainly concerned, Neurath still does *not* recommend treating the wide-ranging programme of a "modern ethnology" (Neurath [1944], 39) separately from the ethnology of "'preliterate' tribes" (Neurath [1944], 27).[99] However, he does not draw a parallel between the anthropology of money calculation and the anthropology of magic and religion. At the beginning of the thirties, he could have argued that money calculation has the same function of social coordination as magic and religion: In modern capitalism, prices have taken over the steering function that magic had in remote ages. In the writings of his late Viennese years, he was, in a few aspects, still closer to Frazer's evolutionary speculations about primeval magic than to new anthropological approaches, even if these had not entirely escaped his attention. The reason why Neurath did not see a great divide between the ethnology of extra-European peoples and the 'ethnology of our age' was provided by the genetic theory that primitive magic and modern science are closely akin, even if each of them goes back to a different and specific mode of production.

In 1944, however, Neurath is better acquainted with the rapidly evolving, mostly English-speaking anthropologic scholarship,[100] and the argument in favour of subsuming the only seemingly disparate fields of 'modern' and 'preliterate' anthropology under the umbrella of a single discipline no longer needs to take

[99]"Since flexibility and adaptability of modern human communities reduce their predictability, many anthropologists seem to prefer the analysis of 'preliterate' tribes, which they think are now relatively more stable. Our own 'ethnology' (with our jurisprudence, economics, etc.) has usually been treated separately – a strange way of forming scientific branches. All reforms in phraseology will not help much, in this field, as long as one does not try to reduce the tendency to create watertight compartments and the tendency to avoid overlapping disciplines" (Neurath 1966 [1944], 27).

[100]In *Foundations*, Neurath explicitly mentions social functionalism and Alfred R. Radcliffe-Brown (Neurath 1944, 17, note 7) as well as anthropologists of a younger generation, among others, Raymond Firth and Edward E. Evans-Pritchard (Neurath 1944, 47, note 8). There are no hints that Neurath knew of Radcliffe-Brown in 1931.

the form of an evolutionary philosophy of history. Rather, it merely assumes a methodological form: Neurath simply warns against watertight compartments and proposes creating overlapping disciplines, perhaps even an encompassing discipline with human communities as its general object – without separating 'their' and 'our' anthropology. In the decades following his death, we have definitely gone in this direction.

References

Ackerman, Robert. 1987. *J. G. Frazer. His Life and Work*. Cambridge: Cambridge University Press.

Beth, Karl. 1914. *Religion und Magie bei den Naturvölkern: ein religionsgeschichtlicher Beitrag zur Frage nach den Anfängen der Religion*. Leipzig: Teubner.

Brusotti, Marco. 2011. "Ethnologie unseres Zeitalters". Otto Neurath über Magie und Technik im Vergleich mit Wittgensteins "Bemerkungen über Frazers *Golden Bough*". In Religionen nach der Säkularisierung. Festschrift für Johann Figl zum 65. Geburtstag, ed. Hans Gerald Hödl and Veronica Futterknecht, 331–354. Vienna: LIT.

———. 2014. *Wittgenstein, Frazer und die "ethnologische Betrachtungsweise"*. Berlin/Boston: de Gruyter.

———. 2016a. Ethnologische Betrachtungsweisen: Wittgenstein, Frazer, Sraffa. *Wittgenstein-Studien* 7: 39–63.

———. 2016b. "That they point, is all there is to it." Wittgenstein, Frazer, eine "Tatsachensammlung" und ihre "übersichtliche Darstellung". In *Wittgenstein's Remarks on Frazer: The Text and The Matter*, ed. Lars Albinus, Josef G.F. Rothhaupt, and Aidan Seery, 311–338. Berlin/Boston: de Gruyter.

———. 2017. Mentalità e stili di pensiero. Note sulla ricezione di Lucien Lévy-Bruhl nella filosofia di lingua tedesca (Jerusalem, Cassirer, Husserl, Fleck). In *Dal pensiero alla formazione*, 2 vols., ed. Gabriella Armenise, vol. 1, 385–393. Lecce: Pensa.

Carnap, Rudolf. 1978. Autobiography. In *Ludwig Wittgenstein: The man and his philosophy*, ed. Kuang Tih Fann, 33–30. Atlantic Highlands: Humanities.

Connor, Steven. 1990. The Birth of Humility: Frazer and Victorian Mythography. In Fraser 1990b, 61–80.

Cunow, Heinrich. 1912. *Zur Urgeschichte der Ehe und Familie*. Stuttgart: Dietz.

———. 1929. *Liebe und Ehe im Leben der Völker*. Berlin: Der Bücherkreis.

Engels, Friedrich. [1884.] *Der Ursprung der Familie, des Privateigentums und des Staats*. In Karl Marx and Friedrich Engels: *Gesamtausgabe* (MEGA), vol. I/29, 1990. Berlin: Dietz Verlag.

Fabian, Johannes. 1983. *Time and the Other. How Anthropology Makes its Object*. New York: Columbia University Press.

Figl, Johann. 2003. *Handbuch Religionswissenschaft. Religionen und ihre zentralen Themen*. Innsbruck/Wien/Göttingen: Vandenhoeck & Ruprecht.

Fleck, Lola. 1991. Otto Neurath's Contribution to the Theory of the Social Sciences. In Uebel 1991, 203–208.

Fraser, Robert. 1990a. *The Making of the Golden Bough. The Origins and Growth of an Argument*. Basingstoke et al.: Macmillan.

———. Ed. 1990b. *Sir James Frazer and the Literary Imagination*. Basingstoke et al.: Macmillan.

Frazer, James George. 1911–1915. *The Golden Bough. A Study in Magic and Religion*. 3rd ed., 12 vols. London: Macmillan.

———. 1922. *The Golden Bough. A Study in Magic and Religion*. Abridged Edition. London: Macmillan.

———. 1928. *Der goldene Zweig. Das Geheimnis von Glauben und Sitten der Völker*. Abgekürzte Ausgabe. Leipzig: Hirschfeld.

———. 2005. *Selected Letters*, ed. Robert Ackerman. Oxford/New York: Oxford University Press.
Graebner, Fritz. 1924. *Das Weltbild der Primitiven: Eine Untersuchung der Urformen weltanschaulichen Denkens bei Naturvölkern*. Munich: Reinhardt.
Haller, Rudolf. 1991a. History and the System of Science in Otto Neurath. In Uebel 1991, 33–40.
———. 1991b. On the Historiography of Austrian Philosophy. In Uebel 1991, 41–50.
Harris, Marvin. 1968. *The Rise of Anthropological Theory: A History of Theories of Culture*. New York: Crowell.
Hegel, Georg Wilhelm Friedrich. 1965. *Sämtliche Werke. Jubiläumsausgabe in zwanzig Bänden*, Vol. 15: Vorlesungen über die Philosophie der Religion. Stuttgart – Bad Cannstatt: Frommann-Holzboog.
Hegselmann, Rainer. 1979. Otto Neurath – Empiristischer Aufklärer und Sozialreformer. In Otto Neurath: *Wissenschaftliche Weltauffassung, Sozialismus und Logischer Empirismus*, 7–78, Frankfurt/M.: Suhrkamp.
Hviding, Edvard and Cato Berg. Eds. 2014. *The Ethnographic Experiment: A. M. Hocart and W. H. Rivers in Island Melanesia, 1908*. New York/Oxford: Berghahn.
Josephson-Storm, Jason Ā. 2017. *The Myth of Disenchantment: Magic, Modernity, and the Birth of the Human Sciences*. Chicago/London: University of Chicago Press.
Kaplan, Leo. 1927. *Das Problem der Magie. Eine ethnopsychologische und psychoanalytische Untersuchung*. Heidelberg: Merlin.
Keyserling, Hermann. 1919. *Reisetagebuch eines Philosophen*, München/Leipzig: Duncker & Humblot.
Kraus, Johann Baptist. 1930. *Scholastik, Puritanismus und Kapitalismus. Eine vergleichende dogmengeschichtliche Übergangsstudie*. Munich/Leipzig: Duncker & Humblot.
Lévy-Bruhl, Lucien. 1921. *Das Denken der Naturvölker*. Vienna/Leipzig: Braumüller. [German translation of *Les fonctions mentales dans les sociétés inférieures*, Paris 1910.]
Lynd, Robert S. with Helen M. Lynd. 1929. *Middletown: A Study in Contemporary American Culture*. New York: Harcourt, Brace and Company.
Lynd, Robert S., and Helen M. Lynd. 1937. *Middletown in Transition: A Study in Cultural Conflicts*. New York: Harcourt, Brace and Company.
Madey, Johannes. 1992. Wilhelm Koppers. In *Biographisch-Bibliographisches Kirchenlexikon* (BBKL), vol. 4, 509-511. Herzberg: Bautz.
Malinowski, Bronisław. 1914. Soziologie der Familie. *Die Geisteswissenschaften* 1, 883–886, 911–914. [Abridgment: Forschungen über die Entwicklungsgeschichte der Familie. In *Schriften in vier Bänden*, ed. Fritz Kramer, vol. 4/2 *Schriften zur Anthropologie*, 1986, 106–123. Frankfurt/M.: Syndikat.]
———. 1922. *Argonauts of the Western Pacific. An Account of Native Enterprise and Adventure in the Archipelagoes of Melanesian New Guinea*. London: Routledge.
———. 1924. *Mutterrechtliche Familie und Oedipus-Komplex. Eine psychoanalytische Studie*. Leipzig/Wien/Zürich: Internationaler Psychoanalytischer Verlag.
———. [1925.] Magic, Science and Religion. Now in Bronislaw Malinowski: *Magic, Science and Religion and Other Essays*, 1998 [1948], 17–92. Prospect Heights: Waveland.
———. 1927a. *The Father in Primitive Psychology*. New York: Norton.
———. 1927b. *Sex and Repression in Savage Society*. London et al.: Kegan Paul et al.
———. 1929a. *The Sexual Life of Savages in North-Western Melanesia. An Ethnographic Account of Courtship, Marriage, and Family Life among the Natives of the Trobriand Islands, British New Guinea*. New York: Readers League of America.
———. 1929b. *Das Geschlechtsleben der Wilden in Nordwest-Melanesien: Liebe/Ehe und Familienleben bei den Eingeborenen der Trobriand-Inseln/Britisch-Neu-Guinea; eine ethnographische Darstellung*. Vorwort von Havelock Ellis. Dt. von Dr. Eva Schumann, n. d. Leipzig/Zürich: Grethlein.
———. 1932. *The Sexual Life of Savages in North-Western Melanesia. An Ethnographic Account of Courtship, Marriage, and Family Life among the Natives of the Trobriand Islands, British New Guinea*. Third Edition, with a Special Foreword. London: Routledge.

———. 1935. *Coral Gardens and Their Magic*. Vol. 1: *A Study of the Methods of Tilling the Soil and of Agricultural Rites in the Trobriand Islands*. Vol. 2: *The Language of Magic and Gardening*. London: Allen & Unwin.

———. [1938.] Introduction. In Jomo Keniatta: *Facing Mount Kenya. The Tribal Life of the Gikuyu*, 1965, vii–xiv. London: Mercury.

Malinowski, Bronisław. 1979. *Das Geschlechtsleben der Wilden in Nordwest-Melanesien: Liebe, Ehe und Familienleben bei den Eingeborenen der Trobriand-Inseln, Britisch-Neuguinea* [= Schriften in vier Bänden, ed. Fritz Kramer, vol. 2], Frankfurt/M.: Syndikat.

Morgan, Lewis H. [1877]. *Ancient Society, or Researches in the Lines of Human Progress from Savagery, through Barbarism to Civilization*. London: Macmillan, Chicago: Kerr 1907.

Müller-Lyer, Franz. [1906]. Phasen der Kultur und Richtungslinien des Fortschritts. Soziologische Überblicke, 1948. Munich: Freitag.

Nadel, Siegfried F. 1957. Malinowski on Magic and Religion. In *Man and Culture. An Evaluation of the Work of Bronislaw Malinowski*, ed. J. Raymond Firth, 189–208. London: Routledge.

Nemeth, Elisabeth. 2007. Logical Empiricism and the History and Sociology of Science. In *The Cambridge Companion to Logical Empiricism*, ed. Alan Richardson and Thomas Uebel, 278–302. Cambridge: Cambridge University Press.

Nemeth, Elisabeth, Stefan W. Schmitz, and Thomas Uebel. Eds. 2007. *Otto Neurath's Economics in Context*. Dordrecht: Springer.

Neurath, Otto. 1909. *Antike Wirtschaftsgeschichte*. Leipzig: Teubner. Now in GS 4, 137–217. English: *Economic History of Antiquity* [excerpts], in Neurath 2004, 120–152.

———. [1917.] Das Begriffsgebäude der Wirtschaftslehre und seine Grundlagen. In GS 1, 103–129.

———. 1921a. Menschheit. *Die Wage*, August 20, 1921, 374–376. Now in GS 1, 197-201.

———. 1921b. *Anti-Spengler*. Munich: Callwey. Now in GS 1, 139–196. English: Anti-Spengler. In Neurath 1973, 158–213.

———. 1928a. *Lebensgestaltung und Klassenkampf*. Berlin: E. Laubsche Verlagsbuchhandlung. Now in GS 1, 227–293. English: Personal Life and Class Struggle. In Neurath 1973, 249–298.

———. 1928b. Kolonialpolitische Aufklärung durch Bildstatistik. *Arbeit und Wirtschaft* 15: 673–680.

———. 1929a. *Wissenschaftliche Weltauffassung. Der Wiener Kreis*. Wien: Artur Wolf. Now in GS 1, 299-336. English: The Scientific Conception of the World: The Vienna Circle. In Neurath 1973, 299–318.

———. 1929b. Bürgerlicher Marxismus. *Der Kampf* 23: 227–232. Now in GS 1, 349–356.

———. 1930a. Wege der wissenschaftlichen Weltauffassung. *Erkenntnis* 1: 106–125. Now in GS 1, 371–391. English: Ways of the Scientific World-Conception. In Neurath 1983, 32–47.

———. 1930b. Kulturbeschreibung auf materialistischer Basis. *Arbeit und Wirtschaft* 18: 717–720.

———. 1930c. Einheitswissenschaft und Marxismus. *Erkenntnis* 1: 75. Now in GS 1, 369–370.

———. 1931a. *Empirische Soziologie. Der wissenschaftliche Gehalt der Geschichte und Nationalökonomie*. Berlin/Heidelberg: Springer. Now in GS 1, 423–527: English (partly): Empirical Sociology: The Scientific Content of History and Political Economy. In Neurath 1973, 319–421.

——— 1931b. Magie und Technik. *Erkenntnis* 2: 82-84. Now in GS 2, 529-531.

———. 1931c. Mutterrecht und Magie. *Arbeit und Wirtschaft* IX/3: 99–102.

———. 1931d. Marxismus eines Jesuiten. *Der Kampf* 24: 271–274. Now in GS 1, 401–405.

———. 1931e. Soziologie im Physikalismus. *Erkenntnis* 2: 393–431. Now in GS 2, 533–562. English: *Sociology in the Framework of Physicalism*, in Neurath 1983, 58–90.

———. 1932. Sozialbehaviorismus. *Sociologus* 3: 281–288. Now in GS 2, 563–569.

———. 1935. *Le développement du cercle de Vienne et l'avenir de l'empirisme logique*. Paris: Hermann.

———. 1939. *Modern Man in the Making*. New York/London: Knopf.

———. 1941. Universal Jargon and Terminology. In Neurath 1983, 213–229.

———. [1944.] Foundations of the Social Sciences [= International Encyclopedia of Unified Science, vols. I and II – Foundations of the Unity of Science, Vol. II – Number 1], [7]1967. Chicago: University of Chicago Press.
———. 1973. *Empiricism and Sociology*, ed. Marie Neurath and Robert S. Cohen. Dordrecht/Boston/Lancaster: Reidel.
———. 1981. *Gesammelte philosophische und methodologische Schriften*, ed. Rudolf Haller and Heiner Rutte, 2 vols. Vienna: Hölder – Pichler – Tempsky [=GS 1, 2].
———. 1983. *Philosophical Papers 1913–1946*, ed. Robert S. Cohen and Marie Neurath. Dordrecht/Boston/Lancaster: Reidel.
———. 1991. *Gesammelte bildpädagogische Schriften*, ed. Rudolf Haller. Vienna: Hölder – Pichler – Tempsky [=GS 3].
———. 1998. *Gesammelte ökonomische, soziologische und sozialpolitische Schriften*, ed. Rudolf Haller, 2 vols. Vienna: Hölder – Pichler – Tempsky [=GS 4, 5].
———. 2004. *Economic Writings. Selections 1904–1945*, ed. Thomas E. Uebel and Robert S. Cohen. Dordrecht/Boston/London: Kluwer.
Neurath, Paul. 1991. Sociological Thought with Otto Neurath. In Uebel 1991, 209–222.
Reisch, George. 1997. Economist, epistemologist... and censor? Otto Neurath's *Index verborum prohibitorum. Perspectives on Science* 5 (3): 452–480.
Rössler, Martin. 2007. Die deutschsprachige Ethnologie bis ca. 1960: Ein historischer Abriss [= Kölner Arbeitspapiere zur Ethnologie, No. 1]. Köln: Institut für Völkerkunde.
Russell, Bertrand. 1929. *Marriage and Morals*. New York: Liveright.
Sandner, Günther. 2007. Economy, Ideology and Culture: Otto Neurath's Approach to a Precarious Relationship. In Nemeth/Schmitz/Uebel, 141-155.
Stadler, Friedrich. 1997. *Studien zum Wiener Kreis. Ursprung, Entwicklung und Wirkung des logischen Empirismus im Kontext*. Frankfurt/M: Suhrkamp.
Stocking, George W., Jr. 1987. *Victorian Anthropology*. New York: The Free Press.
———. 1995. *After Tylor. British Social Anthropology 1888–1951*. Madison: The University of Wisconsin Press.
Tambiah, Stanley J. 1990. *Magic, Science, Religion, and the Scope of Rationality*. Cambridge: Cambridge University Press.
Turk, Michael H. 2018. *Otto Neurath and the History of Economics*. London/New York: Routledge.
Uebel, Thomas E., ed. 1991. *Rediscovering the Forgotten Vienna Circle. Austrian Studies on Otto Neurath and the Vienna Circle*. Dordrecht/Boston/London: Kluwer.
———. 1997. Sozialwissenschaft auf physikalistischer Grundlage im Wiener Kreis. In *Philosophie und Wissenschaften. Formen und Prozesse ihrer Interaktion*, ed. Hans Jörg Sandkühler, 169–192. Frankfurt/M: Peter Lang.
———. 2007. Philosophy of Social Science in Early Logical Empiricism: The Case of Radical Physicalism. In *The Cambridge Companion to Logical Empiricism*, ed. Alan Richardson and Thomas Uebel, 250–277. Cambridge: Cambridge University Press.
Westermarck, Edvard. 1891. *The History of Human Marriage*, 2 vols. London: Macmillan.
Wittgenstein, Ludwig. 1993. Bemerkungen über Frazers *Golden Bough* / Remarks on Frazer's *Golden Bough*. In *Philosophical Occasions 1912–1951*, ed. James Klagge and Alfred Nordmann, 115–155. Indianapolis/Cambridge: Hackett.
Young, Michael W. 1998. Malinowski's Kirivina. In *Fieldwork Photography 1915–1918*. Chicago/London: The University of Chicago Press.

Chapter 5
"God Never Does Mathematics" – Hans Hahn on Religion

Julia Schäfer

It may surprise the reader to find in this collection an essay on Hans Hahn. Even though his interests were varied, he is mainly known for his contributions in the fields of mathematics and philosophy of science. Together with Otto Neurath and Moritz Schlick, Hahn was one of the founding members of the Vienna Circle and, according to Karl Menger, the one who turned the circle's interest towards logic with an eye on related philosophical problems (Menger 1980, 10). How is it that he is now associated with the topic of religion?

The most obvious answer to this question is that Hahn's association with religion arises from the Vienna Circle's concern with the topic of religion in general – and this is what this volume intends to discuss. Hahn is often described as a key figure of the Circle – not only with regard to his long-time affiliation with the group, but also in terms of his intellectual influence. "One can say," Frank wrote in his obituary of Hahn in *Erkenntnis,* Vol. 4, "that in a certain sense Hahn was always a center of the group. He always represented its central ideas without entering into differences of opinion on side issues. No one knew as well as he how to present those leading ideas in such a simple as well as thorough way, in such a logical as well as suggestive form" (Menger 1980, 11).

Even though Hahn eventually became the principal signer of the Circle's famous pamphlet *Wissenschaftliche Weltauffassung* in 1929, it is unclear to what extent he can be said to have actively contributed to this text. Earlier that same year, he had been asked to deliver its inaugural address at the First Congress of the Epistemology of Science in Prague, where he presented a kind of 'creed' for logical positivism and

J. Schäfer (✉)
Institut für Philosophie, Universität Wien, Wien, Austria
e-mail: julia.schaefer@univie.ac.at

E. Ramharter (ed.), *The Vienna Circle and Religion*, Vienna Circle Institute Yearbook 25, https://doi.org/10.1007/978-3-030-76151-6_5

the so-called scientific world-view[1] (see also Sect. 5.1 of this paper). The pamphlet, however, was mainly written by Otto Neurath, while Hahn allegedly received merely its final draft. Not only was it said to lack the succinctness and precision of Hahn's address, it also introduced political ideas and tendencies as common views that were not entirely shared by all members of the highly individualistic group (Menger 1980, 14).

Discussing his view on religion in the following two sections, I will therefore essentially stick to Hahn's own writings. It must be noted that Hahn's concern was mostly with metaphysics in general, rather than with religion in particular, and that he refers primarily to metaphysics within the domain of science. Nonetheless, Hahn, being a politically involved socialist, reflects on the metaphysical world-view also in terms of a superstructure based on practical life. Discussing practical dimensions of what he calls "faith", he provides a psychological and sociological explanation of the "scientific" and "metaphysical" world-view, both on the individual level as well as in terms of society.

In the first section, I will address Hahn's remarks on metaphysics and science. When cited explicitly, religion is referred to as an area where metaphysics traditionally appears. Hahn, very much like other members of the Vienna circle, claims that metaphysical elements such as religious beliefs belong to a different domain than science, namely that of faith. In certain places in his argumentation, it is, however, not clear whether the "scientific world-view" and the "metaphysical world-view" essentially differ, and thus belong to entirely separate domains or whether the scientific world-view is a kind of faith as well, and thus in competition with religious belief[2] for example.

In the second part of the paper (Sects. 5.2, 5.3, and 5.4), I will examine Hahn's remarks on religious belief as "faith", its association with "taste", as well as his conception of two competing paradigms of thinking: world-affirming and world-denying philosophy, which have "as their ultimate cause", according to Hahn, a certain psychological attitude. As I will show, there is a severe problem with his reasoning: On the one hand, the analogy between "matters of faith" and "matters of taste" suggests a kind of relativism: having a specific faith would be as ungrounded and legitimate as having a different one. From this point of view, believing in the scientific world-view would then be no better than believing in another metaphysical faith, such as religion. On the other hand, Hahn, in the further explication of his thought, expresses more and more clearly an (unjustified) preference for what he calls "world-affirming" philosophy over "world-denying" religion, ending up in a missionary plea for liberation of humanity from "the nightmare" of world-denying metaphysics.

[1]Even though the Vienna Circle insisted in their writings on the non-metaphysical nature of their "world-*conception*" (as opposed to the metaphysical "world-*view*"), the distinction is highly questionable. Furthermore, English translations of Hahn's works consistently use "world-view". For these reasons, I will use "world-view" in this essay.

[2]On this topic, see also Romizi (2012, 2013).

5.1 No Place for Metaphysics in Science

Similar to other members of the Vienna Circle, Hahn's concern was with metaphysics in general. In what follows, I will further elaborate on why there should be, according to Hahn, no place for metaphysical elements in science. In his essay *The Significance of the scientific world-view, especially for mathematics and physics*, he states

> The name 'scientific world view' is intended both as a confession of faith and as a delimitation of a subject: It is to *confess our faith* in the methods of the exact sciences, especially mathematics and physics, faith in careful logical inference (as opposed to bold flights of ideas, mystical intuition, and emotive comprehension), faith in the patient observation of phenomena, isolated as much as possible, no matter how negligible and insignificant they may appear in themselves (as opposed to the poetic, imaginative attempt to grasp wholes and complexes, as significant and as all-encompassing as possible). (Hahn 1980b [1930–1], 20)

This so-called "faith" Hahn confesses to (I will come back to the ambivalence about the concept of "faith" below) is based on his understanding of how science ought to be practised, as best displayed by contemporary mathematics and physics. His commitment is, however, to any kind of studious inquiry that methodologically relies on careful logical inference and the rigorous observation of phenomena. This is how logical empiricism is opposed to metaphysics, art, mysticism and any other kind of enterprise, which tries to gain knowledge by building on holistic theories and seeking truth beyond the empirically observable.

Nevertheless, there was and still is – much to Hahn's regret – quite a large amount of such elements to be found in science and especially in philosophy, which he attributes partly to their socio-historical context; the main reason he sees, however, in scientists' erroneous expectations on the epistemic potential of thought and language.

> The intrusion of metaphysical elements into science has been aided and abetted by two tendencies: a tendency to overestimate thought and to assume – while mistaking the tautological character of thought – that thought could lead all by itself to something new, and a tendency to overestimate language. (Hahn 1980b [1930–1], 28)

According to Hahn, science, understood as acquisition of knowledge, can only draw from experience and logical thought. Logical thought, however, is tautological[3] in the sense that it does not "discover" any being on its own, but only transforms empirical information into terms of language. Language, especially ordinary language, is, however, a "highly inaccurate" and "extraordinarily imperfect" instrument. We must not, therefore, make the mistake of inferring the structure of the world from the structure of language.

---

[3]Ludwig Wittgenstein's understanding of logical sentences as tautologies was from the very beginning judged as one of the most important concepts of the *Tractatus* (for example, by Bertrand Russel and Frank Ramsey). The fact that Hahn seizes this idea exemplifies the influence of Wittgenstein on the Vienna Circle.

In *Superfluous entities or Occam's razor* (1980a [1930]), Hahn applies this critique to so-called proofs of God's existence and religious truths:

> Thought is [...] perfectly incapable of leading us from the sensible world into a world of different kinds of entities; there can be no logical chain of thought from the sensible world to other-worldly entities, so [...] thought cannot lead to God or to the dogmas of the church; these are left to a different domain, that of faith. (Hahn 1980b [1930], 4)

Even though the scientific language of symbolic logic "is much less imperfect than the various word languages", a logical proposition by itself, Hahn says, states nothing at all about the world – it allows us to say the same thing in different ways. In his view, the sense and usefulness of such transformations arise only from the fact that we have very fragmentary knowledge of the world:

> And the occasion for introducing a symbolism whose structure deviates from what is to be depicted, a symbolism which allows us to say the same thing in different ways, is that we are not omniscient, that we can know the states of affairs that make up the world only in a very fragmentary way. [...] An omniscient subject needs no logic, and contrary to Plato we can say: God never does mathematics. (Hahn 1980b [1930–1], 23)

To sum up his argument so far: science is about the acquisition of knowledge. Neither thought nor language lead to (new) knowledge. Both have been overestimated in this respect. The main reason why metaphysical elements came into science was this very overestimation. Estimating their potential correctly, there can and should be no metaphysics in science.

Nevertheless, as we have seen at the beginning of this section, Hans Hahn was – in a certain sense – also a man of faith. His use of religious language/rhetoric ("confession", "faith") seems to suggest that he was aware that the scientific world-view lacked final justification, and in this sense contained metaphysical elements. Thus, we can see that, on the one hand, Hahn argued against the presence of metaphysics in science, whereas on the other, he presented the scientific world-view as if it were a faith, and thus metaphysics. It is not clear how this paradox can be solved.

In the following, I will examine Hahn's claim that metaphysical elements such as religious beliefs belong to a different domain than science, namely that of faith.

5.2 Varieties of Faith

Now, what about the domain of faith? Hahn illustrates his argument with the use of the following analogy:

> While [some] believed in [God or in the dogmas of the church], many nowadays do not believe in them; but this difference is, epistemologically speaking, not a big one, just as it makes no difference, epistemologically speaking, that one person loves narcotics and another one does not. (Hahn 1980b [1930], 7)

As argued in the first section, religious beliefs must belong to a domain different from science, for they are not a matter of knowledge. Hahn stresses this point by now claiming that there is little or no epistemological difference between believers and non-believers, meaning that what he calls "faith" has little or no rational grounds. The example he gives in form of analogy is that believing in God or not is like loving narcotics or not.

This comparison is irritating in two ways: First, we remember Hahn calling his commitment to the 'scientific world-view' a "confession of faith": would this "faith" be comparable to love for narcotics as well? Second, "loving narcotics or not" seems to be a very odd example of what we usually understand by "faith". It becomes a little clearer what Hahn has in mind, when he later states:

> But – and now we are returning to our point of departure – people have different tastes. One person enjoys multiplicity, the variegation and capriciousness of change, the confusion in the sensible world which is so difficult to unravel, and he therefore continues to affirm this world, whereas another who has been denied the pleasures of the sensible world denies this world and tries to make up other worlds behind it. (Hahn 1980b [1930], 19)

"Faith" is now construed as taste, which points once again to the lack of conceptual clarity identified before: Hahn's association of faith with taste is misleading because it suggests a relativist concept that conceives of all kinds of faiths as equally legitimate. However, Hahn himself infers an – at this point still largely implicit – valuation of tastes. If we take a brief look back at the analogy between "belief in God and the dogmas of the church or not" and "loving narcotics or not", we see now that the connotation of "loving narcotics", brought up as a counterpart of "belief in God", shows the latter in an extremely negative light. Furthermore, it evokes the association with the Marxist idea of religion as the 'Opium of the people', which Hahn will later cite explicitly ("a means, employed over and over again, of consoling the masses") (Hahn 1980b [1930], 1).

This resonates well with such remarks as 'world-denial' and 'making-up other worlds', depicting the religious 'taste' or 'faith' as a product of psychological distress (mental illness, addiction) and contrasting it with the self-confident, unpretentious manner of the sophisticated modern person who faces life's challenges and continues to affirm this world.

This asymmetry is reinforced by Hahn's Nietzschean-inspired differentiation between two competing paradigms of thinking that have forged the history of philosophy: The so-called world-affirming and world-denying philosophy. He states:

> It seems to me that in the bewildering variety of philosophical systems it is possible to distinguish two main types: those systems of philosophy that affirm the world and those that deny it. The ultimate cause for such [differences] would seem to be certain basic psychological attitudes: a certain optimism on the one hand and a certain pessimism on the other. (Hahn 1980b [1930], 1)

Here, Hahn develops a psychological perspective, which traces 'world-view' as "faith" back to a certain, 'existential mind-set': Whilst a "certain pessimism" directs one's eyes and thoughts away from the "mere appearances" of this world and makes one search for 'true being' behind it, a "certain optimism" promoting a set of assumptions, methods, or notions that "build wholly upon the world how it presents itself" tries to find its way about and come to terms with perceptions inconsistency, irregularity and variety.

This is the "faith" Hahn feels the need to "confess" to: Relying on a "certain optimism" as a basic psychological attitude towards the world, he commits himself to certain methods affiliated with the "scientific world-view". However, this renders the scientific world-view as a concept no different from what Hahn rejects as "the metaphysical" and it invalidates his argument why there should be no metaphysics in science. In the following, I will argue that Hahn did not really think of different world-views, "faiths", or "tastes" in relative terms, but that he assumes a sort of hierarchy.

5.3 Relativism and Faith as Superstructure

Even though Hahn seems to be aware of the fact that his own world-view does not come without preconditions, another instance of his shift from relativism to value judgement can be identified: When he transfers his ideas about faith from the individual level of analysis to the level of society as a whole, Hahn seems to adopt a positivist idea of intellectual progress and social evolution. In classical Marxist fashion, he draws a parallel between practical and intellectual life-conditions, taking the former to determine the latter. As he points out with historical examples, certain "forms of living" give rise to certain existential mind-sets which, on their part, entail certain ways of thinking.

> The Germans are known after all as the nation of thinkers and poets. But gradually a new day is dawning, and the liberation comes from the same land that gave birth to political liberation, viz. England: the English are known after all as the nation of shopkeepers. [...] And it is certainly no accident that the same nation gave the world democracy and the rebirth of the world-affirming philosophy, and no accident that the land that saw the beheading of a king also witnessed the execution of metaphysics. For all those other-worldly entities of metaphysics – Plato's ideas, the Eleatics' One, Aristotle's pure form and first mover, the gods and demons of the religions, and the kings and princes of the earth – all share a common fate, and when the emperor falls the duke must follow suit. (Hahn 1980b [1930], 4)

The forms of living Hahn is here respectively associating with metaphysics and world-affirming philosophy as ways of thinking refer to what we would basically call 'a socio-cultural way of life', addressing work and the economy (a "nation of poets" vs. a "nation of shopkeepers"), on the one hand, and (state) polity (monarchy vs. democracy), on the other. It is, however, not only the business-minded English, to whom he attests "free-thinking", but also defenders of secularism and papal opponents whom he recognises as world-affirmers.

> [T]he struggle between the Ghibellines, who favoured the temporal power of the Emperor, and the Guelphs, who favoured the spiritual power of the Pope, this centuries-long struggle between Ghibellines and Guelphs is only part of the eternal struggle between world-affirmers and world-deniers. (Hahn 1980b [1930], 6)

As illustrated by his historical analyses, Hahn seems to assume different psychological and socio-cultural 'states of development' aligned with different faiths. His concepts of faith, as I showed before, is, however, not a relativistic one, but value-laden. Nonetheless, it remains difficult to say whether Hahn is a positivist also with regards to social progress in a Comtean sense.[4] While he speaks of "eternal struggle between world-affirmers and world-deniers", on the one hand, he states, on the other, that religion, as a means of consoling the masses, has "only in our time", namely through political and intellectual liberation, "been seriously undermined".

Be that as it may, in terms of faith, Hahn has clearly been open to missionary activity and the idea of conversion, as illustrated by the ending paragraph of his essay:

> But we, the adherents of the world-affirming philosophy, regard this as a side-track, a blind alley, a disease from which mankind has suffered for millennia, and it is our intention to free mankind from this nightmare. And if a man emerges from the mystical gloom of the world-denying philosophy and embraces the simple, transparently clear teaching of the world-affirming philosophy, we greet him with these words:
>
> You emerge from death, uncertain sufferings,
> All that was faint and mean and vague,
> And learn with unobstructed vision to distinguish
> The sickly dawn from the clear day!
> (Hahn 1980b [1930], 19)

5.4 Conclusion

It seems Hans Hahn was not much concerned with religion; nonetheless, he was, in a certain sense, a man of faith. The scientific world-view he confesses to provides an understanding of his idea of and attitude towards science. However, it does not give reason to his intention to overcome metaphysics in science. His reasoning contains, as I have shown, a paradox: On the one hand, he argues against the presence of metaphysics in science by saying that it belongs to a "different domain", namely that of faith, whereas on the other, he presented the scientific world-view as if it were a faith as well. Hahn's psychological account of faith, construing it as a sort of taste, illustrates this point even more clearly, as Hahn introduces relativism, for one thing, and a hierarchy of tastes, on the other. The implicit value-judgement becomes evident especially in his sociological treatment of faith where he draws a parallel between practical and intellectual life-conditions, taking the former to determine the latter. While world-affirming philosophy and the "scientific world-

[4]Auguste Comte presents in his work *The Course in Positive Philosophy* the idea that all social development underlies the law of three stages, arguing that the growth of the human mind progresses in stages, and so must societies. He claims the history of both society and science could be divided into three different stages, each of which can be aligned with new philosophical ideas: theological, metaphysical, and positive. (Bourdeau 2021)

view" are associated with facing life's challenges, "free-thinking", "democracy" and secularism, "the metaphysical world-view" and religious faith are characterised as a disease, world-denial and as a blind alley. From Hahn's point of view, people who believe in God do not simply see the world differently, they haven't even come to look at it.

References

Bourdeau, Michel. 2021. Auguste Comte. In *The Stanford Encyclopedia of Philosophy* (Spring 2021 Edition), ed. Edward N. Zalta. Forthcoming. https://plato.stanford.edu/archives/spr2021/entries/comte/

Hahn, Hans. 1980a. Superfluous Entities or Occam's Razor. In *Hans Hahn, Empiricism, Logic and Mathematics: Philosophical Papers*, ed. Brian McGuinness, vol. 13, 1–19. Dordrecht/Boston/London: D. Reidel Publishing Company.

———. 1980b. The Significance of the Scientific Worldview, Especially for Mathematics and Physics [First Published in *Erkenntnis* I, 1930/31]. In *Hans Hahn, Empiricism, Logic and Mathematics: Philosophical Papers*, vol. 13, 20–30. Dordrecht/Boston/London: D. Reidel Publishing Company.

Menger, Karl. 1980. Introduction. In *Hans Hahn, Empiricism, Logic and Mathematics: Philosophical Papers*, ed. Brian McGuinness, vol. 13, 9–18. Dordrecht/Boston/London: D. Reidel Publishing Company.

Romizi, Donata. 2012. The Vienna Circle's "Scientific World-Conception": Philosophy of Science in the Political Arena. *HOPOS: The Journal of the International Society for the History of Philosophy of Science* 2 (2): 205–242.

———. 2013. War die ‚wissenschaftliche Weltauffassung" des Wiener Kreises nicht doch auch eine Weltanschauung? In *Die europäische Wissenschaftsphilosophie und das Wiener Erbe*, ed. Elisabeth Nemeth and Friedrich Stadler, 127–151. Wien: Springer.

Chapter 6
A Rabbi Among the Apostates? Josef Schächter and Religion in the Vienna Circle

Malachi Hacohen

It is not a coincidence that this volume, and the October 2019 Vienna conference that gave rise to it, have been the first to address the subject of religion in the Vienna Circle. The circle was the citadel of secular philosophy, and its members, staunchly anticlerical, were vested in scientific philosophy that endeavored to overcome metaphysics and theology. Nowadays, much of the modern confidence in science is gone. In the twenty-first century, religion has made a surprising and powerful return as a force shaping contemporary society. Scholars surmise that religion never went away in the secular age, and discern secularized religious structures and vocabularies in modern ideologies. Now that postmodernity has relativized modern discourses and deconstructed Enlightenment narratives, we have become more attentive to religion. We live in a post-secular society.

Vienna Circle protégés would not be comfortable in the post-secular society. The 1929 manifesto *Wissenschaftliche Weltauffasung* (scientific worldview), which proclaimed the Vienna Circle, attacked relentlessly metaphysics and theology. It envisioned reforming philosophy by excluding metaphysics, so that philosophy may become the handmaiden of science. (Neurath 1929; Stadler 2001) Progressive Viennese anti-clericalism had a long lineage, and, by the interwar years, the crusade against metaphysics had been several decades old. In fin-de-siècle years, Ernst Mach's Positivism served as the groundwork for the Viennese *Spätaufklärung* (late enlightenment). Among its major projects were a secular school network for adult education (*Volksbildung*), and a controversial private school for children, the *Freie*

Dedicated to the memory of my mother, Zehava (Chaya Golda) Hacohen, née Zucker (1933–2018). She was Josef Schächter's student in the *Realgymnasium* in Haifa in the late 1940s. It was from her that I first heard about the Vienna Circle.

M. Hacohen (✉)
Department of History, Duke University, Durham, NC, USA
e-mail: mhacohen@duke.edu

E. Ramharter (ed.), *The Vienna Circle and Religion*, Vienna Circle Institute Yearbook 25, https://doi.org/10.1007/978-3-030-76151-6_6

Schule. (Boyer 1995; Stadler 1981, 1982) Non-Jewish Jews were prominent in Viennese progressivism and the circle's networks. They left traditional Jewish life behind, sought social integration, viewed religion as a relic, and the Catholic Church as an enemy of enlightenment. They dreamed of an open secular society that would overlook their Jewishness and accept them as German Europeans. Religion and the Vienna Circle appear to be antithetical.

Yet, this volume, and the conference from which it sprang, suggest that if we rethink this antithesis, a more nuanced and complex view of religion in the Vienna Circle emerges. If Philipp Frank still appears as an indefatigable warrior for science, expecting the scientific worldview to overcome, at one and the same time, religion and value-relativism, Hans Hahn reveals at least rudimentary awareness that the crusade against metaphysics claimed for science a religious and metaphysical status. Otto Neurath was relentless in attacking metaphysics and theology but he regarded magic as prescientific technology. For Rudolf Carnap, early Protestant influences may have set the course for his transformation of the Kantian legacy, and disenchantment with religion during the First World War may have triggered his intellectual transformation. Moritz Schlick, less dismissive of religion than most of his colleagues, sought to develop a new ethics. He saw ethical motives as grounded in religious emotions and heralded a new religion of humanity. Ludwig Wittgenstein joined contempt for theology with relish for the mystical. Karl Popper thought himself an agnostic rather than an atheist. He considered radical anticlericalism politically imprudent, the crusade against metaphysics misguided, and Wittgenstein's mysticism a catastrophe. In the postwar years, he allowed metaphysics and theology back into Critical Rationalism. Even more, Kurt Gödel remained a Christian all his life and delved into theology, exploring proofs for God, the afterlife, and free will. The Vienna Circle was not, then, uniform in its rejection of religion, metaphysics and theology. (Hacohen 2000; Ramharter 2022a)

This modification of the old view has been largely confined, however, to finding religious traces in the scientific worldview or discovering openings for religion in the circle's new philosophies. With the exception of Gödel, no circle member saw traditional religion as viable, and none sought the convergence of traditional religion and the scientific worldview. None, that is, except for Josef Schächter (1901–1994), whom the conference has transformed, after almost a century of neglect, from the margins of the circle to the center. Schächter was a Jewish rabbi and Schlick's assistant, and participated regularly in the circle's meeting. Together with Friedrich Waismann, he taught Schlick's courses when Schlick was assassinated in 1936. After the *Anschluß*, Schächter escaped to Mandatory Palestine (the Land of Israel). There he began a new career as an influential educator, who sought to reconcile modernity and traditional Judaism, and forge a new path for the modern Jew. His linguistic philosophy in interwar years was informed by religious queries, and he credited his exploratory path in the postwar years, in part, to the circle's inspiration. He provides a unique example of religion and philosophy coming together in the Vienna Circle. He is the subject of this essay.

6.1 Bridge-Builder and Transmitter: From Central Europe to Israel

Schächter grew up in a ḥasidic Jewish family in Eastern Galicia and received both Jewish and German education. He arrived in Vienna in 1921 in the aftermath of the dislocations occasioned by the First World War and the pogroms of the immediate postwar years in the Ukraine. He taught Talmud in the Jewish seminary (Hebräischen Pädagogium) from 1922 to 1938 and, in his final years in Vienna, he also taught at the Maimonides Institute for adult education. In 1926, he was ordained by a well-known ḥasidic authority as an Orthodox rabbi. Beginning in 1925, he studied logic and philosophy at the university with Moritz Schlick, and completed his doctorate under him in 1931 with a thesis on Nikolai Hartmann. He participated regularly in the Vienna Circle meetings, and, in 1935, published his *Prolegomena to a Critical Grammar* in the Vienna Circle's book series. The book, and his articles from 1937 to 1938, reflect his engagement with the circle's language philosophy and with Schlick's ethics, and his effort to reconcile them with his personal religious quest. He would later describe intimate conversations he had with Schlick on philosophy, religion and ethics. (Asher Schaechter 2009, appendix 6) At the same time, he was writing essays in Jewish studies, logic and philosophy for Hebrew periodicals published in Palestine/Israel. The relationship between science and religion constituted the major challenge he addressed in his life and work. The rabbi and the philosopher, the Talmud and logic, ethics and science converged in his work from start to end.

In Mandatory Palestine, and later in the State of Israel, Schächter quickly became a widely respected public intellectual, director of pedagogical training in seminaries, and coordinator of Jewish studies curricula for Israeli schools. He served on the boards of leading educational and cultural foundations and on juries for literary and scholarly prizes. A prolific writer on Jewish and moral education, he published hundreds of essays, pamphlets, and compilations and lexica of rabbinic material for teaching. Beginning in 1940, he taught Bible, Talmud, literature, and philosophy for more than a decade in Haifa's famed *Realgymnasium*, the Hebrew Reali School. He became a legendary teacher whose students established a new settlement in the Galilee in 1959, Yodfat, dedicated to the pursuit of ethical life and spiritual perfection. The only Zionist member of the Vienna Circle, Schächter was convinced of the necessity of a Jewish state. However, national life was never an end for him, but rather a means to spiritual perfection – moral and religious, universal and Jewish.

Schächter was a major transmitter of the Vienna Circle's legacy to Israel. While still in Vienna, he offered in Hebrew in 1937 an overview of logic and syntax. In Palestine/Israel, he continued to report in essays, pamphlets, and short books on the circle's philosophy and major figures and trends in Viennese culture. He deployed them to advance his pedagogical agenda and moral vision. Israeli school students were among the first in the world to learn about the Vienna Circle (החוג הווינאי). My mother ז"ל was Schächter's student in the Reali School in the late 1940s in Haifa and was full of admiration and respect for him. Growing up in Israel, I heard from her about Karl Popper and his interlocutors well before I would encounter them in an academic setting. I found among her school papers thin pamphlets put together by Schächter on Adolf Loos and the Vienna Circle, published in typescript on low quality yellowish paper. (Engelmann and Schächter 1946; Schächter 1948) Schächter returned periodically to the circle in the postwar years as he was developing his moral and religious vision. In the 1970s, he brought out, in his own press, a pamphlet on *Moritz Schlick: His Personality and His Philosophy*. The memory of the old friendship, and the circle's legacy, remained alive.

6.2 Religion and Philosophy

Even at the height of his engagement with the circle in the 1930s, Schächter harbored serious reservations about the scientific worldview, and felt tormented by the tensions between positivist philosophy and his religious sensibilities. In an autobiographical account, he described his commitment to eliminating metaphysics, the difficulties he encountered praying at the time, and his joyful escapes to the Jewish study halls to delve into the Talmud. (Schächter 1943, 77–82; 1964, 91–4) He thought logical positivism risked impoverishing human life by eliminating transcendence, interiority, and the religious experience. He was critical of Carnap's logical syntax and Neurath's physicalism as confining language and flattening reality, and felt closer to Schlick and Waismann. He admired Schlick's epistemology and sympathized with his explorations into ethics. He regarded Schlick, like Waismann, as a poetic soul. (Schächter 1970)

Schächter endorsed the circle's effort to reach philosophical clarity and accepted science as constitutive of the modern experience. He considered the elimination of metaphysics as a religious duty. Clearing away myth and idol worshiping would make religious experience purer and an authentic encounter with the divine possible. (Schächter 1943) But the scientific worldview had dangerous ramifications: It fetishized materiality, ignored emotions and values, and delegitimized desires and dreams. Formal language and the effort to reduce language to data reports were chimerical. Language had to be expansive enough to accommodate ethics, poetry, and religion. In his *Prolegomena* (1935; 1973, Part 2, I, 10, 8), Schächter gave examples of statements of thoughts, feelings, value judgements, commands, and prayers that were "not perceivable by the senses," yet were not meaningless. He suggested a grammar of ordinary discourse to complement Carnap's syntax of formal language and accommodate religion.

A series of short essays on religion, science, and language that Schächter wrote in 1937–38 reveal his innovative view of language emerging from the dilemmas of a religious believer. (1937a, 1937b, 1938, 1943) The scientific search for truth was, for Schächter, religious in character, but it was limited to reality. Religious language, expressing deep human desires, always pointed to an ideal beyond reality. Normative, emotive, and aspirational language was not referential, and not verifiable, but it pointed out the direction, in which the speakers wished reality to turn. "God help me!" was meaningful prayer, and the idea of immortality reflected a profound aspiration, a human project. Language pointed to ideal worlds and opened space for morality and religion.

In "The Meaning of Pessimistic Sentences" (1938), Schächter highlighted the pronouncements about the vanity of the world in the biblical book of Ecclesiastes, such as, "there is nothing new under the sun (1:9)." Pessimistic sentences represented to Schächter a double challenge: They were meaningless – they felt true but were unverifiable, and they reflected cultural despair. Schächter suggests that they testify to an imagined state of affairs, a desired alternative, a perfect world, and this alone gave them meaning. Language was multilayered and its different strata represent interrelated spheres of life. Physicalist sentences amenable to verification constituted the sphere of science, but emotive, moral, poetic and religious language pointed to a reality beyond them. The relationship between the two would preoccupy him his entire life but, after his departure from Vienna, his focus gradually shifted to religion and the world beyond science.

Schächter may have intuited a potential opening to ordinary language in Wittgenstein's formal language in the *Tractatus,* uses of which could not be understood by a universal grammar, apart from their local cultural contexts. The late Wittgenstein (1958, e.g.) lay hidden in the *Tractatus*, and Schächter appeared to have discovered ordinary language and its shifting contexts before him.[1] (Schächter 1964, 23–27; Ramharter 2022b) Like Wittgenstein, Schächter thought that silence may be appropriate in configuring God: This much of the elimination of metaphysics remained with him throughout life. God was not a theological construct but an interior experience. But, unlike Wittgenstein, Schächter did not relegate the world beyond science and referential language to the mystical. He insisted that language expressed norms, commands, wishes and dreams, vindicating the viability of religious life.

6.3 The Modern Crisis

Schächter regarded the scientific worldview as reflecting both the achievements and the crisis of modernity. A world confined to science and the reality at hand, and devoid of moral and religious meaning, was one where humanity had lost its way. He felt that the world cried out for a new vision that would restore meaning. Traditional religion was in crisis. In a scientific universe, many of its hallowed concepts could no longer be maintained. New moral and religious projects were needed. Re-envisioning Judaism was Schächter's main course for healing this ill, but

[1]This is the way Schächter rendered the *Tractatus* after the publication of *Philosophical Investigations*. (Schächter 1964, 23–27) His account registered also the impact of conversations with Wittgenstein's friend, Viennese architect Paul Engelmann, who became a close friend in Palestine.

he drafted resources from other traditions, from European philosophy and literature and, later, Eastern religions, to shape a new worldview.

Schächter began charting the modern crisis in late 1930s Vienna. He found urban culture, especially the lax interwar sexual norms, alienating. The rise of Fascism and National Socialism signaled a profound cultural and social upheaval, the like of which humanity had never before encountered. The scientific worldview, which Schächter initially found exhilarating, increasingly looked to him as a major cause of the crisis. Enlightenment hopes for revealing truth and authentic humanity were ending in frustration. Science was dissolving the myths grounding society, undermining morality and religion, and creating a sensuous and soulless human devoid of interiority. The integrity of the human soul, which Schächter identified with monotheism, was gone, and human destiny and eternity had vanished. Wonderment about nature and the inherent mystical relationship to the unknown disappeared, and religious rites appeared fossilized. Technological mastery replaced spiritual quest. The existential disaster was both individual and collective in character: "It is essential for a culture that its members have a common conception of life and find a common meaning in life." (Schächter 1937b, 27) But the scientific worldview emptied life of meaning, and humanity experienced a "concealment of the divine" (הסתר פנים). Man in God's image vanished. (Schächter 1937 b, 1943, 1953, 11–26, 129–130, 143–154; 1963a, 9–18)

Dire as the human condition was, Schächter would not allow despair to set in. The house stood in ruins, and there was no way of rebuilding the original structure, but the ruins themselves testified to myth's continued appeal – to a spiritual and religious quest. (Schächter 1943, 77–82) Schächter made it his life mission to seek reconstruction. There was no going back to the old myth and religion. In his old age, Schächter would regard his youthful enthusiasm for scientific language as heretical, but he recognized that the Vienna Circle had rendered the old metaphysics and theology obsolete. Those engaging in them risked an apostasy of their own. (Schächter 1943, 77–82; Asher Schaechter 2009, text to note 269) All the same, there were true metaphysical motifs that reflected the deepest wishes of the human soul, and its internal search and needs. Divinity and other metaphysical ideas were not cognitive but the product of internal experiences, of the search and desire for transcendence. To deny them was to deny the essence of humanity. (1937a, 1937b, 1953, 11–26) In Israel, national projects seemed to sideline the spiritual quest for a time, but Schächter would not relent and kept reminding his growing audience of the acuteness of the universal crisis. In later years, the increasing popularity of psychoanalysis, modernist literature and drug culture reflected, Schächter thought, Israeli youth's unanswered spiritual craving. (Schächter 1976, 64; Margolin 2014). Throughout his life, he was convinced that social suffering would produce a new morality and the awaited if tardy spiritual awakening would take place. His life was a life of searching and seeking.

6.4 Spiritual Renewal

The search for transcendence and moral perfection was imperative for replenishing meaningful life in a scientific universe. The true philosopher searched not only for knowledge, as scientists did, but for reconstituting the integral person, rebuilding interiority, and recovering the transcendental dimension of human life. The philosopher was the "essential human," who no longer had access to God, yet endeavored to protect the divine in life. The essential human cared little for the transient dimensions in life – for a career, possessions, or technological control. Like Socrates, the true philosopher sought eternity and perfection, and was willing to defy injustice even at the cost of his or her life. Socrates became the model: Having refused the Sophists' joyful criticism of tradition and religion, he sought to reconnect his contemporaries with the eternal and divine, without ever having the security of metaphysics and theology. Schächter's true philosopher represented the Vienna Circle as it could have been, should its members have complemented their critical scientific project with Schächter's proposed spiritual quest. (1937b, 31–32; 1953, 217–37; 1964, 69–75)

Without religious confidence or support, the philosopher, deprived of the primal sources sustaining humanity, remained vulnerable. Schächter posed this essential human as a model for his Viennese colleagues, but he himself remained a religious person and focused on redirecting the philosopher to religion. He sought a renewed encounter with the divine, and restored rituals that would anchor the transcendental in daily practice. He redefined Jewish principles of faith as phenomenological experiences of interior human life. He thereby drew them away from their former metaphysical articulation, which posed them as external to humanity, and left them open to the critique of theology. It was not God's uniqueness in the universe that was central, but the unity of the believer with God and the world. God was manifested immanently, and belief in God was belief in the meaning of life. (Schächter 1963a) The heavenly Torah implied the human encounter with God. In contrast, apostasy was cynicism and contempt for the ethical life. As the search for the divine revealed itself in language, Schächter availed himself of literature and philosophy to exemplify it. He progressively turned to existentialism to show freedom, dependence, love, fear and loneliness operating as "existentialist" forces at the core of the human soul. (Schächter 1966, 300) Later, in the 1960s and 1970s, he would turn to Yoga and Eastern religions in search of paths for the divine. Jewish religion became the laboratory for universal religious renewal.

6.5 Jewish Religious Renewal

More than most Jewish thinkers, Schächter viewed the modern Jewish crisis through the prisms of the universal human crisis. Yet, he was both a traditional Jew and a Zionist, and he regarded the Hebrew revival in Palestine/Israel as the culmination of a millennia-old Jewish civilization. Surrounded in the Vienna Circle by secular and assimilated Jews who broke with Judaism, he could not but sense the depths of the Jewish predicament. In Palestine/Israel, he was taken aback by the secularism of mainstream Zionism and the growing rift between Orthodox and secular Jews. He wanted to invest the national project with universal and Jewish values and build a bridge between religious and secular Jews. He was himself a practicing, if not strictly, observant Jew, who covered his head when teaching Jewish Studies in the secular Reali School. He was convinced that a Hebrew national culture must draw on Judaism, but equally persuaded that a major reconstruction of Judaism was necessary to make it speak to Israeli youth. He deployed all his energy and intellectual resources in an effort to make Judaism relevant. He did so as an educator and coordinator of Jewish Studies programs in secular schools; as an interpreter of biblical and talmudic literature who developed an original midrashic mode; and as a public intellectual promoting "education for values." He engaged New Age religions and international trends to make Judaism speak to the present. (Schächter 1963; Margolin 2014)

Moses' Torah, said Schächter, is the peak of human achievement. It has brought together myth, history and wisdom, and has found a place for the one (God and the individual), the many (community), and the unique (God's uniqueness and that of human marriage; in Hebrew: יחד, ייחוד, יחיד). (Schächter 1976, 71–72) Elucidating the Torah's contemporary meaning was not a scientific project: Schächter rejected historical and literary criticism in biblical studies. He doubted the scholar's ability to resist modern prejudices, and, in any case, biblical criticism missed the point: Making the Bible speak to the modern age required elucidating its timeless motifs, not its historicity. (Schächter, 1953, 125–134; Schächter 1963b) Yet, he himself sought rapprochement between scientific and religious motifs by showing their cosmic and historical durability and their convergence in the human soul. The biblical creation myth interwove, he said, divine, cosmic and human: Sun and moon were incorporated into a monotheistic universe, the human body and home were formed and protected, the life cycle was noted and marked, the seasons were celebrated in annual festivals, and marriage was instituted and sacrilized. (Schächter 1963, 238–267) Schächter often used Eastern religions to expound biblical motifs, and deployed literature and psychology to advance his spiritual experiential approach, seeking to create an encounter between the Bible and the reader. He saw himself as mimicking biblical and talmudic interpretive modes and continuing the Jewish project by creating a modern *Midrash.* (Schächter 1963; Schächter 1963b; Schächter 1972)

Schächter had a wide audience and a group of devoted followers – the Schächterites – but his Jewish religious renewal never took hold. None of his followers had his masterful command of rabbinic literature, and most did not remain loyal to traditional Judaism. In line with his ḥasidic upbringing, Schächter viewed ritual as symbolic practice evoking spiritual transcendence, and beseeched his followers in Yodfat to retain Jewish practice. They did not, and he never moved his residence there. (Margolin 2014, 547–48) His "New Age" Orthodox Judaism, which joined Jewish observance with Yoga, Midrash with Karl Jung, talmudic study with physical labor, Zionism with ecological consciousness, and sexual puritanism with philosophical iconoclasm was *sui generis*. Even in old age, he paid homage to Schlick as inspiring his search for divinity, but denounced the scientific worldview. (Schaechter 2009, appendix 6) He preached tirelessly "education for values" and experimented with new modes to enhance Jewish renewal, as if sensing that the project was failing to take root.

In the decades after his death, Israeli scholars showed considerable interest in Schächter's work, generating a small field of academic literature. (Margolin 2014) Work on Schächter continues today, a steady but gentle stream. As with many remarkable philosophers, and with most of the Jewish *maskilim* (proponents of enlightenment), who sought to reconcile modernity and Jewish tradition, Schächter's search was distinguished by its originality and brilliance, and his dedication and commitment, rather than by its success.

6.6 Conclusion: A Rabbi Among the Apostates

A religious Jewish *maskil*, Schächter was an outlier in the Vienna Circle. He shared with them only part of his world. None of his fellow Jews in the circle understood, let alone shared, his Jewish renewal project: They had broken with Judaism. Schächter's Jewishness was a polar opposite to Popper's: Popper was an avowed assimilationist, who refused a Jewish identity for much of his life, whereas Schächter professed loyalty to rabbinic Judaism and sought its renewal. This author has followed Schächter's path of a religiously educated Jewish *maskil*. I began my academic journey with the Vienna Circle and the young Popper, but have since written a Jewish European history grounded in rabbinic sources. Schächter was no more a guide to Jewish Studies than Popper, however: Inspired by the critical "Litvak" and maskilic traditions rather than by Ḥasidism, my *Jacob & Esau* hopes for a Jewish Studies renewal via the very historical scholarship, which Schächter deemed detrimental to his modern Midrash (Hacohen 2019).

I find most attractive the young Schächter, the ambivalent scientific philosopher. He revealed the circle's softer side: If the interlocution with Popper exposed a measure of dogmatism in the circle, their tolerance of a rabbi – who took over instruction at the university from Schlick, under a protofascist government – showed their accommodation of religious pluralism. Schächter seems to me at his best as a language critic. Moved by rabbinic Judaism, he brilliantly exposed the limits of scientific language, opening up language philosophy to the diversity of human cultural endeavors. His critical philosophy seems more interesting than his Jewish spiritual renewal, his life in-between philosophy and Jewish practice more appealing than his commitment to their unification. The path for the modern Jew is best blazed not by the rabbi among his disciples but by the rabbi among the apostates.

References

Boyer, John W. 1995. *Culture and Political Crisis in Vienna: Christian Socialism in Power, 1897–1918*. Chicago: University of Chicago Press.
Engelmann, Paul and Josef Schächter. 1946. *Style of Truth: In Memory of Adolf Loos* (in Hebrew). Tel-Aviv: Mafyil.
Hacohen, Malachi. 2000. *Karl Popper – The Formative Years, 1902–1945: Politics and Philosophy in Interwar Vienna*. New York: Cambridge University Press.
———. 2019. *Jacob & Esau: Jewish European History Between Nation and Empire*. Cambridge: Cambridge University Press.
Margolin, Ron. 2014. Was Josef Schächter the Prophet of New Age Spirituality in Israel? (in Hebrew). In *Beyond the Halakhah*, ed. Yaacov Yadgar, Gideon Katz, and Shalom Rtzbi, 529–554. Sde Boker: Ben-Gurion University.
[Neurath, Otto.] 1929. *Wissenschaftliche Weltauffassung. Der Wiener Kreis*. Wien: Artur Wolf. English: The Scientific Conception of the World: The Vienna Circle. In Neurath 1973. *Empiricism and sociology*, eds. Marie Neurath and Robert S. Cohen, 299–318. Dordrecht – Boston – Lancaster: Reidel.
Ramharter, Esther. 2022a. *The Vienna Circle and Religion*. Vienna: Springer.
———. 2022b. Josef Schächter on Religion and the Philosophy of Language: Preferring Beginnings to Ends. In Ramharter 2022a
Schächter, Josef. 1935. *Prolegomena zu einer kritischen Grammatik* (Schriften zur wissenschaftlichen Weltauffassung 10). Vienna: Springer. English: Schächter 1973.
———. 1937. *A Short Introduction to Logic* (in Hebrew). Vienna: Victoria.
———. 1937a. Relegie en Wetenshap. *Synthese* 2 (5): 159–167.
———. 1937b. Notes on Problems of Ethics and the Philosophy of Culture. Vienna (unpublished). Published in *Ethics and the Will*. Essays (Vienna Circle Collection 21), eds. Brian McGuinness and Joachim Schulte, 1994. Dordrecht: Springer, 19–32.
———. 1938. Der Sinn pessimistischer Sätze. *Synthese* 3(5): 223–233. English: Schächter 1994. The Meaning of Pessimistic Sentences. In *Ethics and the Will. Essays*, Vienna Circle Collection 21, ed. Brian McGuinness and Joachim Schulte, 9–17. Dordrecht: Springer.
———. 1943. Torah u-Minut (teaching and false teaching למודי היהדות). *Gilyonot* 14: 77–82.
———. 1948. *On Understanding: A Dialogue* (in Hebrew). Tel-Aviv, Mordechai Newman.
———. 1953. *From Science to Faith* (in Hebrew). Tel-Aviv: Devir.
———. 1963. Pathways in the Education of this Generation (in Hebrew). Tel-Aviv: Aleph. http://www.daat.ac.il/daat/vl/tohen.asp?id=493
———. 1963a. *The Superhuman in the Human* (in Hebrew). Tel-Aviv: Aleph
———. 1963b. *Jewish Studies and High-School Education* (in Hebrew בחינוך העל יסודי). Tel-Aviv: press of the author's students.
———. 1964. *Essays and Notes* (מסות ורשימות). Jerusalem: Kiryat Sepher.
———. 1966. The Task of the Modern Intellectual. In *An Anthology of Hebrew Essays*, eds. Israel Cohen and B. Y. Michali, 2 vols., 2:299–310. Tel Aviv: Institute for the Translation of Hebrew Literature and Massada Publishing.
———. 1970. *Moritz Schlick: His Personality and His Philosophy* (in Hebrew). N.p: n.p.
———. 1972. *Weekly Torah Portion* (in Hebrew פרשת השבוע). Tel-Aviv: Yaḥdav.
———. 1973. *Prolegomena to a Critical Grammar*. Dordrecht: Springer (eBook).
———. 1976. *Pathways in the Education for Values* (in Hebrew). Tel-Aviv: Yaḥdav.
Schaechter, Asher 2009. From Science to Faith: The Early Thought of Joseph Schaechter in the Land of Israel (in Hebrew). Ph.D. Dissertation, The Hebrew University of Jerusalem.
Stadler, Friedrich. 1981. Spätaufklärung und Sozialdemokratie in Wien 1918–1938. In *Aufbruch und Untergang: Österreichische Kultur zwischen 1918 und 1938*, ed. Franz Kadrnoska, 441–474. Vienna: Europaverlag.
———. 1982. *Vom Positivismus zur "Wissenschaftliche Weltauffassung"*. Vienna: Löcker.

———. 2001. *The Vienna Circle. Studies in the Origins, Development, and Influence of Logical Empiricism*. Dordrecht: Springer.

Wittgenstein, Ludwig. 1958. *Philosophical Investigations, translated by G.E.M. Anscombe*. Oxford: Blackwell.

Chapter 7
Josef Schächter on Religion and the Philosophy of Language: Preferring Beginnings to Ends

Esther Ramharter

7.1 Introduction

Even though Josef Schächter (1901–1994) is certainly not the most prominent member of the Vienna Circle, he should not be seen as a marginal figure. Not only was he important for the Circle's activities, he also developed remarkable philosophical ideas on his own. In this paper, I will not be able to do justice to his philosophy in general, but rather focus on one particular point of intersection between his considerations about language and his thoughts concerning religion—a point that I consider nonetheless as characteristic for his philosophy.

Schächter was born in Galicia, he grew up being educated in a Hassidic tradition and was ordained for the rabbinate. He went to Vienna in 1921 (see Even 2014, 82) and studied at the University of Vienna, where he became a member of the Vienna Circle. At the same time, he began to give courses at the Jewish Teachers Seminary (see Diamond 1964, 17). In 1938, he emigrated to Israel (then Palestine). This means that he—the rabbi!—was one of the last members of the Vienna Circle having stayed in Vienna. After the murder of Moritz Schlick, whose assistant Schächter was, he took over Schlick's seminars for a short period (see Stadler 2015, 720). In Israel, he taught in secondary schools: first in Tel Aviv, later on in Haifa (see Schoneveld 1976, 67). From 1951, he worked as a supervisor of secondary schools in Haifa (see

E. Ramharter (✉)
Institut für Philosophie, Universität Wien, Wien, Austria

Institut Wiener Kreis, Universität Wien, Wien, Austria
e-mail: esther.ramharter@univie.ac.at

E. Ramharter (ed.), *The Vienna Circle and Religion*, Vienna Circle Institute Yearbook 25, https://doi.org/10.1007/978-3-030-76151-6_7

Kressel 2007, 101). Furthermore, he inspired the founding of Kibbutz Yodephet (see Weiner 1961, 261–268; Schoneveld 1976, 71).[1]

As already suggested by his biography, there are two important backgrounds for his works: Hassidism and the Vienna Circle (see Diamond 1964, 17; Benyoëtz 2003, 378). Schächter himself states this:

> Almost fifty years ago I thought that the logical elucidation of thought and language is a primary means among the ways of correcting humankind... Over the years I came to realize that the inner work, of which the Hassidic literature and the literature of the Far East speak, is preferable to it... I must note that the clarifications I received in the "Vienna Circle" helped me much in my studies in religion to separate the wheat from the chaff. (Schächter 1977, 176–177, as cited and translated in Even 2014, 82)

Schächter's writings can be divided chronologically into his early works, written in Vienna, before the war, and those, which he composed later in Israel. A corresponding distinction can be drawn on a contentual level: his publications after the emigration from Austria were, with a few exceptions, dedicated to the Bible, the Talmud, and (religious) education in general; his earlier publications on science and language, however, were very much in the spirit of the Vienna Circle. Nonetheless, Schächter always maintained some independence in his thinking. This can especially be seen from the three books that he wrote during these years: his doctoral thesis *Kritische Darstellung von N. Hartmanns „Grundzüge einer Metaphysik der Erkenntnis"* in 1931 (which has not been published), *Prolegomena zu einer kritischen Grammatik*[2] in 1935, and מבוא קצר ללוגיסטיקה (A Short Introduction to Logic) in 1937. In the following, some passages of the earlier publications will be studied in detail, whereas later publications will only be referred to in terms of tendencies.

7.2 Schächter's Philosophical and Religious Background

7.2.1 Schächter's Involvement with Wittgenstein and the Vienna Circle

Schächter attended the meetings of the Vienna Circle very regularly and so became an important constant in the life of the Circle. Nevertheless, the relation between him and others participating in the discussions of the Vienna Circle has been constantly

[1] According to Weiner 1961 (263), the group was sometimes called "Schechterists".

[2] The book did not receive much attention. The reason for this might be that it was published at the same time as Carnap's *Logische Syntax der Sprache* and Popper's *Logik der Forschung*, although a review was published stating that "[t]his volume constitutes a kind of complement to Dr. Carnap's *Logische Syntax der Sprache*, and to some extent accomplishes for the syntax of ordinary discourse what Dr. Carnap's book does for scientific languages" (Weinberg 1937, 334; for a list of further reviews see Reitzig 1978, 278).

described—both in the secondary literature and by Schächter himself—as one-directional: he was influenced by them, not the other way round. Although a critique of this view—in particular, a study of a potential influence of Schächter on Ludwig Wittgenstein[3] seems promising to me, this has to be left to another paper. Besides Wittgenstein, the most important persons for Schächter's philosophical development were Moritz Schlick[4] and Friedrich Waismann (see Stadler 2015, 720). Schlick was the advisor of his doctoral thesis, which clearly associates it with the Vienna Circle, even though the content might give the impression that Schächter was distancing himself from Logical Empiricism (see Reitzig 1978, 272). In both a footnote of this dissertation (Schächter 1931, 83) and in the Author's Preface of *Prolegomena*, Schächter says that he owes much to Waismann, who told him a lot about his unpublished work. Waismann's *Logik, Sprache, Philosophie* had not appeared yet; however, he discussed his ideas—which he took from Wittgenstein—within the Circle (see Reitzig 1978, 273). Wittgenstein himself was another influence that reached Schächter via the Vienna Circle. It seems, however, unlikely that Schächter would have known Wittgenstein personally (see Benyoëtz 2003, 379), as they saw each other at most once or twice, in the lectures of Brouwer. It must have been therefore mainly other members of the Vienna Circle, especially Schlick and Waismann, by whom Schächter was made familiar with Wittgenstein's ideas.[5] What exactly Schächter heard of Wittgenstein or which of his texts he read can not be reconstructed with certainty. He could, for example, have been acquainted with a version of the Big Typescript.[6] Even though it is not clear what he knew of Wittgenstein's writings, there can be no doubt that his thoughts expressed in *Prolegomena* are essentially Wittgensteinian.[7] He himself declared that he intended to keep the book "in the kind of philosophic spirit for which, I believe, the enquiries of Ludwig Wittgenstein are the model" (Author's Preface to Schächter 1973).

What Schächter took over from both the Vienna Circle and Wittgenstein was *clarifying thoughts* as his method (see the quotation in Sect. 7.1) and a strong and lasting interest in structure and functionality of languages.

---

[3]In the Translator's Foreword to the English Edition of *Prolegomena* (Schächter 1973), Paul Foulkes explicitely addresses this issue.

[4]Even his later book העל אנושי באנושי (*The Superhuman in the Human*, Schächter 1963) was written under Schlick's influence, see Benyoëtz 2003, 378.

[5]Paul Engelmann later played a major role in Schächter's relationship to Wittgenstein's philosophy, see Benyoëtz 2003, 379.

[6]Or, he could have known *Philosophical Remarks*, as Beckmann 2001 (8) claims—without further explanation.

[7]In *Prolegomena* one finds comparisons of matters of grammar and chess (Schächter 1973, Part 1, II, 0, 5), considerations concerning the grammar of "I know" (Schächter 1973, Part 1, IV, 3, 5), of "I feel your pain" and "it is good in itself"' (Schächter 1973, Part 2, I, 5, 7), the conception of thoughts as "sentences not perceivable by the senses" (Schächter 1973, Part 2, I, 10, 8) and many other similarities with Wittgenstein's considerations.

7.2.2 *Schächter's Hassidic Background*

As we have seen (compare Sect. 7.1), commentators on Schächter usually name Hassidism and the Vienna Circle as the two main influences on Schächter's life and work.[8] The importance of Hassidism, in particular, becomes manifest in the understanding of teaching that Schächter took from Hassidism (see Schoneveld 1976, 68; Weiner 1961, 268). Schächter was a passionate teacher. Living in Israel after World War II, he developed a conception of education corresponding to the needs of young people in Israel at that time. J. S. Diamond points out that the situation of society in Israel played an important role in Schächter's thinking:

> [T]he bitter struggle between an entrenched, politically strong Orthodoxy and an equally militant secularism has, psychologically at least, polarized Israeli society, and has resulted in two separate school systems—'religious' and 'secular'. [...] To overcome this problem, a philosophy is needed, "which will enable these texts [Tanakh, Talmud, and Midrash] to be taught as sources of Jewish religious inspiration. One attempt to develop such a philosophy has been that of Dr. Yosef Schechter [...]. (Diamond 1964, 17)

Schächter's rejection of the radical separation of religion and the secular domain has also been noticed by other authors. J. Schoneveld, e.g., writes: "Schächter does not attach much importance to the common separation of teachers into the 'religious' and the 'non-religious'." (Schoneveld 1976, 68) In the context of education Schächter wants to contribute to the overcoming of to the dichotomy of religion and the secular:

> By developing ideas that are essentially Hasidic, Schechter is able to translate the tradition into terms that are intelligible to the secular Israeli youth, and thus he is able to bridge the unfortunate gap between "religious" and "secular". (Diamond 1964, 23)

This potential to overcome the separation of religion and the secular domain is the most important aspect of Hassidism for the purposes of this paper.

7.3 A Preference of Beginnings to Ends: Schächter's Approach to Religion and Language

7.3.1 *Religion and the Meaning of Life*

How does Schächter think that he could bridge the gap between the religious and the non-religious?

> We do not see ourselves as progressive people who look back on our fathers as the primitive, albeit interesting, composers of the Tanakh; we see ourselves rather as pupils coming to our

[8] Eastern sources are certainly also very relevant for him, see Weiner 1961, 265.

> fathers to learn how they found meaning for their lives.[9] (Schächter 1960, 20, as cited and translated by Schoneveld 1976, 172)

> Disturbed by the gap between Orthodoxy and secularism in Israel society, Schaechter tried to turn Israel non-Orthodox youth back to its Jewish heritage. He sought to achieve this by approaching the Bible and Talmud as a philosophy of life. Schaechter felt that modern man was uprooted and cynical in the technological world, because he had forgotten to invest life with meaning and to look within himself. This inner search was the heritage of Israel. (Kressel 2007, 102)

Finding meaning in life is, hence, the link between religion and secularity. Schächter holds that it "is essential for a culture that its members have a common conception of life and find a common meaning in life." (Schächter 1937b, 27) Or, exalted in his conception of the essential man: "The essential man strives to give his life a higher meaning." (Schächter 1937b, 31)

Finding a meaning in life roughly corresponds to what Schächter calls "religion in a broader sense"[10] in a paper published in 1937 and entitled "Religie en wetenschap". Additionaly, he defines "Religion in the narrower sense": "One can say that religion tries to 'answer' the question of the meaning of life in a certain way and that an understanding of the supernatural, providence, and more such concepts play an important role in religion." (Schächter 1937a, 166, transl. E.R.) Religion understood in this narrower sense is therefore a special case of the search for meaning in life.

This "inner search" (see quotation above) for meaning cannot yield a meaning that comes from outside, from an external goal, aim, or endpoint:

> For example, the Aggadah stresses the notion that "beloved is man" (Hebrew: "haviv") this in the face of a persuasive fatalism. Such an attitude may today be called "optimism"; and it is imperative that it and its assertion of the worth of human existence should be communicated to modern youth, so imperiled by the cynicism and despair that prevail today. Schechter finds in the Aggadah other touchstones on which adolescents today may build their inner lives: that existence is only meaningful when it is *intrinsically* meaningful. (Diamond 1964, 22)[11]

[9]Note how far Schächter has veered away from the Vienna Circle.

[10]"'Belief' in the religious sense of the word means [...] to be narrowly bound to certain concepts (or objects or behaviour) and to examine their value. Wherever there is such a binding, which influences the motives and decisions of men, one can speak of religious belief in a general sense." (Schächter 1937a, 163–164, transl. E. R.)

[11]Schächter says: "The structure of man's soul is such that the so-called 'existentialist' forces, such as freedom, dependence, love, fear and loniliness operate at its core, whereas in the circumference of the soul external motives such as social, economic and political considerations are at work. For some inconceivable reason, human beings tend to disregard the inner life, and there is always a need to remind them [...] that it is the inner life which is important and that only the person whose external existence radiates from the inner core, can be said to be truly alive." (Schächter 1966, 300) Compare also: "In Schächter's view, the belief that the act of elucidation is valuable is the main thing, not the satisfaction of the action. [...] In his view, there is no external way to grasp or talk about the divine; the work has to be internal and to focus on self-transformation [...]." (Even 2014, 83)

Accordingly, religious education should not start with God, but within one's soul; that means, it does not start with the end, one could say, but with trying and searching:

> A basic ingredient of Schechter's thinking is the Hasidism he absorbed in his boyhood, and if this thinking may be said to have a starting-point it is the dictum of the Hasidic Rabbi Zusya that in seeking to create a Weltanschauung one begins with the human condition. (This is opposed to Rabbi Elimelech, who begins with God.) (Diamond 1964, 18)[12]

Education should not start with a specific aim or goal, but at the only point where one can actually start: oneself. As a consequence, it is not so important whether someone is religious or not, it is much more relevant to be in search for something, to start a process of finding a meaning:[13]

> [H]e rejects attempts [...] that go to great pains to differentiate between Israelite monotheism and paganism. Instead, the common denominator between the two must be stressed—that both spring out of the same source, e.g. the sincere attempt of ancient man to invest his life with meaning. (Diamond 1964, 20–21)

What counts is not whether one falls under a category such as "believer" or "non-believer", nor the sort of attitude or belief-system one finally *ends up* with,[14] but a sincere search for the meaning of life. Diamond repeatedly writes about "this attempt of a people to fashion a meaningful way of life" (Diamond 1964, 21).

This preference of beginnings and attempts in contrast to ends and endpoints is integrated into every day life by his appeal to "regard each day as the first":

> Drawing on Kierkegaard, Schechter takes seriously the latter's series of sketches on "The Rose and the Bird." In one of these, the rose, knowing it is destined to wither, yet blossoms fulfillingly; and the bird, in spite of its apportioned time on earth, still sings unceasingly. This is what man must strive for in his life-time. The secret in achieving this is not by pursuing entertainment and trivia, but by mastering the art of repetition. For life, as Kierkegaard pointed out, is repetition. When man regards each day as the first, when he performs each routine activity not for ulterior motives but because it is to him inherently worthwhile, then he begins to fill his life with meaning, then is he really living. (Diamond 1964, 18)

7.3.2 *Meaning in Language*

In his philosophy of language, Schächter essentially drew from the ideas of the Vienna Circle, but he also—partly under the influence of Wittgenstein, who was on his way to become "the late Wittgenstein"—went beyond them. He turned to

---

[12]This view is also ascribed to the Maggid of Mesritz in a Hassidic tale from a collection of Martin Buber, see (Schoneveld, 1976, 68).

[13]In order to motivate the initiation of such a process, Schächter wants to teach optimism (Diamond 1964, 22); for similar thoughts that can be found with Schlick, see Siegetsleitner 2014, 282.

[14]A consequence is that "[t]he dead-end state is not infrequent with [the true philosopher]"—as Schächter has to admit (Schächter 1963, cited and transl. as in Even 2014, 83).

a broader understanding of language, which was also the basis of the linguistics that emerged at that time. J. F. Staal notes in the Forword to the English version of *Prolegomena*:

> It is instructive to witness how an avowedly positivist philosopher discovers facts of language which so-called positivistically-oriented linguists—albeit obsessed by almost mythological matters of procedure and methodology—had not been able to detect. More recently linguists and logicians, unhampered by positivist dogma and methodological paranoia, have been able to go further. (Schächter 1973, Forword to the English Edition)

In the 1930s, Schächter developed for the first time the idea that there are elements of language that could be said to have meaning although they do not possess a reference and that there are sentences which are neither true nor false.[15] This idea occurs in several forms in his writings: in what he calls "poetic philosophy" (see Schächter 1936, 11–12), in "religion in the narrower sense" (Schächter 1937a, see above) and in ethics (see Schächter 1937b, 22). In retrospect, Schächter himself puts it like this:

> I discovered that the test of meaning, which was a fixed asset in this Circle [the Vienna Circle, E. R.] [...] was not efficacious regarding some types of statements [...] that there are statements in human language in which it is evident that the meaning is not dependent on verification [...]. We understand these statements, because the understanding of the statements consists of finding its place in the fabric of language. (Schächter 1963, 9, as cited and transl. in Even 2014, 82)

As a consequence, there are, according to Schächter, at least some cases of linguistic expressions in which one has to search for the meaning within language, not somewhere out there, where the expression points to. In fact, it is not only that there are such cases, Schächter is not interested in the other cases. To be more precise: he is not interested in the referencing function of these other cases. Although some propositions can certainly said to be true or false and some expressions do have a reference, Schächter is hardly ever paying attention to this fact. For all expressions—whether they possess a reference or not—his focus is other than a focus on reference. He states:

> Knowing the meaning we can use the sign. By stating its use we exhaust the meaning of the sign. Thus a person knows the meaning of a sign, when he knows how to operate with it and how to place it in its linguistic context. (Schächter 1973, Part 1, I, 1, 4)

The emphasis on the linguistic context and the absence of any occurrence of "truth" or "reference" in the surrrounding paragraphs become particularly conspicious by comparing the quotation with a similar passage by Moritz Schlick, written at about the same time:

> Thus, whenever we ask about a sentence, "What does it mean?", what we expect is instruction as to the circumstances in which the sentence is to be used; we want a description of the conditions under which the sentence will form a *true* proposition, and of those which

[15]According to a list of books at the end of *Prolegomena* which collects all books that Schächter read before writing *Prolegomena*, he only knew Frege's *Grundlagen der Arithmetik* and *Grundgesetze der Arithmetik*, but not "Über Sinn und Bedeutung".

> will make it *false*. The meaning of a word or a combination of words is, in this way, determined by a set of rules which regulate their use and which, following Wittgenstein, we may call the rules of their grammar, taking this word in its widest sense. (Schlick 1936, 341; emphasis not by me)[16]

The referral to verification clearly distinguishes Schlick's account from that of Schächter.

As in his understanding of religion, Schächter is interested in the origin: the language. He does not care whether the use of expressions ends up at a point outside language, which can be called reference, or not.

When it comes to religious language in particular, two things should be noted: First, Schächter does not make any difference between religious and other expressions. Consider, for example, the following two passages on religious matters in *Prolegomena*:

> We sometimes say "this linguistic structure lies between concept and sentence". This can mean only that it is sometimes the one and sometimes the other, since both at once is logically impossible. For example, the word "God" may occur as concept or as fragment of the sentence "God help us!" and so on (but it may be simply a way of reacting, that is neither concept nor sentence; this holds from the point of view both of the exclaimer and of the hearer who can interpret it only as symptom). (Schächter 1973, Part 2, I, 10, 3)

> A prayer is a request to God or other beings satisfying certain conditions: (i) they understand our language (ii) they are more powerful than we are and can make our wishes come true (iii) they are occasionally influenced by our requests to fulfil our wishes. But other combinations that resemble these only in certain respects are often called "prayer" too; to resolve conflicts, to struggle free from mental distress, a person often prays without expecting this to produce fulfilment of a wish. For the linguistic average, his sentences are prayers, but for him they are to be compared with music, song and so on. Music and lyrical poetry too can express ideas, and can be understood and translated. To that extent we are concerned with signs, sentences and language; but for the rest, music and poetry do not consist of signs or sentences. (Compare also "mechanical" prayers, such as the Buddhist "omani pad me um" and the like.) (Schächter 1973, Part 2, IV, 7, 10)

For a member of the Vienna Circle this expresses an attitude astonishingly open towards religion. Schächter does not treat religious expressions any differently from other expressions: no caveats, no exceptions, no hints to any differences.[17]

Second, in a book from 1970, he says with respect to religious expressions something similar to what he has said about elements of language in general:

> I did not want (nor do I want) to revive dead carcasses (of superstitions and of the creatures of university metaphysics), but on the other hand I did not want to ignore the existence of real religious and existential elements in thought and in language. (Schächter 1970, 68–69, cited and transl. as in Even 2014, 82)

---

[16]Compare also: "Stating the meaning of a sentence amounts to stating the rules according to which the sentence is to be used, and this is the same as stating the way in which it can be verified (or falsified). The meaning of a proposition is the method of its verification." (Schlick 1936, 341)

[17]In contrast, Carnap, in his well-known paper "Überwindung der Metaphysik durch logische Analyse der Sprache" (Carnap 1931), discusses the question if there is a reference of the word "God"—and wants to proof that there is none –; Schächter does not care if there is one.

Religious matters are situated *within* language here. Again, Schächter does not care if there is any external endpoint of the search for meaning.

7.3.3 The Preference of Beginnings to Ends as a Common Element of Schächter's Philosophy of Language and His Philosophy of Religion

So far, we have observed that Schächter, on the one hand, focusses on the search for meaning in religion, and, on the other hand, concentrates on the search for meaning in language. In both cases, he directs his attention mainly to the beginnings and the process, and much less to the result. That these are not two separate observations, i.e., that there is indeed a connection between him focussing on the starting points rather than on the end points in both the search for meaning in life and the meaning within language—and that Schächter himself sees it this way—becomes clear if we look at the development of his thought after the papers he published in 1936 and 1937. In his papers of 1936 and 1937, his considerations ended up in a position concerning religious language that one could call emotivistic (see Schächter 1936, Schächter 1937a, 164). This outcome obviously did not satisfy him anymore when he wrote the paper "Der Sinn pessimistischer Sätze", which was first published 1938 in German (Schächter 1938), and in 1994 in an English version. In this article, he elaborated on a theory of layers or strata ("Schichten") of language.[18] His idea is, roughly, that language has many layers, each of which is constituted by its own rules. That these strata have developed in the course of time seems an unavoidable fact, but this does not matter for Schächter's considerations, as he is "looking at strata exclusively under their linguistic aspect" (Schächter 1994, 14). The meaning of an expression can (sometimes) be found in a deeper or different layer.[19]

One example for such a layer is the "ideal language stratum": It is built up by the expressions that we use when we idealize something that we have experienced or heard of. Schächter gives the example of a man who has been told many nice things about the Alps and thereby makes himself an ideal picture of the Alps (see Schächter 1994, 16). If this person expresses his desire to see the Alps or his admiration for the Alps, his utterances have a meaning. Even if one day he manages to see the real Alps and feels disappointment, his words do not become meaningless. A characteristic of this language stratum is that "certain concepts and grammatical categories display a curious anomaly, which makes it more difficult to describe.[...] I will try to

[18]Note that Friedrich Waismann later wrote two articles on language strata, too; see Waismann 1968.

[19]Sometimes the empirical ground breaks away (e.g. for some presumed teleological relations, the Natural Sciences have shown that they do not hold); nevertheless, the meaning can still be found in the respective layer. This is the reason why we can make sense of propositions like, e.g., "everything is void".

elucidate it by means of a grammatical category which assumes a different meaning in it, namely the concept of 'value'. What is regarded as valuable in this area is not things and states, but 'directions'." (Schächter 1994, 15) If someone is asked "What is valuable?", the question "is not answered by *describing* a state of affairs, but by pointing in the *direction* of a wish." (Schächter 1994, 16) The word "directions" already indicates a process that is to be understood from aiming at something, but not necessarily from its goal. This idea is made explicit by Schächter:

> The word "direction" is meant figuratively in this connection; for example, a stone can be thrown in two ways: (I) we see the target clearly and try to hit it; (2) we merely guess that there is a target there and throw in that direction, but without seeing it clearly. In the same way, our striving for "absolute" correctness corresponds to aiming at a target which is only known to lie in a certain direction. (Schächter 1994, 16)

The conception of "direction" that Schächter has in mind here can be explained by the difference between searching in archeology and searching for one's key. In the latter case, one knows exactly what one is looking for, in the former one only needs to have a very rough idea. The end of the search, its objective, is not decisive for the meaning here: "Man just entertains some wishes without being able to state what exactly their fulfilment looks like." (Schächter 1994, 16)

By covering such areas as values, ideals, wishes, etc., the ideal language stratum is obviously one that is central to any search of meaning in life, and it is also one that is relevant to explaining an attitude towards the meaning of linguistic expressions: "From the perspective of the rest of language, to idealize value terminology is, in every case, *to go beyond* the limits of language." (Schächter 1994, 16) The expressions of the ideal language stratum are underdetermined not only empirically, but also within the other language strata, yet have a meaning. Therefore, they exemplify what Schächter has stated (see Sect. 7.3.2): that there is a meaning of linguistic expressions that does not depend on verification.

Another language stratum important for the search of meaning in life is the "language stratum of immortality". In this stratum, "[a]n individual human being appears [...] as having always existed and existing for ever; the individual 'I' is dominant" (Schächter 1994, 14). Schächter explains:

> This "I" is defined in the "language stratum of immortality". It does not begin with year a and end with year z. In the sentence "It is better for one not to have been born", the subject exists before his birth; we are therefore dealing with the latter meaning of "I". This meaning can be given in two ways: as a real description, i.e. as a soul which first exists in and for itself and only then enters a body, after which it leaves it again, but without therefore ceasing to exist; or as the direction of a wish which is not even realized in a description. Such concepts, which are given only by their direction, are very characteristic of modern thought. We have banished the mystical way of thinking from our world view (which consists of assertoric sentences), but without eliminating it from our "language of interests" (which consists of sentences expressing wishes and values). (Schächter 1994, 17)

Both layers—the ideal language stratum and the language stratum of immortality—are constitutive for the searching of meaning in life, *and* they are focussed on some starting points rather then on definite aims: Both depend strongly on "concepts, which are given only by their direction", where direction is understood in the above described way, and in particular not via the relation to a definit reference. The fact that Schächter considers them as "very characteristic of modern thought" ascribes to them a prominent position in his philosophy.

Schächter's "bias" for beginnings, which connects his philosophy of religion (the search of meaning in life) to his philosophy of language, also manifests itself in the metaphors that Schächter uses, for example:[20]

> There is a grave danger that the modern soul has been so baldy damaged that the ray of light radiating from within towards the external circumference will be obscured or extinguished, and only the shell of externality will remain. (Schächter 1966, 300)

Another example does not stem from Schächter himself but is a poem by Karl Kraus entitled "Zwei Läufer" (Two Runners). Nevertheless, the fact that Schächter cites it in his introduction to the correspondence between Wittgenstein and Engelmann (Schächter 1965/2006, 113) makes it an example of Schächter's predilection. Engelmann resembles, according to Schächter, the runner, who comes from the origin:

> Zwei Läufer laufen zeitenlang,
> der eine dreist, der andere bang:
> Der von Nirgendher sein Ziel erwirbt;
> der vom Ursprung kommt und am Wege stirbt.
> Der von Nirgendher das Ziel erwarb,
> macht Platz dem, der am Wege starb.
> Und dieser, den es ewig bangt,
> ist stets am Ursprung angelangt.

The implicit preference of the one who "has always been arriving at the origin" is not only Kraus', but also Schächter's.

7.4 Schächter, the Vienna Circle, and Wittgenstein

Before concluding, I want to answer to one possible objection: One might argue that Schächter underwent a development, whereas I treat his publications as if he held them true all together at the same time. Although there are certainly shifts in his views and different questions becoming more or less important to him in the course

[20]It also manifests itself in the fact that Schächter declares some important questions to be matters of terminology, e.g.: "We do not know how and why the supreme principles of ethics have arisen and been accepted. To express this in religious terms, one could say that such laws are the result of man's contact with God. To avoid such a form of expression, one could say that they have their origin in elementary living conditions common to all men. This is a terminological question (though one terminology may have a stronger influence on people than the other)." (Schächter 1937b, 26) Here, he prefers the way of speaking in which the starting point—man—is in the focus, and he does so by reasons that he finds in language.

of time, I do not see any real contradictions or even tensions between his earlier and later writings, *as far as the issues of this paper are concerned.*

An overall estimation of the relation between Schächter's philosophy and the views of the members of the Vienna Circle goes far beyond the scope of this paper. Even with restriction on religion, such a comparison is not possible within this framework. Perhaps the most important point to be made concerning Schächter and the Circle is that Schächter's obviously much more sympathetic attitude towards religion did not hinder him from being an esteemed member of the Circle—which says a lot about the tolerance practiced by the group. With respect to the specific point that I wanted to make in this paper, one concrete theoretical tension can, however, be identified: The Circle's interest in theories is opposed to Schächter's focus on beginnings and origins of intellectual projects—theories can be seen as endproducts of science, which turn their back on beginnings.[21]

On the other hand, everybody being familiar with the "late Wittgenstein" should be able to see from the above considerations that there are strong affinities between his and Schächter's thoughts.[22] If we call a Wittgensteinian a person who adheres to the ideas of the late Wittgenstein, Schächter is one already in the 1930s—far more than other members of the Vienna Circle ever became. Even four decades later, in the 1970s he devoted a chapter of a book (Schächter 1966a) to Wittgenstein. Perhaps, Schächter was the only real Wittgensteinian among the members of the Vienna Circle after all.

References

Beckmann, Susanne. 2001. *Grammatik der Metapher*. Tübingen: Niemeyer.

Benyoëtz, Elazar. 2003. Paul Engelmann, Der Andere. Ein Teppich, aus Namen geknüpft, zu seinem Gedenken aufgerollt. In *Wittgenstein-Jahrbuch 2001/2002*, eds. Wilhelm Lütterfelds, Andreas Roser, and Richard Raatzsch, 369–427. Bern: Peter Lang.

Carnap, Rudolf. 1931. Überwindung der Metaphysik durch logische Analyse der Sprache. *Erkenntnis* 2: 219–241.

Diamond, James S. 1964. Yosef Schechter. An Approach to 'Jewish Consciousness'. *The Reconstructionist* 30 (Dec. 25): 17–24.

Even, Rachel. 2014. Josef Schächter's turn: From the Vienna Circle to Non-Dual Religiousness via Wittgenstein. In *Analytical and Continental Philosophy: Methods and Perspectives. Contributions of the Austrian Ludwig Wittgenstein Society* Vol. XXII, eds. Sonja Rinofner-Kreidl and Harald A. Wiltsche, 82–84. Kirchberg am Wechsel: The Austrian Ludwig Wittgenstein Society.

Kressel, Getzel. 2007. Entry Josef Schächter. In *Encyclopedia Judaica*, 2nd Edition, Vol. 8, 101–102. Detroit: Thomson Gale.

[21]Even Carnap's later interest in formal reconstruction, which, in a sense, tries to do justice to the origins of scientific ideas, considers these ideas from a unifying point of view with hindsight.

[22]This can be said, I think, in spite of the fact that there are also considerable differences between the two philosophers. E.g., Wittgenstein, contrary to Schächter, saw a deep gap between the religious and the non-religious person.

Reitzig, Gerd H. 1978. Nachwort. In Reprint (1978) of [Schächter, Josef. 1935. *Prolegomena zu einer kritischen Grammatik* (Schriften zur wissenschaftlichen Weltauffassung 10). Vienna: Springer. (Reprint 1978 by Reclam)], 271–279.
Schächter, Josef. 1931. *Kritische Darstellung von N. Hartmanns „Grundöge einer Metaphysik der Erkenntnis"*. Vienna: Dissertation.
Schächter, Josef. 1935. *Prolegomena zu einer kritischen Grammatik* (Schriften zur wissenschaftlichen Weltauffassung 10). Vienna: Springer. (Reprint 1978 by Reclam)
Schächter, Josef. 1936. Sull'essenza della filosofia. *Rivista di Filosofia* 27: 1–13.
Schächter, Josef. 1937. מבוא קצר ללוגיסטיקה [*A Short Introduction to Logic*]. Vienna: Victoria.
Schächter, Josef. 1937a. Relegie en Wetenshap. *Synthese* 2(5): 159–167.
Schächter, Josef. 1937b. Notes on Problems of Ethics and the Philosophy of Culture. Vienna 1937 (unpublished). Published in *Ethics and the Will. Essays* (Vienna Circle Collection 21), eds. Brian McGuinness and Joachim Schulte, 1994, 19–32. Dordrecht: Springer.
Schächter, Josef. 1938. Der Sinn pessimistischer Sätze. *Synthese* 3(5): 223–233.
Schächter, Josef. 1960. פרקי הדרכה בתנ"ך [*Chapters of Guidance in the Bible*]. Jerusalem – Tel Aviv: M. Newman Publishers Ltd.
Schächter, Josef. 1963. העל אנושי באנושי [*The Superhuman in the Human*]. Tel Aviv: Alef Publishers.
Schächter, Josef. 1965/2006. Vorwort. In *Wittgenstein – Engelmann. Briefe, Begegnungen, Erinnerungen*, ed. Ilse Somavilla, 12–13. Innsbruck – Wien: Haymon Verlag.
Schächter, Josef. 1966. The Task of the Modern Intellectual. In *An Anthology of Hebrew Essays* II, eds. Israel Cohen and B. Y. Michali, 299–310. Tel Aviv: Institute for the Translation of Hebrew Literature and Massada Publishing.
Schächter, Josef. 1966a. יהדות וחינוך בזמן הזה [*Judaism and Education during These Times*] Tel Aviv: Dvir.
Schächter, Josef. 1970. פרקי עיון לנבוכי זמננו [*Theoretical Studies into the Perplexities of Our Times*]. Tel Aviv: Dvir.
Schächter, Josef. 1973. *Prolegomena to a Critical Grammar*. Dordrecht: Springer (eBook).
Schächter, Josef. 1977. עיונים במחשבת זמננו [*Studies in the Thought of Our Times*]. Tel Aviv: Yachdav Press.
Schächter, Josef. 1994. The Meaning of Pessimistic Sentences. In *Ethics and the Will. Essays* (Vienna Circle Collection 21), eds. Brian McGuinness and Joachim Schulte, 9–17. Dordrecht: Springer.
Schlick, Moritz. 1936. Meaning and Verification. *Philosophical Review* 45(4): 339–369.
Schoneveld, Jacobus. 1976. *The Bible in Israeli Education. A Study of Approaches to the Hebrew Bible and its Teaching in Israeli Educational Literature*. Assen: Van Gorcum.
Siegetsleitner, Anne. 2014. *Ethik und Moral im Wiener Kreis. Zur Geschichte eines engagierten Humanismus*. Wien – Köln – Weimar: Böhlau.
Stadler, Friedrich. 2015. *The Vienna Circle. Studies in the Origins, Development, and Influence of Logical Empiricism*. Dordrecht: Springer.
Waismann, Friedrich. 1968. Language Strata. In *How I see Philosophy*, 91–121. London: Palgrave Macmillan.
Weinberg, Julius R. 1937. Review of J. Schächter: *Prolegomena zu einer kritischen Grammatik. The Philosophical Review* 46(3): 334–335.
Weiner, Herbert. 1961. *Wild Goats of Ein Gedi*. Garden City (New York): Doubleday.

Chapter 8
Philipp Frank and the "Conference for Science, Philosophy and Religion", 1940–1968

Friedrich Stadler

8.1 The Vienna Circle and Religion: Some General Remarks

It is a well-known fact that religion and theology hardly figured in the discussions and publications of the classical Vienna Circle – the so-called Schlick Circle (1924–1936). However, already before and also later there were individual members who took religious and theological thinking under scrutiny, this being related to an attitude critical of metaphysics. In the *Manifesto* of 1929 we can read the following in connection with the method of logical analysis á la Carnap (*Wissenschaftliche Weltauffassung* 1929/2012, 142–43):

> If someone asserts "there is a God", "the primary basis of the world is the unconscious", "there is an entelechy which is the leading principle in the living organism", we do not say to him: "what you say is false"; but we ask him: "what do you mean by these statements?". And then it becomes apparent that there is a sharp boundary between two kinds of statements. To one belong statements of the sort made in empirical science; their meaning can be determined by logical analysis or, more precisely, through reduction to the simpler statements about the empirically given. The other kind of statements, to which those cited above belong, reveal themselves as empty of meaning if one takes them in the way that the metaphysician intends. One can, of course, often reinterpret them as empirical statements; but then they lose the emotional content which is usually essential to the metaphysician. The metaphysician and the theologian believe, thereby misunderstanding themselves, that their statements say something, that they denote a state of affairs. Analysis, however, shows that these statements say nothing but merely express a certain feeling of life. To express such attitudes can be a significant task in one's life. But the proper medium for doing so is art, for instance lyric poetry or music. It is dangerous to choose instead the linguistic garb of a

Thanks go to Ádám Tuboly for his critical reading and Camilla Nielsen for translations.

F. Stadler (✉)
Institut Wiener Kreis, Universität Wien and Vienna Circle Society, Vienna, Austria
e-mail: friedrich.stadler@univie.ac.at

E. Ramharter (ed.), *The Vienna Circle and Religion*, Vienna Circle Institute Yearbook 25, https://doi.org/10.1007/978-3-030-76151-6_8

> theory: a theoretical content is simulated where none exists. If a metaphysician or theologian wants to retain the usual medium of language, then he must himself realize and bring out clearly that he is giving not description but expression, not theory or communication of knowledge, but poetry or myth.

Psychology (Freudian psychoanalysis), sociology (Marx) and logic (Frege, Russell) are then cited as ways of explaining these "aberrations of metaphysics". Most of the members of the Vienna Circle were of Jewish origin, and they saw themselves as enlightened and assimilated, placing themselves in the political and ideological area between socialism and liberalism. They represented a cosmopolitanism which was embedded in the social-reformist Viennese modernism – analogous to the literary modernism around Hermann Broch, Elias Canetti and Robert Musil. In any case, they all felt obliged to a separation of church and state, as well as the corresponding distinction between faith and knowledge. They were more or less representatives of the "late enlightenment" (with agnostic monists, freethinkers, and proponents of ethical and youth movement) since the turn of the century around 1900 (cf. Stadler 1981, 2015).

This attitude also sealed their personal fate and that of the Vienna Circle as a whole, with dismissals and expulsion taking place no later than the "Anschluss" in 1938. The public reactions to Schlick's assassination in 1936 already produced the image of the "Jewish-oriented" Vienna Circle, although the Protestant Schlick himself was not of Jewish descent. Here, too, the exception was the Catholic and colleague Dietrich von Hildebrand (Stadler 2001/2015, 597–632).

Most of the Vienna Circle members took a critical stance of the role of religion and church in history and society and identified with the tradition of a (radical) French Enlightenment and its emancipatory development, which rejected all dogma and transcendental thinking. On the other hand, the negative accounts of the Vienna Circle by political Catholicism and in National Socialism and the struggle against anti-Semitic discourse confirmed this image of the enemy on the part of a "Black Vienna" (Wassermann 2014; Silverman 2009, 2012). These antagonisms were also an expression of political and cultural conflicts in the interwar period.

A more differentiated look at individual members of the Vienna Circle reminds us of the saying that exceptions confirm the rules. On the one side, there is the example of the trained rabbi and Schlick disciple Josef Schächter (see Esther Ramharter, Chap. 7, in this volume), on the other, mention must also be made of the emancipation of Rudolf Carnap from his pietist youth period, via the German youth movement to the leftist "scientific world view", and the scientific humanism of his late years in the USA (Damböck and Wolters 2020). And Edgar Zilsel dealt very early and critically in Vienna with the religion of the genius (1918) and the concept of genius (1926), for which he was ultimately denied a habilitation at the University of Vienna. The well-known "Zilsel thesis" on the emergence of modern science also takes into account the formative role of medieval scholasticism in the interaction with the "artist-engineers", of brainwork and manual work.

Influenced by his intensive discussions with Wittgenstein, Friedrich Waismann explored the relationship between faith and knowledge, conjecture and evidence, will and motive, which was expressed above all in his later writings. In addition to Wittgenstein, who himself was very interested in the relationship between philosophy, religion and faith, the proximity to pragmatism was also an occasion for a discussion of faith and knowledge. This family resemblance certainly goes back to William James and his *Varieties of Religious Experience* (cf. Maddalena and Stadler 2019). Otto Neurath is also one of the little-noticed exceptions: He had studied theology before the First World War and with his thesis of Austrian philosophy ("Neurath-Haller-Thesis") he seriously investigated the (positive) role of Aristotelian scholasticism, which was also connected with the discovery of the Catholic enlightener and social reformer Bernard Bolzano and the Catholic dissident Franz Brentano. In his philosophical-historical treatises, there are countless depictions of the social and political role of religions in comparison to the "scientific world-conception" as an expression of social developments and conditions. In any case, he was very specific about the role of magic, religion and science in the development of human thought (see Marco Brusotti, Chap. 4, in this volume). Last but not least, we should not forget Neurath's preference for ancient Chinese philosophy, mythology and religion, in the tradition of which he himself contributed poetic parables under a pseudonym (Neurath 1973). In the unofficial discussion circle on Marxism in the 1930s Neurath spoke out more directly about religion, which is evident from the minutes, located in the Archives of the Austrian Republic (Neurath 1930, 20–21).

After his second escape to England, Neurath, in addition to his activities related to the International Unity of Science Encyclopedia and the Isotype movement, worked intensively on a book project on "Persecution and Tolerance", which we can only reconstruct in part from his archives. In any case, we know from his exile library at the Vienna Circle Society how much this preoccupied him up until his death (Stadler 2019). In this context, his reading of Stefan Zweig in British exile is remarkable: in the English translation of Zweig's *The World of Yesterday* (1943), he notes that the Albigenses were not taken into account, and in the English edition of Zweig's book *The Right to Heresy. Castellio against Calvin* (1936) (a parable about the rise of Hitler), we can find critical comments on the history of religious wars and heresy in Christianity. Most recently, the transcription of Kurt Gödel's diaries has revealed his intensive occupation with religion and theology including the ontological proof of God (see Tim Lethen's paper). So far some fragmentary comments on the general topic of this volume as a first approach to its context.

8.2 Philipp Frank (1884–1966) and his Contributions to the "Conference on Science, Philosophy and Religion" (CSPR), 1940–1968

Who was Philipp Frank (cf. Cohen and Wartofsky 1965; Nemeth 2003, 2007; Stadler 2001/2015, 2007; Tuboly 2017; Schuster 2019)?

He was born in Vienna on 20 March 1884, and after attending secondary school he studied mathematics and physics at the Universities of Vienna and Göttingen, advised by Ludwig Boltzmann, Felix Klein and David Hilbert among others. In 1906 he received his PhD with a dissertation entitled "On the criteria for the stability of the motion of a material point and its relation to the principle of least action". After only three years (1909) he completed his habilitation in physics at the University of Vienna with the thesis "Die Stellung des Relativitätsprinzips der Mechanik und Elektrodynamik" (The significance of the principle of relativity in mechanics and electrodynamics) and worked from 1910–1912 as a private lecturer in physics at the University of Vienna. Subsequently, he was appointed Associate Professor of Theoretical Physics from 1912–1916 as successor to his colleague and friend Albert Einstein at the German University of Prague, where he finally served as full professor and director of the Institute of Theoretical Physics from 1917–1938 until his forced emigration to the USA in 1938. As early as 1918 Frank acquired Czechoslovakian citizenship and in 1920 he married Hania Frank from Poland, who facilitated his access to the scientific scene there and in Russia (e.g., Mach and Lenin, and Marxism-Leninism) through her knowledge of the Russian language. Philipp's brother was the famous architect Josef Frank, also an employee and friend of Otto Neurath's Social and Economic Museum in Vienna (Gesellschafts- und Wirtschaftsmuseum in Wien). In the Viennese years before the First World War from 1907 onwards, Philipp Frank, together with Hans Hahn and Otto Neurath (possibly also Richard von Mises), was one of the participants in a regular café-house round, which has come to be known as a proto-circle of the Vienna Circle ("Erster Wiener Kreis") in research. In Prague, Frank acted as the intellectual and organizational center of a circle with representatives of philosophy, natural science, and with contacts to the "Prague Circle" around Max Brod (Brod 1979; Frank 1947, 84–85). Following Rudolf Carnap's appointment in 1931, he acted together with him as the central figure of the "Prague Branch" of Logical Empiricism. In addition, he regularly came to his hometown Vienna for discussions, among others with the Schlick Circle and the Ernst Mach Association (Verein Ernst Mach).

Not least because of his Jewish origins and political views Philipp Frank followed Carnap (who left in 1936) to escape the threatening conditions in Prague, and in 1938, after a lecture tour on modern physics at American universities, he remained at Harvard University, initially as a lecturer in physics and mathematics. An offer from Istanbul University, where his friend and collaborator Richard von Mises had been working since 1933, was turned down by Frank because of a contract with an American publisher for a book on physics, but both met again at Harvard, where they worked together until the end of their lives. From 1939 to 1953, Frank was a Hooper Fellow, where he played a major role in shaping the Science Teaching Program on behalf of President James B. Conant, while at the same time he was a visiting professor at numerous American universities.

As early as 1934, Einstein's biographer Frank, together with Otto Neurath and Rudolf Carnap and Charles Morris, was theoretically and practically involved in the Unity-of-Science movement during and also after the Second World War. From 1948 to 1965, he founded and directed the "Institute for the Unity of Science" at

Harvard, the institute which became the intellectual center for the continuation of the Vienna Encyclopedia Program in exile, which had considerable impact on his collaborators and followers (among others Gerald Holton, Robert S. Cohen, Ernest Nagel). However, the "Zeitgeist" during the Cold War was hostile to this initiative of leftist immigrants, as George Reisch convincingly demonstrated in his book on the philosophy of science during the Cold War period (2005).

As the only representative of the Logical Empiricism and the *International Encyclopedia of Unified Science*, Frank participated actively until his death from 1940 on in the annual "Conferences on Science, Philosophy and Religion", additionally in the "Harvard Science of Science Discussion Group" and the "Boston Colloquium for the Philosophy of Science". Frank, who became also a member of the American Academy of Arts and Sciences, died on July 21, 1966 at Harvard (Stadler 2007).

Although Frank was an internationally renowned physicist, he spent his life studying the philosophy, history and sociology of science, and also addressed the relationship between philosophy, science and religion, arguing historically and theoretically in accordance with Einstein in the sense of his relativistic position. If we take a look at his writings, we recognize this knowledge interest that makes his late involvement in the CSPR plausible: already in his Prague days he wrote popular scientific articles on science and theology in newspapers on the relation of science and theology ("Wissenschaft und Theologie", in: *Freier Gedanke* 40, 1920, or "Theologische Elemente in den exakten Wissenschaften", in: *Prager Tagblatt*, 8. December 1926). Here already the proximity to the Free Thinkers' Movement is visible, which was strengthened by his co-operation with the Ernst Mach Association (Verein Ernst Mach), which from 1928 to 1934 as "external organ" of the Schlick Circle manifested itself as late enlightenment institution in the context of "Red Vienna". In his book *Das Kausalgesetz und seine Grenzen* (1932/1998) Frank addressed the speculative and transcendental misinterpretations of the concept of causality against the background of relativity theory and quantum physics (including a critique of the vitalism of a Hans Driesch or the philosophy of wholeness of an Othmar Spann). Against the accusation of scientistic reductionism, he opposed the horizon of a limit to the sciences that had to be rejected, as an expression of the so-called "school philosoph": the "ignorabimus" turned out to be senseless. In the book with his collected articles (1949) Frank defended the modern scientific world view against the attacks of Catholic *and* Marxist sides (e.g. Mach vs. Lenin), emphasizing the positive role of neo-Thomism in the formation of modern philosophy of science, where he wrote ("The New Positivism and Duhem's 'Thomism'", 15): "By studying Duhem thoroughly we gained a more subtle understanding of the relations among science, metaphysics, and religion than has been customary among empiricists." – seen as an integration of Mach and Poincaré. And Frank concluded remarkably clearly:

> As a matter of fact, Duhem was an advocate of Aristotelian and Thomistic metaphysics and a faithful believer in the whole Catholic theology. When I noted this merging of the most advanced type of 'new positivism' with Thomistic metaphysics into a coherent system it impressed me strongly. Duhem interpreted on the basis of this hybrid philosophy the historic conflict between the Roman Church and the Copernican system. Although our group did not

> follow Duhem's metaphysical predilection, his doctrine became for us a frame of reference to which we could relate all the conflicts that have raged between science and religion and, more generally, between science and political ideologies. (16–17).

About this theory dynamics, Frank in his autobiographical introduction to the same book entitled *Modern Science and is Philosophy* (1949), wrote in the preface:

> If we want to evaluate precisely and critically how firmly this philosophy is anchored in the ground of science, we must not ignore the extra-scientific factors, but must analyze carefully the social, ethical and religious influences. Every satisfactory philosophy of science has to combine logic of science with sociology of science.

And at the end he refers to the related book of his friend Richard von Mises on "positivism" (1939/1951), which also deals with non-scientific areas (498–499) in an optimistic and at the same time cautious summary:

> The course of development of knowledge as hitherto observed shows a tendency toward ever clearer and more determined separation between religious-metaphysical and scientific lines of thought; the range of applicability and validity of the latter expands continually at the expense of the former. One may recognize in the further progress of this trend a characteristic of Western civilization. To what extent, however, this trend may yet spread over the earth can hardly anticipated. (1951, 363).

This attitude, as an attempt to bridge the dualism of context of discovery and context of justification, had also been developed in the conference volume *The Validation of Scientific Theories* (1956), which he edited with an introduction and where he wrote about "The Variety of Reasons for the Acceptance of Scientific Theories". In addition, he included a separate section on "Science as a Social and Historical Phenomenon" (with essays from Henry Guerlac, Alexandre Koyré, Edwin Boring and Robert S. Cohen). The challenge of Einstein (1921) to the traditional philosophy of Neo-Thomism or Neo-Kantianism (e.g., with J. Maritain and Ernst Cassirer) was, according to Frank, shrouded by a metaphysical discourse:

> This kind of language has been in use to present cherished types of science, ethics, or religion. If we disregard temporarily this background there is a large common ground between neo-Thomism, neo-Kantianism, and the new positivism. In extreme cases this common ground may be so extensive that one can read hundreds of pages of a neo-Thomist or a neo-Kantian without recognizing that he is not a positivist of the new type. The most outstanding example is the French physicist and philosopher, Pierre Duhem. (Frank 1949, 24–25)

Accordingly, a new "Iron Curtain" between science and philosophy was created – this, as opposed to filling "New Wine into New Bottles" (Frank 1949, 25). For Frank, this process was strongly linked to the emergence of new republics and democracies after 1918, its disintegration and destruction, thus, also with the emergence of totalitarian regimes.

Against this background, it is not surprising that Philipp Frank, after his emigration to the US, was willing to participate in the regular CSPR in order to represent the standpoint of modern sciences against traditional philosophy, ethics and theology. He himself published his many contributions in a book under the telling title *Relativity – A Richer Truth* (1950), with a foreword by Albert Einstein.

The German translation followed two years later with the different title *Wahrheit – relativ oder absolut?* (Truth – Relative or Absolute?, 1952).

Frank himself wrote about this interaction:

> In 1940 Professor Howard Shapley, Director of the Harvard College Observatory, introduced me into the conference of Science, Philosophy and Religion, which meets every year. This conference is a group of philosophers, educators, social workers, and ministers of all denominations, and includes a sprinkling of scientists. They are interested particularly in the contributions that each of these fields can make to understanding and supporting of the democratic way of life. The leaders in this group have been, besides Dr. Shapley, Dr. Louis Finkelstein, President of the Jewish Theological Seminary of America, and Dr. Lyman Bryson, Director of the Educational Department of the Columbia Broadcasting System. From these meetings I have learned more about the attitudes of different groups of people towards science, and what points in the philosophy of science support or are believed to support a certain way of life. The large majority in these meetings has been rather critical of the scientific outlook and its contribution to human welfare. (Frank 1949, 52)

And to this cautious description of the context Frank added in defense of relativism:

> I addressed this group several times between 1940 and 1947. My contributions centered around the question of whether the "relativism" of modern science is actually harmful to the establishment of objective values in human life. I made an argument about human behavior. "Relativism" is not responsible for any deterioration of human conduct. What one calls "relativism" is rather the attempt to get rid of empty slogans and to formulate the goals of human life sincere and unambiguously. (Frank 1949, 52)

What were these interdisciplinary and intra-confessional conferences, which were already dealt with by George Reisch (2005, 2017) in connection with the philosophy of science and unity of science movement in the USA: In view of the rise and victorious communism, fascism and NS in Europe, leading intellectuals and philosophers were worried about the continued existence of Western liberal democracies. Among these about eighty persons it was above all Rabbi Louis Finkelstein of the "Jewish Theological Seminary" (JTS) in N.Y.C., who together with like-minded people (Harold D. Lasswell, Mortimer J. Adler, Paul Tillich, Robert MacIver, Franz Boas, Enrico Fermi, Pitirim Sorokin), also including Philipp Frank initiated these annual conferences. Originally, Horace Kallen, Sydney Hook and John Dewey had opposed the inclusion of religion in the title.

Against the background of an alleged loss of values and faith cited as the reason for the increasing anti-democratic currents, "value-free science" and godlessness were severely criticized by the majority. Yet this conflict between religion, theology, metaphysics and science had already erupted at the beginning of these conferences with Mortimer Adler's programmatic speech on "God and the Professors" (1941) in response to Einstein's "Science and Religion" (1941), which triggered fierce critical reactions from John Dewey and others and was also criticized by Sidney Hook, John Dewey and Ernest Nagel in the leftist *Partisan Review*. Adler concluded his declaration of war on liberal secularized scientific culture with the provoking paragraphs:

> For all these reasons I say we have more to fear from our professors than from Hitler. It is they who have made American education what it is, both in content and method: in content an indoctrination of positivism and naturalism; in method, and exhibition of

> anarchic individualism masquerading as the democratic manner. Whether Hitler wins or not, the culture which is formed by such education cannot support what democracy we have against interior decay. (1941, 12)

> Certainly if it is part of the Divine plan to bless man's temporal civilization with the goodness of Democracy, that civilization must be rectified. It is probably not from Hitler, but from the professors, that we shall ultimately be saved. (1941, 13)

By the way, an echo in the left spectrum was also Horace Kallen's criticism of his relative Otto Neurath, whom he accused of a kind of complicity with totalitarian thinking through the Unity of Science movement, which ended dramatically in 1945 with a response from Neurath. (cf. Neurath 1945/46).

The attack by Mortimer Adler was obviously prompted by Einstein, who explained in his article on "Science and Religion" (1941, 3–4):

> The further the spiritual evolution of mankind advances, the more it seems to me that the path to genuine religiosity does not lie through the fear of life, and the fear of death, and blind faith, but through striving after rational knowledge. In this sense I believe that the priest must become a teacher if he wishes to do justice to his lofty educational mission.

In this turbulent and emotionally highly charged scenery, Philipp Frank, with admirable composure (which he had probably already practiced in Europe), subsequently attempted with his contributions to save the honor of science in a delicate balancing act between science, ethics and religion, pleading for a fundamental compatibility of scientific relativism and objectivism. He began with a lecture on "Science and Democracy" and continued his regular contributions until the end of his life, which he published in *Relativity – A Richer Truth?* In his foreword to this book, Einstein, whose cosmic religiosity Frank had benevolently explained to a broad audience here and in his Einstein biography, wrote that logic and ethics are compatible, although

> scientific statements of fact and relations, indeed, cannot produce ethical directives. However, ethical directives can be made rational and coherent by logical thinking and empirical knowledge. If we can agree on some fundamental ethical propositions, then other propositions can be derived from them, provided that the premises are stated with sufficient precision. Such ethical premises play a similar role in ethics to that played by axioms in mathematics. (Einstein, in Frank 1950, 9)

And in the sense of a meta-ethical non-cognitivism (emotivism) against all apriorism and absolutism, he concludes on the origin of the ethical axioms:

> For pure logic all axioms are arbitrary, including the axioms of ethics. But they are by no means arbitrary from a psychological and genetic point of view. They are derived from our inborn tendencies to avoid pain and annihilation, and from the accumulated emotional reaction of individuals to the behaviour of their neighbours. ... Ethical axioms are found and tested not very differently from the axioms of science. *Die Wahrheit liegt in der Bewährung*. Truth is what stands the test of experience. (10)

In his introduction to the same book, Frank then goes directly into the cultural struggle for "relativism", which he, almost heroically, as one of the few natural scientists, offensively defended as an essential characteristic of research and science. He was confronted with the question: science as a means or as an end in itself?

> In the beginning I soon saw that by the fight against "relativism" and its "totalitarian affiliations", the Conference was about to "jump out of the frying pan into fire". I saw that the fight against "relativism" could easily degenerate into a fight against the spirit of modern science. [...] I have not presented any excuses for the "relativism" in science, but rather have launched a frontal attack against its enemies. I have attempted to show that so-called 'relativism' has not the slightest thing to do with agnosticism and scepticism, that *it is in no way hostile to the belief in ethical or democratic values,* that it is accompanying every advance in science and is nothing but a significant representation of the enrichment of human expression which is inseparably connected with our gradually increasing experience. (Frank 1950, 12)

In order to make his position more understandable, Frank included his conference presentations, with two additional sections only in the German edition ("Metaphysical Interpretations of Science", "The Logical and Sociological Aspects of Science"). He was particularly concerned with how "truth" is actually used as practice in scientific research, and with the fact that inner-scientific criteria "are not the exclusive motives for the acceptance or rejection of general principles. Besides the 'logical' and 'empirical' criteria of the scientific worker, general principles are often accepted or rejected for 'metaphysical reasons'" (Frank 1952, 11).

In addition to this, Frank has in the section on "Metaphysical Interpretations of Science" thematized the connection between metaphysics and faith in science and referred to the language of "commonsense" as an explanation for this, which, however, in his view would not be a sufficient justification (cf. Tuboly 2021). In the third and last added part on "The logical and sociological aspects of science" Frank tried to show "how closely metaphysical interpretations of science are connected with the practical life of teachers and students", because also in areas separated from life like astronomy or physics "one has tried to formulate principles of science in such a way that they are useful or at least not harmful to ethical and religious advice" (11–12).

With reference to politics, society and religion, Frank notes that any development towards liberalism was linked to a development of semantics, which brought with it a pragmatic and operationalist view of the meaning of slogans, as was the case, for example, with the liberal theologian Theodor Parker on the "natural sacrament" in accordance with James and Peirce. This connection is important for Frank because by Parker's sermon "the 'creeds' and 'rites' of Christianity are *accepted even by the worst criminals, if no operational meaning is attached to these creeds and rites*" (1950, 45).

Subsequently, Frank presents Einstein's position in the chapter on "The 'Rationality of Nature' and Einstein's 'Cosmic Religion'", and also in the last chapter of his posthumously published last book (Frank 2021), as an alternative to dogmatic and aprioristic doctrines: according to him, the existence of a rational nature is independent of the interpretation of its usefulness. The belief in this reason as "cosmic religiosity" is belief in a certain interrelation between nature and spirit in the context of the creation myth and its specific way of speaking. In any case, it was clear to Frank that thereby the connection or compatibility between science and religion need not be problematized, because – as he literally states –

> [a]ccording to Einstein, the fundamental logical structure of science has not changed. Cosmic religiosity [. . .] has its basis in the existence of simple mathematical laws which describe the immense complexity of phenomena. In this sense every advance in science is an advance in our understanding of cosmic religion, since every advance of science consists in the establishment of more adequate laws. The amazing simplicity of principles compared with the complexity of observable phenomena has become more and more conspicuous; in this sense, and only in this sense, can modern physics be seen from the point of view of religion. Einstein, however, has never claimed and never admitted that the specific content of modern physics would be a contribution to peace between science and religion. The achievements of modern physics confirm our belief in the "rationality of nature". Since this belief is, according to Einstein, the basis of cosmic religion, modern physics is supporting this cosmic religion in the sense as does every other advance in our knowledge of the general laws of nature. (117–118)

And Frank continues, speaking of the "rationality of nature" is therefore a fundamental characteristic of the human relationship to the environment, although it itself remains "incomprehensible", formulated by Einstein: "*The most incomprehensible thing about the world is that it is understandable*" (118), whereby the mysterious and mystical is given from the psychological point of view without any practical goal and orientation. This is provided only by the institution of the Churches. However, the ultimate goals are derived from life and tradition itself, which can only be realized by a scientific goal-means relation with truth as basic regulative value, or in Einstein's dictum in variation of Kant, later modified by Imre Lakatos: "Religion without science is blind; science without religion is lame" (120). To which Frank dryly adds this: "Mankind needs religious leaders who face with open eyes the results of science, and mankind needs scientists who have the moral courage to act according to their knowledge" (120). The corresponding literary recommendations by Frank once again clearly define the frame of reference for his lectures, once more confirming the convergence of logical empiricism and American pragmatism, including references to the works of James B. Conant, John Dewey, Charles Morris, C. L. Stevenson, but also to the ethics of Moritz Schlick.

Incidentally, this also corresponds to the further development and justification of his ideas by Frank's former assistant and colleague Gerald Holton, who critically examined Paul Tillich's position with his plea for a modernized Unity-of-Science: In his 16th Vienna Circle Lecture, the renowned physicist and historian of science once again provided a "thematic analysis" on the topic of unity or multiplicity in science and summarized his experiences with Frank at Harvard in connection with the heated debate on relativism. Among them we read the episode about Einstein's meeting with the Archbishop of Canterbury in 1921 on the influence of relativity on religion, which the creator of relativity theory promptly commented: "None, relativity is a purely scientific matter and has nothing to do with religion" (Holton 2010, 251). But the criticism of the church held until Einstein's exile. The Archbishop of Boston, Cardinal O'Connell, argued against the relativity theory as questioning a personal God and his creation. At Harvard, Paul Tillich, who emigrated from Germany, argued against Einstein's "cosmic religiosity" in a more academic discourse. His manifesto "Science and Theology: A Discussion with Einstein" (1940) ended with a strong plea for a personal God. Still in his late work

My Search for Absolutes (published in 1967, one year after Frank's death) he turned against what he considered to be the dominant relativism in everyday life, but also in science, philosophy, religion, ethics – as "the great spectacle of scientific relativism" vs. "Absolutes" (252–253) – as a manifestation of the cultural struggle, scientific dispute and methodological dispute on the dualism of "absolute" and "relative" that continues to this day. In this history of problems, Holton finally presents Philipp Frank's book *Relativity – A Richer Truth* (1950), as a discussion of relativism within and outside the sciences. He argues that this discussion is an essentially illegitimate transfer into the life-world, with the exception of a necessary operational "linguistic act" of expressions in the context of a skeptical rationalism and empiricism. And that remained – as Holton explains – up to the present an anti-thesis to the official Catholic doctrine in the beginning twenty-first century, when Cardinal Ratzinger warned of a "dictatorship of relativism" (actually a self-contradiction), which was literally repeated by George W. Bush three years later on the occasion of the visit of Pope Benedict XVI (and flared up in philosophy with Hilary Putnam vs. Richard Rorty and Paul Feyerabend).

An answer to this conflict is provided by Frank himself in his book, when he refers to the pragmatic and historical-social aspect of every science, worldview and religion: "The idea of a super-science, called philosophy or metaphysics, only arises when we understand 'science' to be a single science, instead of meaning the system of all sciences" (Frank 1952, 66). And he refers in this connection to Jean Piaget and John Dewey, with a quotation from Dewey's *Reconstruction in Philosophy* (1920):

> The history of philosophy will acquire a new meaning. What appears as a loss from the point of view of a supposed science will turn out to be a gain from the point of view of pure knowledge of humanity. The history of philosophy will then no longer be the scene of disputes about the "essence of reality", but will be characterized by the clash of opposites in the objectives and desires for human coexistence. (Frank 1952, 166)

This closes a thematic circle (including a paradoxical return to the "Science Wars"), which, based on a historical and theoretical analysis of science in human history, defends the openness, incompleteness and plurality against any dogmatic (Reisch/Tuboly 2020) belief with ultimate justification, not least as a contribution to a democratic culture (Reisch and Tuboly 2020). Perhaps as a variation of Churchill's saying on democracy: "Science is incomplete and imperfect, but there is no better form of human thought and action [. . .]."

References

Adler, Mortimer. 1941. *God and the professors: Conference on science, philosophy and religion (1941)*. http://www.ditext.com/adler/gp.html. Accessed 8 Apr 2016.

Brod, Max. 1979. *Der Prager Kreis*. Mit einem Vorwort von Peter Demetz. Frankfurt/M.: Suhrkamp.

Cat, Jordi, and Adam Tamas Tuboly, eds. 2019. *Neurath Reconsidered. New Sources and Perspectives*. Cham: Springer.

Cohen, Robert S., and Marx Wartofsky, eds. 1965. *In Honor of Philipp Frank*, = Boston Studies in the Philosophy of Science, Vol. II. New York: Humanities Press.
Damböck, Christian, and Gereon Wolters, eds. 2020. *Young Carnap in a Historical Context 1918–1935*. Cham: Springer.
Einstein, Albert. 1921. *Geometrie und Erfahrung*. Berlin: Springer.
———. 1941. Science and religion. In *Science, Philosophy and Religion, A Symposium. The Conference on Science, Philosophy and Religion in Their Relation to the Democratic Way of Life*. New York: Harper.
Frank, Philipp. 1932. *Das Kausalgesetz und seine Grenzen*. Wien: Springer. English edition: Philipp Frank, *The Law of Causality and its Limits*, ed. Robert S. Cohen. Dordrecht-Boston-London: Kluwer.
———. 1947. *Einstein. His Life and Times*. New York: Knopf.
———. 1949. *Modern Science and its Philosophy*. Cambridge: Harvard University Press.
———. 1950. *Relativity – A Richer Truth*. Foreword by Albert Einstein. Boston: The Beacon Press.
———. 1952. *Wahrheit – Relativ oder Absolut?* Mit einem Vorwort von Albert Einstein. Zürich: Pan-Verlag.
———, ed. 1956. *The Validation of Scientific Theories*. Boston: The Beacon Press.
———. 2021. *The Humanistic Background of Science*, eds. George Reisch and Ádám Tamás Tuboly. New York: SUNY Press (forthcoming).
Holton, Gerald. 2010. On Unity and Disunity in the sciences: Variations of ancient Thematas. In *The Vienna Circle in the Nordic Countries. Networks and Transformations of Logical Empiricism*, ed. Juha Manninen and Friedrich Stadler, 245–262. Cham: Springer.
Maddalena, Giovanni and Friedrich Stadler. Eds. 2019. *Symposium European Pragmatism*. (= *European Journal of Pragmatism and American Philosophy*. XI-1). ISSN 2036-4091.
Nemeth, Elisabeth. 2003. Philosophy of science and democracy. Some reflections on Philipp Frank's *Relativity – A Richer Truth*. In *Philosophy of Science and Politics*, ed. Michael Heidelberger and Friedrich Stadler, 119–138. Wien/New York: Springer.
———. 2007. Logical empiricism and the history and sociology of science. In *The Cambridge Companion to Logical Empiricism*, ed. Alan Richardson and Thomas Uebel, 278–304. Cambridge: Cambridge University Press.
Neurath, Otto. 1930. Diskussionsprotokolle, Inoffizieller Diskussionszirkel über Marxismus, Archiv der Republik Österreich. Nr. 1433: 1–11.
Neurath, Ottto. 1945/46. For the discussion: Just annotations, not a reply. *Philosophy and Phenomenological Research* 6 (4): 526–528.
Neurath, Otto. 1973. *Empiricism and Sociology, With a Selection of Biographical and Autobiographical Sketches*, eds. Marie Neurath and Robert S. Cohen. Dordrecht/Boston: Reidel.
Reisch, George. 2005. *How the Cold War Transformed Philosophy of Science. To the Icy Slopes of Logic*. Cambridge: Cambridge University Press.
———. 2017. Pragmatic Engagements. Philipp Frank, and James Bryant Conant on Science, Education, and Democracy. In Tuboly 2017, 227–244.
Reisch, George, and Ádám Tamás Tuboly, eds. 2020. *The Humanistic Background of Science*. Albany: SUNY Press.
Schuster, Radek, ed. 2019. *The Vienna Circle in Czechoslovakia*. Cham: Springer.
Silverman, Lisa. 2009. Reconsidering the margins: Jewishness as an analytical framework. *Journal of Modern Jewish Studies* 8 (1): 103–120.
———. 2012. *Becoming Austrians. Jews and Culture between the World Wars*. Oxford: Oxford University Press.
Stadler, Friedrich. 1981. Spätaufklärung und Sozialdemokratie in Wien 1918–1938. Soziologisches und Ideologisches zur Spätaufklärung in Österreich. In *Aufbruch und Untergang. Österreichische Kultur zwischen 1918 und 1938*, ed. Franz Kadrnoska, 441–473. Wien/München/Zürich: Europa-Verlag.
———. 2001/2015. *The Vienna Circle. Studies in the Origins, Development, and Influence of Logical Empiricism*. Vienna/New York/Cham: Springer.

———. 2007. History of the philosophy of science. From *Wissenschaftslogik* (logic of science) to philosophy of science: Europe and America, 1930–1960. In *Handbook of the Philosophy of Science – Focal Issues*, ed. Theo Kuipers, 577–658. Amsterdam: Elsevier.

———. 2015. Wiener Kreis. In *Enzyklopädie jüdischer Geschichte und Kultur 6*, ed. Dan Diner, 401–404. Stuttgart/Weimar: Metzler.

———. 2019. A Viennese Library in Exile: Otto Neurath and the Heritage of Central European Culture in the Anglo-Saxon World. In Cat/Tuboly 2019: 23–44.

Tuboly, Ádám Tamás. Ed. 2017. *The Life and Work of Philipp Frank* (= *Studies in East European Thought* 69(3): 200–276).

———. 2021 (forthcoming). Understanding Metaphysics and Understanding through Metaphysics: Philipp Frank on Scientific Theories and their Domestication. In *Logical Empiricism and the Physical Sciences: From Philosophy of Nature to Philosophy of Physics*, eds. Sebastian Lutz and Ádám Tamás Tuboly. New York/London: Routledge.

Von Mises, Richard. 1939/1951. *Kleines Lehrbuch des Positivismus. Einführung in die empiristische Wissenschaftsauffassung*. Den Haag: Van Stockum & Zoon. New edition and transl. Friedrich Stadler. Frankfurt/M.: Suhrkamp 1990. English edition: *Positivism. A Study in Human Understanding*. New York: Dover.

Wassermann, Janek. 2014. *Black Vienna. The Radical Right in the Red City, 1918–1938*. Ithaca: Cornell University Press.

Wissenschaftliche Weltauffassung. 1929/2012. *Wissenschaftliche Weltauffassung. Der Wiener Kreis* (1929, hrsg. vom Verein Ernst Mach). Wien: Artur Wolf Verlag. Reprint der Erstausgabe, mit Übersetzungen ins Englische, Französische, Spanische und Italienische. Hrsg. mit Einleitungen und Beiträgen von Friedrich Stadler und Thomas E. Uebel, Wien/New York: Springer.

Zilsel, Edgar. 1918/1990. *Die Geniereligion. Ein kritischer Versuch über das moderne Persönlichkeitsideal, mit einer historischen Begründung*. Leipzig – Wien: Braumüller. New edition, introduction by Johann Dvorak, Frankfurt/M.: Suhrkamp.

———. 1926/1972. *Die Entstehung des Geniebegriffes. Ein Beitrag zur Ideengeschichte der Antike und des Frühkapitalismus*. Tübingen: Mohr. New edition, preface by Heinz Maus. Hildesheim/New York: Georg Olms Verlag.

Chapter 9
Kurt Gödel's Dogmatic Theology

Tim Lethen

9.1 Introduction

Well known as a logician, mathematician, and physicist, Kurt Gödel (1906–1978) also devoted a great amount of time and effort to philosophy and—by far less known—to theology. In philosophy, his main aim and focus was to draw inferences with mathematical rigor and precision and, as Hao Wang points out, he anticipated "philosophy as an exact theory" which "should do to metaphysics as much as Newton did for physics" (Wang 1987, 192).

A closer look into Gödel's *Nachlass* shows that the aim to look at philosophy as an exact theory has to be extended to theology as well, an opinion for example explicitly expressed in one of the items describing his general "philosophical viewpoint", reading as follows:[1]

> Es gibt eine wissenschaftliche (exakte) Philosophie (und Theologie), welche die Begriffe der höchsten Abstraktheit behandeln.

Whereas literally nothing has been published in this direction so far, the purpose of this article is to give an example of Gödel's work in theology by analyzing notes connected to the topic of *dogmatic theology*, taken as "the scientific exposition of the whole theoretical doctrine revealed by God about God Himself and His activity and which we accept on the authority of the Church" (Ott 1957, 3). These notes clearly

[1] Kurt Gödel Papers, Box 11b, Folder 15, item accessions 060168, on deposit with the Manuscripts Division, Department of Rare Books and Special Collections, Princeton University Library. Used with permission of Institute for Advanced Study. Unpublished Copyright Institute for Advanced Study. All rights reserved.

T. Lethen (✉)
Department of Philosophy, History and Art Studies, University of Helsinki, Helsinki, Finland
e-mail: tim.lethen@helsinki.fi

E. Ramharter (ed.), *The Vienna Circle and Religion*, Vienna Circle Institute Yearbook 25, https://doi.org/10.1007/978-3-030-76151-6_9

show that Gödel was concerned with the idea of treating dogmatic theology as an exact theory, comparing it to the field of theoretical physics and developing different but strongly interconnected logics based on the notion of the catholic dogma.

After a general introduction into Gödel's idea of the formalization of metaphysics and theology in Sects. 9.2 and 9.3 we present Gödel's lecture notes on the subject of dogmatic theology, thus uncovering an early scholastic influence on his work in this direction. Section 9.4 will then gather source material from Gödel's *Nachlass* which is in some respect connected to the catholic dogma and which does clearly reflect his thoughts and ideas connected with this topic. Based on these notes, Sects. 9.5 and 9.6 develop two formal logical systems, one being purely propositional, the other being modal in character. Whereas Sect. 9.7 draws a close connection between different grades of *certainty* of the catholic dogma on the one hand, and the notion of *evidence* on the other, the final section examines Gödel's thoughts about the dogma of free will. Rounding off, an appendix gathers the literature on dogmatic theology as listed by Gödel himself.

Whereas Gödel published very little about philosophy during his lifetime, literally nothing can be found about his theological views. This article therefore heavily relies on unpublished material found in Gödel's *Nachlass* and uses the opportunity to quote from it in an extended manner. Transcriptions from the Gabelsberger shorthand system were made by the present author as part of the GODELIANA project based in Helsinki, Finland, led by Jan von Plato. Additions made by the present author are marked by angled brackets.

9.2 Formalizing Metaphysics and Theology

Gödel's aim of formalizing philosophy is probably best expressed in his own words found in his notebook series *MaxPhil*. In *MaxPhil III* (on page 41)[2] Gödel writes:

> Max⟨ime⟩: (d.h. Formalisierung der Phil⟨osophie⟩) Ich sollte mich einmal entschließen, welches philosophische, psychol⟨ogische⟩ System ich acceptieren ⟨*sic*⟩ will und in welchem Sinn ich es akzeptieren will. Dann habe ich bei einer Disk⟨ussion⟩ zuerst die Frage in mein System zu übersetzen [eventuell erst durch Fragen den genauen Sinn feststellen, wenn mehrere Übersetzungen möglich sind], und dann die möglichen Wege zur Antwort durch Kenntnis von Ax⟨iomen⟩ und Theoremen und Beweismethoden dieses Systems oder die entgültige Antwort zu produzieren, oder, wenn jemand etwas behauptet, die Widerlegung aus den Axiomen meines Systems (nach Feststellung des Sinns) zu bringen. Bei Fragen der Sinnklärung muss man aber nur solche Terme verwenden, die allgemein und intersubjektiv gleich verstanden werden.[3]

[2]Kurt Gödel Papers, Box 6b, Folder 66, item accession 030089.

[3]Here Gödel adds the following footnote: "Dabei ist wichtig, dass nicht einmal der Sinn der Worte 'wenn', 'wahr', '∨' klar ist."

There are several hints in the same series of notebooks that Gödel finally made the decision which worldview to choose and accept as his own, clearly tending towards theology. For example, in *MaxPhil III* (21) he writes:

> Meine geänderte Weltanschauung (Glaube an Theologie) hat zur Folge, dass die Ereignisse sich viel mehr nach den "Rechtsverhältnissen" richten müssen (als ich früher glaubte).

In his notebook *MaxPhil IX*[4] (78–79), written between November 1942 and March 1943, he again describes his general methodology, pointing into the same direction:

> Bem. (Phil.): Meine Arbeit in Bezug auf Phil⟨osophie⟩ soll in einer Analyse der obersten Begriffe bestehen (der logischen und psychol⟨ogischen⟩). Das heißt, was letzten Endes zu tun ist, ist eine Liste dieser Begriffe aufschreiben und die möglichen Ax⟨iome⟩, Th⟨eoreme⟩ und Def⟨initionen⟩ für sie überlegen (selbstverständlich samt Anwendung auf die empirisch gegebene Wirklichkeit). Um das aber tun zu können, muss man zuerst durch (halb verstandene) philosophische Lektüre[5] ein "Gefühl" dafür erwerben, was man annehmen kann. Andererseits wieder wird der Versuch einer Axiomat⟨isierung⟩ das Verständnis der philosophischen Schriftsteller erhöhen. [Also Wechselwirkung von "oben" und von "unten", wobei das richtige Verhältnis wichtig ⟨ist⟩.]

According to Hao Wang (1987, 192), Gödel told him that he had "not yet determined the complete list of the primitive concepts" but only mentioned "object, concept, substance, cause, and sometimes a few others". Nevertheless, a clear exposition of such a list is given by Gödel himself on a loose page[6] preserved in his *Nachlass*. It reads as follows:

> reason, cause, substance, accidens, necessity (conceptual), harmony ⇂ value ⇃ (positiveness), God (= last principle), cognition, force, volition, time, form, content, matter, life, truth, class ⇂ = absolute ⇃, concept ⇂ das allgemeine und das einzelne ⇃, idea, reality, possibility, ineducible, Many & one, Essence

As a very clear and striking example, Gödel's ontological proof heavily relies on the formalization of the concepts "God", "positiveness", and "essence", all three appearing on the above list. Already in the earliest known version,[7] compiled in 1941, the predicate of being *God-like* is defined as

$$G \equiv \hat{x}\,[(\varphi)\,[\varphi\,\varepsilon + \rightarrow \varphi(x)]]$$

[4]Kurt Gödel Papers, Box 6b, Folder 69, item accession 030095. This passage (slightly corrected here) has already been published by Crocco and Engelen (2016).

[5]Here Gödel adds the following footnote: "und das andere Schreiben von philosophischen 'Bemerkungen' "

[6]Kurt Gödel Papers, Box 11b, Folder 15, item accession 060168. The list is headed "Phil. Bem." and is written in longhand English. Phrases between harpoons are added above the proceeding word.

[7]For an account on the existing variants of Gödel's ontological proof, see Kanckos and Lethen 2019.

closely following the writings of Anselm of Canterbury in his *Proslogion*.[8] The notion of *essence* entered the stage in a third (undated) version of the proof, where it is defined by the equivalence[9]

$$\varphi\ \varepsilon\ Ess_x \equiv (\psi)\{\psi(x) \supset .N(x)[\varphi(x) \supset \psi(x)]\}\ .\ \varphi(x)$$

Finally, *positiveness* has been decribed as being captured only vaguely in Gödel's writings,[10] but the exact opposite is the truth: Many writings in his *Nachlass*, some of them still unpublished, present a detailed axiomatization as well as a collection of careful interpretations.

Whereas Gödel certainly would count his ontological proof as a metaphysical investigation, it is not quite clear whether he would also consider it a piece of rational theology. In a conversation with Rudolf Carnap in 1940, Gödel says:

> Man könnte ⟨ein⟩ exaktes Postulatensystem aufstellen mit solchen Begriffen, die gewöhnlich für metaphysisch gehalten werden: "Gott", "Seele", "Ideen". Wenn das exakt gemacht würde, wäre nichts dagegen einzuwenden. (Gierer 1997, 9)

Carnap himself then mentions the analogy to theoretical physics and uses the term "metaphysical theology" which underlines the smooth and floating border between metaphysics and theology.

On the last pages of his notebook *MaxPhil XV*,[11] written after 1955, Gödel carefully describes the just mentioned border, now referring to "idealistic philosophy" and "dogmatic theology" (Dogmatik) instead of metaphysics and theology.

> Philosophie: Punkte, wodurch sich die spez⟨ielle⟩ ideal⟨istische⟩ Ph⟨ilosophie⟩ (auch wenn sie nachher axiomat⟨isiert⟩ wird) von der Dogm⟨atik⟩ unterscheidet:[12]
>
> 1.) ⟨Der⟩ Unterschied zwischen Erscheinung und Ding ⟨ist⟩ ungeheuer.
> 2.) Das "Eins-Sein" von allem.
> 3.) Eine spezielle Erkenntnisart (Schauen des Absoluten, Unzeitlichkeit).
> 4.) Es werden Beziehungen zwischen den Fundamentalbegriffen (und Reduktionen von Grundsätzen[13]) vorgenommen, welche dem Dogm⟨a/atiker⟩ nicht mehr möglich sind.

---

[8]The connection between Gödel's proof and Anselm's writings is described by Kanckos and Lethen 2022. Basically, being God-like is defined as possessing every positive property.

[9]The property φ is an essence of the object x, if it necessarily entails all of x's properties, and x has the property φ.

[10]For an example, see (Sobel 2003, 118). Without addressing the material in Gödel's notebooks, Sobel writes: "[The] intended meaning [of positiveness] must be gleaned from cryptic explanatory sentence fragments in Gödel's hand, as there is nothing to it in Scott's hand." (The "fragments" mentioned are those that appear in Gödel's 1970 version of the proof.) And Anderson and Gettings write: "We hasten to add that the idea [of positiveness] is not crystal clear; Gödel's own explanations are extremely terse and somewhat cryptic." (Anderson and Gettings 1996, 168) One has to keep in mind though that the authors cited were obviously limited to published sources.

[11]Kurt Gödel Papers, Box 6b, Folder 72, item accession 030100.

[12]Gödel's footnote: "Letzter Grund für ⟨den⟩ Unterschied: Es liegt ihr eine nicht-pos⟨itivistische⟩ Wissenschaft zu Grunde."

[13]Gödel's footnote: "oder Bestreitungen (wie der Satz vom Widerspruch bei Hegel)"

5.) Man kann sie nicht in demselben Sinn auf dieselbe Weise lernen (verstehen) wie eine pos⟨itivistische⟩ Wissenschaft (Unverständlichkeit).
3'.) Abgeschwächt: Vorhandensein eines universellen erkenntnistheoretischen Subj⟨ekts⟩ in allem.
6.) Wir erkennen das Wesen der Dinge, indem wir diese oder Gott in uns haben.
7.) Vollkommener Apriorism⟨us⟩, aber nicht aus dem Satz vom Widerspruch.
8.) Es wird alles aus Gegensätzen aufgebaut. (Und Gott ist Einheit.)
9.) Neg⟨ative⟩ Theol⟨ogie⟩ und Pantheismus.

It is remarkable that Gödel here again mentions the axiomatization of an (idealistic) philosophy. As we shall see in the following sections, he carefully developed an axiomatization for the concept of 'dogma', clearly belonging to his work in theology. And in a list[14] of about 80 theological questions and remarks (henceforth abbreviated ND), he parallels Carnaps comparison to theoretical physics:

> Kann nicht das Dogma vollkommen exakt sein, ähnlich wie theor⟨etische⟩ Physik?

Although the term "dogma" does not appear on it, it should be worth quoting a remark appearing in another theological list of about 98 items called "Fragen Theologie" (questions theology) now discovered in Gödel's *Nachlass*[15] (henceforth abbreviated FT), in which he transfers the earlier-mentioned program of the compilation of definitions, axioms, and theorems from philosophy to theology, this time immediately naming the uppermost concepts. Here he writes:

> Progr.: Die theologischen Grundbegriffe wie z.B.: Wille, Wunsch, Gebet, Gnade, Verdienst, Lehre, Strafe, Gerechtigkeit, gut, böse, in der Macht jemands, Vernunft, Erbsünde, Sünde, Begierde, Tatsünde, Natur, Ursache, ... nach ihrer logischen Kategorie bestimmen, definieren und ständige Term⟨inologie⟩ dafür (im Anschluss an den Sprachgebrauch) einführen und Grundsätze formulieren. Danach an irgendwelchen Häresien (z.B. Baii Mich., Luther, etc.) prüfen, ob entscheidbar (beweisbar).

Clearly, the expression "Grundsätze formulieren" refers to an axiomatization of the defined concepts, and a maxim in his notebook *MaxPhil V*[16] supports the central role of the logical category when analysing the listed basic concepts:

> Um die Fage "Was ist X" zu beantworten, hat man sich zu fragen:
>
> 1.) Welcher logischen Kategorie gehört X an? [Zunächst, ob es Einheit oder Vielheit ist.]
> 2.) Wie lauten die Atomsätze, in denen X vorkommt? [Sätze der Form $X R a$]
> 3.) Wann sind zwei X miteinander identisch?

[14]This list appears on the first pages of book VII (Kurt Gödel Papers, Box 8a, Folder 65, item accession 040218, entitled "Logik Vorl. N.D. VII") of a series of notebooks preparing Gödel's Notre Dame lecture on elementary logic, published in (Adžić and Došen 2017). An attempt was made by Adžić and Došen to transcribe this list but was restricted to those parts that are not written in the Gabelsberger shorthand system. Several less fortunately transcribed passages appear even in the English longhand part of the list.

[15]Kurt Gödel Papers, Box 6a, Folder 51, item accession 030074.

[16]Kurt Gödel Papers, Box 6b, Folder 67, item accession 030091.

And indeed, in his notebook *MaxPhil II*[17] (156), one can find a short list of seven axioms, the first one reading "Die Strafe gehört derselben Kategorie an wie die Sünde", revealing that Gödel was indeed following his own program.

Whereas the *dogma* is not explicitly mentioned among the aforementioned theological concepts, the closely related term "Häresie" (heresy) seems to play an important role for Gödel, naming Luther and Michael Baius[18] as 'test cases' for the defined concepts and their axiomatization.

Other remarks obviously seem to support the cental role the heresy plays in Gödel's theological point of view, notably the following passage, to be found in his notebook *MaxPhil II* (99).

> Bei ⟨der⟩ Aufstellung theologischer Theorien ist es offenbar eine fruchtbare Idee, immer erst zu fragen: "Ist das eine Häresie?" Daher wäre das Nicht-Stellen dieser Frage eine Sünde.

9.3 Gödel's Theological Lecture Notes

None of Gödel's existing biographies mention that he attended at least 30 theological lectures in Vienna in the winter term of 1937, spanning such topics as the history of scholasticism, epistemology, the relationship between revelation and science, dogmatic theology, and many others. His lecture notes are preserved in the notebook "*Theologie 1 – Nur Vorlesungen*"[19] which has now entirely been transcribed by the present author. As an example, the notes of a lecture given by Joseph Lehner (1868–1949)[20] entitled "*Tractatus de Deo Uno et Trino, de Creatione, de Deo Redemptore sexies*" are reproduced here, which will give an insight into Gödel's first contact to the subject of dogmatic theology.

Mi 6./X., 10–11, Lehner

Zu leugnen, dass Maria in den Himmel erhoben wurde, wäre haeresi proximum.[21]

Der Glaube ist ein unmittelbarer Akt der Tugend. ⟨Er⟩ spricht darüber, was zu glauben Pflicht ist.

Die unmittelbar geoffenbarten Wahrheiten

[17] Kurt Gödel Papers, Box 6b, Folder 65, item accession 030088.

[18] In 1567, Pope Pius V condemned seventy-six propositions (some sources speak of seventy-nine) from Baius' works in the bull *Ex omnibus afflictionibus*. The condemnation was renewed by Pope Gregory XIII in 1580.

[19] Kurt Gödel Papers, Box 7a, Folder 107, item accession 030129.

[20] For a short biography see (Suttner 1984).

[21] Pope Pius XII defined the Assumption of Mary as a dogma in 1950, see (Ott 1957, 208). Note the close relation between 'dogma' and 'heresy', a phenomenon reflected by the modal operators introduced in Sect. 9.6.

fide humana
ist es Pflicht zu glauben {
Scholastiker
Entscheidungen der Bischöfe, denen man als Lehrer Gehorsam Schuldig ist.

z.B. Provinzialentscheidungen? Kann man zu den armen Seelen beten?

Gründe des Glaubens (Gründe, weswegen etwas geglaubt werden muss.)

Scholast⟨ische⟩ Methode:

1. Vernunftsbeweis für Dogmen zum tieferen Verständnis
2. Analogieschluss (filiatio)
3. Vergleichung der Dogmen miteinander

Wenn man auch die Dogmen nicht begreifen kann, so dringt man tiefer ein.

4. Ein System aus diesen Wahrheiten zu machen. Großartiger Aufbau und innerer Zusammenhang
5. Einwände gegen ein Dogma zu entkräften { polemische Methode / apologet⟨ische⟩ Methode } *

Ein Dogma bleibt immer mit einer Dunkelheit des Glaubens behaftet. Aber ein allgemeiner Beweis, dass das Dogma nicht widerspricht der Vernunft, gibt es nicht.

* Diese Methode soll man nicht so hervorkehren.

Das Wesentliche ist die ~~dogmatische~~ Pos⟨ition⟩. Aber ⟨die⟩ apolog⟨etische⟩ ⟨ist⟩ doch notwendig, besonders im Verkehr mit Einzelnen. Es besteht ⟨die⟩ Gefahr, dass Zweifel hineingebracht werden, die ihm sonst nicht gekommen wären.

Das Ziel ist ein ganz Übernatürliches, welches mit den natürlichen Kräften nicht einmal erkannt werden kann. Daher muss Gott dieses Ziel durch Offenbarung bekannt machen. Deswegen ⟨ist⟩ Offenbarung notwendig, sowohl physisch als ⟨auch⟩ moralisch.

~~Theodizee~~ Gerechtigkeiten, Strafgerechtigkeiten können durch natürlichen Verstand erkannt werden (z.B. Theodicee), ⟨wird⟩ aber nicht von allen wirklich erkannt (oder nicht genug rasch). Seit der Offenbarung aber kann es jedes Kind erkennen.

Die Theologie ist im Glauben beschlossen. Man muss in der Theolgie schon vom Glauben ausgehen. Wer keinen Glauben hat, kann gar nicht zur Theologie kommen. Und sie kann nicht über den Glauben hinaus.

Rationalisten meinen, man könne alles verstehen und es werde eine Zeit kommen, wo man die Glaubenswahrheiten einsehen werde. Man kann niemandem den Glauben andemonstrieren.

Gebetet wird nicht, weder am Anfang noch am Ende.

Es sind mehr Hörer als bei Pohl,[22] 140, und fast ausschließlich Theologen (in Tracht).

Beginnt so leise, dass man ihn nicht versteht, und liest eine Zeit während des Gemurmels der Studenten.

Addressed as a rationalist and a logician, Gödel was obviously most impressed by the 4th item of the "scholastic method", the idea that the dogmas as a whole form a system of truths of their own, thus constituting the mentioned "great inner connection". Nevertheless, in item FT26 (see Sect. 9.4) he also touches upon the question if "a time would come" when the dogma itself could be rationally

[22] Wenzel Pohl (1870–1949). Gödel attended his lecture entitled "Erkenntnistheorie".

"understood", thus implementing Lehner's remaining items (1) to (3) and (5). We will consider this point in Sect. 9.8, concentrating on the dogma of free will.

9.4 The Dogma in Gödel's Notes

The lists ND and FT cover a wide range of theological topics. Here we present those items[23] which are in some respect connected with dogmatic theology. In their entirety, they will form a basis for the development of several formal logics in the following sections.

ND1 Every doc⟨t⟩r. prop. is true.
ND2 Every one ~~not~~ (christ., catholic) bel. the neg. of ~~in~~ ⟨a⟩ dogm. commits a mort. sin.
ND3 Every one not bel. in a dogma although he knows that it is a dogm. commits a mort. sin.
ND4 Every one teaching publicly the neg. of a doctr. as the truth (although) commits a mort. sin.
ND5 Every one asserting privately —
ND1' If *A* is ⟨a⟩ dogma at some time, it is a dogma at any later time.
ND2' If *A* is a doc⟨t⟩r. prop. at some time, it is —
ND30 Is every prop. of the bible ~~taken literally~~ which is interp. unanimously a dogma?
ND34 Def. von Dogma und Doctrine aus den vorhandenen Dokum.
ND35 Ist jeder berechtigt, an Rom zu appellieren, was ein Dogma ist?
ND41 Wie wird die Exc. ipso facto behoben? (durch Lesen eines Buches, nicht Grüßen der Heiligenbilder, Verkehr mit dem Teufel, etwas nicht Dogmatisches glauben)
ND46 Martyrolog., Dogmenverz.,[24] wie feststellen, was Dogma oder Lehrmeinung?
ND75 Trinit. dogma: una natura, tres persones
ND76 Homousie: una persona, duae naturae
ND78 Kann nicht das Dogma vollkommen exakt sein, ähnlich wie theor. Physik?
FT26 Conc. Vat.: Können nicht einmal alle Dogmen mittels der Vernunft verstanden werden? (intellegi?)
FT34 Ist es möglich, einen päpstlichen Beschluss (insbesondere bezüglich einer Lehrmeinung) rückgängig zu machen? Und welche Beschlüsse?
FT42 Unterschied zwischen anathema und Excom.; nur das Dogma,[25] was unter Anathem.; Anathem. bezieht sich nur auf letzten Satz; gleichgültig, in welchem Doc. ein solches Anathem. steht; welche Art von Doc., für welche der Papst zeichnet, gibt es (in welchen die Anathem. gültig sind); Kriterien der Gültigkeit; wo sind sie aufbewahrt; kann etwas, was eine Häresie war, aufhören, eine solche zu sein; ist ⟨−⟩ insbesondere in einer Epistel möglich? [Conc. Rom.] oder überhaupt Prov. Synoden?
Constitutio, Decr. Sancta Congreg. S. off. Declaratio, Breve Motu proprio, offizielle Ausgabe, p.339, formala missa, ebenso 337, Instructio
Es sind drei Dinge zu berücksichtigen:

[23]The original order and the item numbers are retained. For easier reference the abbreviations ND and FT are prefixed.

[24]Gödel's footnote: "1. Welches Dogma? 2. Welcher Wortlaut? (? 3. Wo auffindbar?)".

[25]Gödel's footnote: "Wenn es heißt *resistit spir. S., fallitur; diffinimus ut ... ab omnib. Christ. credatur est haeretica, est falsa (erronea, suspecta, temeraria, scandalosa)*".

1. In welchen Dok. ~~ist etwas Dogma~~ kann ein Dogma festgestellt werden?
2. Mit welchen Worten muss es festgestellt sein?
3. Wo sind die Dok. wahrheitsgemäß referiert?

Ist Lehrmeinung ein ebenso fester Begriff? Beispiel einer Änderung; temeritas bezieht sich darauf?

Some of these items seem to suggest that Gödel—at least at the time of writing—was confused about the (relation between the) terms *dogma*, *doctrine*, and *Lehrmeinung*. In item ND2, the word "doctr" has obviously been overwritten by the word "dogm", whereas it remains (the wrongly spelled) "docr" in ND1.[26] In Sect. 9.6 we will assume that ND1 is a proposition about dogmas (i.e. *doctrinae divinae*). Similarly, item ND4 seems to speak of the negation of a *dogma*. Eventually, the items ND34 and ND46 seem to underline that Gödel was trying to clarify this matter.

9.5 Intuitionistic Dogmatic Logic

Contrary to a dogmatic *modal* logic which will be presented in the following section, there is no explicit mention of a dogmatic propositional logic in Gödel's notes.[27] Nevertheless, there are quite a few hints amongst his writings which show that he obviously was fully aware of the fact that a formal logical system can be built upon the notion of the catholic *dogma* that replaces the general term of *proposition*.

A basis for these considerations is formed by models of branching time which were considered by Gödel as early as 1935.[28] Such a model $(\mathcal{M}, \leq, \Vdash)$ consists of a set of *moments* $\mathcal{M}$ which is ordered into a tree-like structure by the partial order $\leq$. If $\mathfrak{D}$ is the set of dogmas then $\Vdash$ is a relation between $\mathcal{M}$ and $\mathfrak{D}$, $m \Vdash d$ being read as 'm forces d'. Figure 9.1 pictures an example of a model involving the set of dogmas $\mathfrak{D} = \{c, d\}$.

Note that in the given example model the following stipulation (*monotonicity*) holds for all $m, m' \in \mathcal{M}$ and all $d \in \mathfrak{D}$.

$$\text{If } m \Vdash d \text{ and } m \leq m' \text{ then } m' \Vdash d \tag{9.1}$$

This condition perfectly reflects the axiom given in ND1', "If A is $\langle$a$\rangle$ dogma at some time, it is a dogma at any later time", and item FT34 clearly shows that Gödel had a particular interest in the question for what kind of papal decisions

[26] The wrongly spelled "docr." also appears in ND2'.

[27] As we shall see, the modal logic considered in the next section turns out to be the *modal companion* of this section's propositional variant.

[28] In 1935, Gödel wrote two successive notebooks in Gabelsberger shorthand on the foundations of quantum mechanics (Kurt Gödel Papers, Box 6b, Folder 78, item accessions 030106 and 030107). In the second of these books, Gödel considers models of branching time in connection with the notion of free will. Details are given in (Lethen 2020b). See also Sect. 9.8.

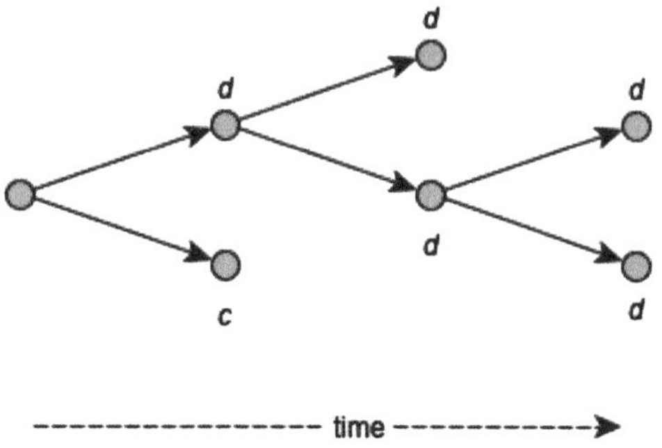

Fig. 9.1 A model involving the set of dogmas $\mathfrak{D} = \{c, d\}$. Note that in the earliest (leftmost) moment $d \vee \neg d$ (*tertium non datur*) does not hold

this rule would hold. In the course of time, new dogmas may appear, and it is interesting to see how Gödel expresses his thoughts about the process of the promulgation of new dogmas, asking in ND35 if everybody has the right to propose new dogmas. As far as a general dogmatic development is concerned, the catholic church distinguishes between a *material* and a *formal* development, the latter now being reflected in the proposed models: Whereas the material development of the dogma concerns the actual divine revelation and is said to have concluded with Christ and the apostles, the formal side concerns the promulgation of the dogma by the teaching authority of the catholic church, raising the material dogma to the status of a formal dogma, formulating clearly defined concepts, and deciding formerly disputed questions. Contrary to the catholic point of view, protestant or modernistic conceptions allow for the possibility of the development of a dogma itself, independent of any ecclesiastic promulgation, following a progression in philosophy and in the sciences in general. The First Vatican Council condemned the view of a development in this sense as heretical. For details, see (Ott 1957, 6).

The models described so far do indeed resemble what is known as *Kripke models* with the catholic view of the dogma (*propositio ecclesiae*) replacing the more general notion of proposition. What is missing up to now is an extension of the forcing relation $\Vdash$ to terms of a logical language build on top of the set of dogmas $\mathfrak{D}$ to be defined by the following grammatical production rule, d ranging over the set $\mathfrak{D}$.

$$\varphi ::= \bot \mid d \mid (\varphi \wedge \varphi) \mid (\varphi \vee \varphi) \mid (\varphi \supset \varphi) \tag{9.2}$$

Here $\bot$ denotes the *absurd dogma*, with $\neg d$ as usual abbreviating the implication $d \rightarrow \bot$. The forcing relation can now be extended to composite terms as follows.

$$m \Vdash \bot \text{ never holds,} \tag{9.3}$$

$$m \Vdash (\varphi \wedge \psi) \text{ iff } m \Vdash \varphi \text{ and } m \Vdash \psi, \tag{9.4}$$

$$m \Vdash (\varphi \vee \psi) \text{ iff } m \Vdash \varphi \text{ or } m \Vdash \psi, \tag{9.5}$$

$$m \Vdash (\varphi \supset \psi) \text{ iff from } m' \Vdash \varphi \text{ follows } m' \Vdash \psi \text{ for all } m' \text{ with } m \leq m'. \tag{9.6}$$

Note that $m \Vdash \neg\psi$ holds iff $m' \Vdash \psi$ does not hold for any moment m' with $m \leq m'$.

Again, absolutely no difference appears between the ordinary Kripke models and the now established dogmatic models and—without going into further details[29]—it follows that a dogmatic (propositional) logic based on any syntactical proof system will be sound and complete with regard to the dogmatic models just developed. It follows immediately that any dogmatic logic based on a set of dogmas $\mathfrak{D}$ is an intuitionistic logic with rules like the *tertium non datur* $(d \vee \neg d)$ or *Peirce's law* $(((c \rightarrow d) \rightarrow c) \rightarrow c)$ not being validated.

As stated above, Gödel himself did not explicitly mention a formal propositional dogmatic logic. In this connection it is worth though to return to his description of the main differences between an idealistic philosophy and dogmatic theology given in Sect. 9.2. Here item 7 ("Vollkommener Apriorismus, aber nicht aus dem Satz vom Widerspruch.") hints at the intuitionistic character of the underlying logic.[30]

Of course, the dogmatic models just introduced suggest a comparison between the terms *dogma* and *knowledge*. As far as intuitionistic logic and knowledge are concerned, Gödel states in his notebook *MaxPhil V* (292):

> In der Brouwer'schen Interpretation der Mathematik werden gar nicht die mathematischen Sätze P interpretiert, sondern die Sätze "Ich weiß P" (oder in einer schwächeren Form "P ist beweisbar").

Transferred to an intuitionistic dogmatic logic this means that a dogmatic proposition d should now be interpreted as 'it is a dogma that d', a fact clearly reflected in the corresponding translation into modal logic given at the end of the following section. It should also be worth mentioning Gödel's own definition of the term 'Wissen [knowledge]' which can indeed be applied without any hesitation to the notion of a dogmatic belief. In unpublished notes[31] concerning his 1951 Gibbs lecture (Gödel 1951), Gödel writes:

> wissen = mit Recht in jedem beliebigen Grad davon überzeugt sein [insbesondere also mit Recht beschließen, es unter keinen Umständen zu revidieren][32]
>
> Theorem: Ich weiß p. $\supset$ Ich bin in unmittelbarem Kontakt mit den Gegenständen der Aussage p.

Note how the theorem is supported by item 6 of Gödel's list given in Sect. 9.2. However, it should be mentioned that the extension of the forcing relation $\Vdash$ and thus of Gödel's notion of knowledge is problematic, at least as far as implicational statements are concerned. Whereas the extensions (9.3)–(9.5) touch only

[29] For details see for example (von Plato 2013, chapter 7), or (Sørensen and Urzyczyn 2006, chapter 2).

[30] Usually the law of noncontradiction is taken to be $\neg(p \wedge \neg p)$ which is valid both classically and intuitionistically. Here it is assumed that Gödel is alluding to a special instance of Peirce's law, $(\neg p \rightarrow p) \rightarrow p$, which draws the conclusion p whenever there is an implicational contradiction between $\neg p$ and p.

[31] Kurt Gödel Papers, Box 12, Folder 43, item accession 060573.

[32] Here, Gödel adds a footnote which is discussed further below.

upon (combinations of) dogmas or propositions given at the same very moment, clause (9.7) draws an intimate relation between the present moment and all possible future moments. Whereas the implication from left to right seems to be very natural, the opposite direction presupposes the opportunity to be able to foresee future situations, which is certainly problematic in connection with both the notion of *knowledge* as well as with the notion of *dogma*, a "direct contact" not being given in either case. Gödel himself comments on this phenomenon in a footnote concerning the just mentioned notes supplementing his Gibbs lecture. Addressing the special case of negation, i.e. the implication of absurdity, he adds to his definition of knowledge:

> Weil sie Erkenntnis möglich machen, also sicherlich ihre Negation niemals anzunehmen wäre, denn das hieße erkennen wollen und gleichzeitig seine Möglichkeit zu negieren.[33] (Für Wissen in diesem Sinne gilt das *Th.* nicht.) Aber woher weiß ich diese *Implic.*? (Zu kompliziert für eine direkte Schau. Außerdem gar nicht evident, denn ein Umlernen vielleicht möglich, und der Satz betrifft eine sehr allgemeine Aussage über unseren Erkenntnis-*Appar.*, der gar nicht mit einem Blick zu übersehen ⟨ist⟩.)

Summing up, the above investigations into an intuitionistic (propositional) dogmatic logic underline the close connection between knowledge and belief based on an isomorphism between their underlying formal logical systems. The mutual simulation is precisely expressed in a theological lecture delivered by Johannes Messner (1891–1984) and attended by Gödel in October 1937. Here Gödel's (heavily underlined) lecture notes concerning the relation between 'Offenbarung und Wissenschaft [revelation and science]' read as follows: "Die Methode ist die gleiche, nur die Quellen sind verschieden. [The method is the same, only the sources are different.]"

9.6 Modal Dogmatic Logic

Contrary to the logic outlined in the previous section, Gödel does explicitly mention the idea of a modal logic interpretation of the terms "Dogma" and "Lehrmeinung" (doctrine). In his philosophical notebook *MaxPhil II* (131), written between 1938 and 1942, he states:

> Dogma, Lehrmeinung, etc., bilden auch zwei Modalitäten. Ferner muss es eine noch stärkere Notwendigkeit als logische oder mathematische Notwendigkeit geben um zu erreichen, dass nicht alle mathematischen Gesetze dasselbe bedeuten. Z.B. $(x)\varphi(x) \equiv (y)\varphi(y)$ ist eine stärkere Notwendigkeit als ⟨eine⟩ logische.

We will denote the corresponding modal operators as D and L, Dp being read as 'it is a dogma that p' and Lp as 'it is a doctrine [Lehrmeinung] that p'. Taking the

[33]The phrase "seine Möglichkeit zu negieren" should be taken taken to mean "seine Negation zu ermöglichen".

operator D in the strictest sense as denoting an immediately revealed truth or fact, the modal logic in question obviously is a *normal* logic that satisfies the axiom K,

$$D(p \rightarrow q) \rightarrow (Dp \rightarrow Dq). \tag{9.7}$$

In this direction, Ludwig Ott states: "If [...] both premisses are immediately revealed truths,[34] then the conclusion also must be regarded as being immediately revealed and as the object of Immediate Divine Faith (*Fides Immediate Divina*)" (Ott 1957, 8). However, relying on the infallibility of the teaching authority of the catholic church, other *veritates catholicae* may be used as premisses, namely historical facts (*facta dogmatica*) as well as what is known as *truths of reason*, i.e. philosophic teachings which do not endanger the heritage of faith through the production of contradictions to a revealed truth.[35] Adressing these contradictions, the First Vatican Council, convoked by Pius IX in 1868, states in its dogmatic constitution on the catholic faith, *Dei Filius*: "But although faith is above reason, there can never be any real discrepancy between faith and reason, since the same God who reveals mysteries and infuses faith has bestowed the light of reason on the human mind, any God cannot deny Himself, nor can truth ever contradict truth. [...] We define, therefore, that every assertion contrary to a truth of enlightened faith is utterly false."[36] (Manning 1871, 215)

Furthermore, Gödel's remark ND1 clearly verbalizes the modal axiom T,

$$Dp \rightarrow p, \tag{9.8}$$

underlining the religious view that any promulgated dogma has to be believed in as an actual fact. In this respect, the First Vatican Council states (*Dei Filius*): "And the Catholic Church teaches that this faith, which is the beginning of man's salvation, is a supernatural virtue, whereby, inspired and assisted by the grace of God, we believe that the things which He has revealed are true; not because the intrinsic truth of the revealed things viewed by the natural light of reason, but because of the authority of God Himself who reveals them, and Who can neither be deceived nor deceive." (Manning 1871, 212)

As far as any further modal axioms are concerned, the preceeding section might suggest a comparison to the notion of *knowledge*, in a modal context usually represented by the operator K. In this connection the axiom $Kp \rightarrow p$ (interpreted as *veridicality*) is often paired with an axiom expressing *positive introspection*, i.e.

---

[34] An example of an immediately revealed implicational truth is "The nature of the angels is spiritual", formally expressed as $D(angel(x) \rightarrow spiritual(x))$. See (Ott 1957, 116).

[35] These circumstances also lead to the possibility to include the rule of necessitation $\frac{p}{Dp}$ into a formal dogmatic proof system.

[36] For example, the constitution condemns *materialism* ("If anyone shall not be ashamed to affirm that, except matter, nothing exists; let him be anathema.") and *pantheism* ("If anyone shall say that the substance and essence of God and of all things is one and the same; let him be anathema."), both being rejected by Gödel himself. See also (Pius XII 1950).

$Kp \to KKp$. Regarding these two axioms, Gödel writes in his notebook *MaxPhil III* (99):

> Das Wissen ist eine Art der Wahrnehmung (Erkenntnis), welche ihrem Wesen nach die Richtigkeit impliziert. (Aber kann man auch wissen, dass man etwas weiß?) Kann man überhaupt etwas wissen?

Adding a kind of *dogmatic introspection*

$$Dp \to DDp \tag{9.9}$$

to the modal system finally leads to a modal logic **S4**. As far as this axiom's evidence is concerned, it should be noted that the first D operator of the conclusion should be taken to express a 'catholic truth' rather than a revealed dogma. Also, the 'plausibility' of the axiom might depend on fact that it 'completes' the modal logic, transforming it into a *modal companion*[37] of the intuitionistic logic developed in the preceding section. (See the discussion on evidence in Sect. 9.7.) *Negative introspection* $\neg Dp \to D\neg Dp$ (or equivalently $L\neg p \to DL\neg p$, see below) is certainly ruled out by the fact that new dogmas may be introduced in the course of time and remain valid ad infinitum. Thus, our logic does not take the step onto the **S5** level.

Up to now, only the operator D has been considered. Its dual, the operator L, captures the notion of *doctrine* [*Lehrmeinung*] in the sense of a *theological opinion*, which is described by Ott (1957, 9) as follows: "Theological opinions are free views on aspects of doctrines concerning Faith and morals, which are neither clearly attested in Revelation nor decided by the Teaching Authority of the Church. Their value depends upon the reasons adduced in their favour (association with the doctrine of Revelation, the attitude of the Church, etc.)" The operator L is put into context by Gödel's remark ND4, the word "doctr." being taken to be used for '*doctrina divina*' or simply 'dogma'. If now 'p is a mortal sin' is formalized as Hp with the operator H being defined by the equivalence $H\varphi \leftrightarrow D\neg\varphi$, also being read as 'it is a heresy that p', then ND4 may be translated into $H(Dp \wedge L\neg p)$, which is equivalent to $D(Dp \to \neg L\neg p)$, which in turn implies

$$Dp \to \neg L\neg p. \tag{9.10}$$

In his Encyclical "Humani generis" (Pius XII 1950, §20), Pope Pius XII substantiates exactly this implication: "But if the Supreme Pontiffs in their official documents purposely pass judgment on a matter up to that time under dispute, it is obvious that that matter, according to the mind and will of the Pontiffs, cannot be any longer considered a question open to discussion among theologians."

[37] Briefly speaking, a modal logic is a modal companion of a superintuitionistic logic if provability on the intuitionistic side corresponds to provability of the Gödel-McKinsey-Tarski translation in the modal system. The translation between the two dogmatic systems is briefly considered at the end of this section.

Unfortunately, in (9.10) one does not get the direction from right to left. This is due to the fact that ND4 refers to an *actual* existing theological opinion or doctrine as opposed to a *possible* doctrine. In the latter case, the implication (9.10) can be extended to an equivalence: If p is not a dogma then $\neg p$ is a possible doctrine. In the former case one ends up with an intuitionistic modal logic **S4**, to be enriched by the axiom $D(p \to q) \to (Lp \to Lq)$ which in this case cannot be deduced from the other axioms. For details see (Simpson 1994, chapter 3). Figure 9.2 summarizes the introduced modal operators. Note that ND2, ND3, and ND5 suggest that the interpretation of the operator L can also be reduced to a possible personal view or belief.

The results of this and the preceding section now offer the possibility to parallel Gödel's translation from intuitionistic propositional logic into a classical logic enhanced by the modal operator B that represents provability, as given in (Gödel 1933). Gödel's main result states that the translation of a formula valid in the propositional system is provable in the modal system. Later, McKinsey and Tarski (1948) were able to show that the opposite direction of this statement is also true, a fact which had already been conjectured by Gödel himself. Here we give a translation τ from a dogmatic propositional logic into a dogmatic modal logic which is based on one of the schemes given by McKinsey & Tarski, substituting the operator D for the provability operator B.

$$
\begin{aligned}
d^\tau &= Dd \text{ for all dogmatic variables } d \in \mathfrak{D} \\
\bot^\tau &= \bot \\
(\varphi \wedge \psi)^\tau &= D\varphi^\tau \wedge D\psi^\tau \\
(\varphi \vee \psi)^\tau &= D\varphi^\tau \vee D\psi^\tau \\
(\varphi \supset \psi)^\tau &= D\varphi^\tau \to D\psi^\tau
\end{aligned}
$$

Whereas this translation is able to emphasize the close relation between two different kinds of logical (dogmatic) systems, it also underlines another very subtle parallel: The notion of *dogma* not only closely resembles the notions of *necessity* and *knowledge* but also reveals an intimate relationship to the notion of *provability*.

operator	**interpretation**	**defined by**
D	It is a dogma that ...	
L	It is a possible doctrine/view that ...	$\neg D \neg$
H	It is a mortal sin/heresy that ...	$D \neg$

Fig. 9.2 Dogmatic modal operators.

9.7 Catholic Grades of Certainty and a Theory of Evidence

Since ancient times, the nature of the *axiom* has been discussed among mathematicians, physicists, and philosophers, firmly related to the notions of *perception* and *evidence*. The scholastic philosopher and theologian William of Ockham (c.1287–1347) for example laid down the following principle, quoted by (Spade and Panaccio 2019): "For nothing ought to be posited without a reason given, unless it is self-evident (*literally*, known through itself) or known by experience or proved by the authority of Sacred Scripture." As far as the "Sacred Scripture" is concerned, Gödel himself in ND30 poses the question if every proposition of the bible was a dogma, which—at least from a formal catholic point of view—has to be answered in the negative as long as these revealed truths have not finally been defined as a dogma by the teaching authority of the catholic church.[38]

Gödel, addressing the notion of evidence, writes in his notebook *MaxPhil III* (54):

> Bem.: Alles, was irgendwie eingesehen werden kann, ist entweder[39]
>
> 1. vollkommen klar (das, was man wissen kann),
> 2. einigermaßen klar (Ersetzungsaxiom),
> 3. plausibel, d.h. annehmbar aus aesthetischen, Vollständigkeitsgründen, etc.
>
> Fra⟨gen⟩:
>
> 1. Kann das Plausible analytisch gemacht werden mittels der Vollkommenheit Gottes?
> 2. Kann alles Erkennbare (auch das synthet. und empirische) plausibel gemacht werden?

Here it becomes clear that Gödel is taking into consideration different grades of evidence and that he is obviously prepared to develop theories on the grounds of aesthetic assumptions. This attitude appears to be in perfect accordance with Bertrand Russell's conception regarding the certainty of an axiom. In his contribution to P. A. Schilpp's volume on Russell's philosophy, Gödel states:

> [Russell] compares the axioms of logic and mathematics with the laws of nature and logical evidence with sense perception, so that the axioms need not necessarily be evident in themselves, but rather their justification lies (exactly as in physics) in the fact that they make it possible for these "sense perceptions" to be deduced; which of course would not exclude that they also have a kind of intrinsic plausibility similar to that in physics. (Gödel 1944, 127)

And he continues:

> Furthermore it seems likely that for deciding certain questions of abstract set theory and even for certain related questions of the theory of real numbers new axioms based on some hitherto unknown idea will be necessary. Perhaps also the apparently unsurmountable difficulties which some other mathematical problems have been presenting for many years are due to the fact that the necessary axioms have not yet been found. Of course, under these circumstances mathematics may lose a good deal of its "absolute certainty"; but, under the

[38]For a detailed answer to Gödel's question, see (Kaulen 1868, chapter IV).

[39]The first two items are marked as "analytisch" at the right margin.

> influence of the modern criticism of the foundations, this has already happened to a large extent. (Gödel 1944, 128)

Addressing Russell's conception of the evidence of axioms, Hao Wang (1987, 180) stresses Gödel's "suggestion of a probable decision of the truth of new axioms of set theory inductively by studying its success", the more easily decidable questions not lying at the beginning of a theory but somewhere "in the middle". This view is certainly supported by an entry in Gödel's notebook *MaxPhil III* (43–44):

> Der tatsächliche Fortschritt in Mathematik und Philosophie besteht darin, dass man zuerst einige wenige unscharfe Begriffe hat. Indem man diese irgendwie gegeneinander wirken lässt und aufeinander anwendet, werden sie schärfer und erzeugen neue unscharfe Begriffe, welche in derselben Weise behandelt werden, etc.[40]

Before we turn to a formal theory of evidence developed by Gödel, it is worth looking at a parallel to the aforementioned different grades of evidence in catholic teaching. The logics developed in Sects. 9.5 and 9.6 both considered a proposition as either being a dogma or not, not allowing for any in between. Nevertheless, a grade of certainty may be ascribed to the dogma, thus imposing a fine gradation on the set of dogmas, paralleling the phenomenon of evidence of axioms. According to (Ott 1957, 8–9, §8), the levels of certainty include: (1) *de fide* (defined by the catholic teaching autority, divided into *de fide divina et catholica definata*, *de fide divina et catholica*, and de *fide divina*); (2) *de fide ecclesiastica*; (3) *sententia fidei proxima*; (4) *doctrina catholica*; (5) *sententia certa*; (6) *sententia communis*; (7) *sententia probabilis* (divided into *sententia probabilior*, *sententia bene fundata*); (8) *sententia pia*; and (9) *sententia tolerata*.

The analogy between the (mathematical) axiom and the (catholic) dogma is therefore twofold. On the one hand, evidence corresponds to a grade of certainty. On the other hand, there is neither a fixed pre-established set of basic mathematical axioms, nor has the act of defining and promulgating new formal dogmas come to an end. The former analogy now seems to suggest the invention of a many-valued logic capturing the different grades of certainty in their corresponding truth values. Gödel himself published such a logic (Gödel 1932), giving the following stipulations based on a logical matrix[41] $\mathfrak{M} = (\mathcal{A}, \mathcal{B}, \vee, \wedge, \supset, \neg)$ with $\mathcal{A} = \{1, 2, \ldots, n\}$ and $\mathcal{B} = \{1\}$.

$$a \vee b = \min(a, b) \tag{9.11}$$

$$a \wedge b = \max(a, b) \tag{9.12}$$

$$a \supset b = \begin{cases} 1, & a \geq b \\ b, & a < b \end{cases} \tag{9.13}$$

[40] Note that this entry appears right after the remark "Die Intuitionisten haben den Begriff der Wahrheit nicht".

[41] A *logical matrix* consists of a set $\mathcal{A}$ of truth values, a subset $\mathcal{B} \subset \mathcal{A}$ of values representing 'absolute truth', and a set of logical operators including negation and implication, satisfying the following axiom: If $a \in \mathcal{B}$ and $a \supset b \in \mathcal{B}$, then $b \in \mathcal{B}$.

$$\neg a = \begin{cases} n, & a \neq n \\ 1, & a = n \end{cases} \tag{9.14}$$

The drawback of this approach can be found in the way negation is handled in stipulation (9.14). To see this, one has to take into consideration that the negation of a dogma is considered as a heresy, which in Gödel's system had to be represented by the value n. Nevertheless, following (Ott 1957, 10, §9), the set of heresies may also be equipped with a fine gradation in exactly the way the dogmas may. Here, the different levels are: (1) *propositio haeretica* (opposed to the formal dogma); (2) *propositio haeresi proxima*; (3) *propositio haeresim sapiens* or *de haeresi suspecta*; (4) *propositio erronea*; (5) *propositio falso* (contradicting a dogmatic fact); (6) *propositio temeraria* (deviating from the general catholic teaching); (7) *proposition piarum aurium offensiva*; (8) *propositio male sonans* (being subject to misunderstanding); (9) *propositio captiosa* (incuding intensional ambiguity); and (10) *propositio scandalosa* (exciting scandal). Gödel's footnote to his remark FT42 clearly testifies that he knew these different theological censures, a fact also suggested by the first sentence[42] of Joseph Lehner's lecture attended by Gödel ("Zu leugnen, dass Maria in den Himmel erhoben wurde, wäre haeresi proximum.").

Returning to many-valued logics again, the idea of different grades of falsity (in addition to different grades of truth) is captured by a variant introduced in 1922 by Jan Łukasiewicz (Łukasiewicz 1922–23). Based on a logical matrix with $\mathcal{A} = \left\{0, \frac{1}{n-1}, \frac{2}{n-1}, \ldots, 1\right\}$ and $\mathcal{B} = \{1\}$, implication and negation are defined as follows:

$$p \supset q = \begin{cases} 1, & p \leq q \\ 1 - (p - q), & p > q \end{cases} \tag{9.15}$$

$$\neg p = 1 - p \tag{9.16}$$

If the set $\mathcal{A}$ is now divided into two partitions comprising (grades of) heresies and dogmas respectively, then definition (9.16) seems to reflect the underlying symmetry. However, as far as the definition of implication (9.15) is concerned, Gödel (*MaxPhil III*, 44) himself expresses his doubts about its adequacy, the central term "Klarheit" possibly being interpreted as evidence or as a theological grade of certainty.

> In der mehrwertigen Logik ist der Wahrheitswert von $q <$ als die Wahrheitswerte von $p, p \supset q$, d.h. (wenn durch "Klarheit" interpretiert), nach endlich vielen Schlüssen hört jeder Grad von Klarheit auf, d.h., man sieht nichts mehr. Das deutet darauf hin, dass, um weitergende Folgerungen zu erzielen, man versuchen muss, die einfachen Dinge vollkommen klar zu sehen. Das gibt vielleicht auch eine natürliche Reihenfolge der Wissenschaften, nämlich: zuerst die Fundamente aller Wissenschaften, dann das Kompliziertere aller Wissenschaften, etc. (harmonische Ausbildung).

[42]Note that the Assumption of Mary was not classified as a formal dogma at the time the lecture was given.

It is interesting to note that the phenomenon decribed does actually hold only for Łukasiewicz's variant:[43] Whereas in Gödel's system q always lies between p and $p \supset q$, in Łukasiewicz's system $q < p$ does indeed imply $q \leq p \supset q$.

Possibly inspired by the idea of a many-valued logic, Gödel soon developed a formal theory of evidence as can be found in his notebook *MaxPhil 0*[44] (29). Here Gödel states:

> Bem.: Die Funktion $E(p)$ [Evidenz von p] genügt anderen Regeln als ⟨die⟩ Wahrscheinlichkeit, nämlich $E(p\,.\,q) = \min[E(p), E(q)]$.
> Daher auch $E(q) = \min[E(p), E(p \supset q)]$.
> Daher ist die Menge $\hat{p}(E(p) > a)$ ein formales System,[45] das mit abnehmendem a zunimmt.
> Ferner $-1 \leq E(p) \leq +1$ und $E(\sim p) = -E(p)$.
> Es wäre interessant, die Klasse $\hat{p}(E(p) > a)$ für empirische Sätze, die für mich gelten, festzustellen.

In the case of an underlying classical logic, the equation

$$E(p \vee q) = \max[E(p), E(q)]$$

is an immediate consequence of the given axioms. Otherwise it may be added itself as an axiom, following the example of equations (9.11) and (9.12) in Gödel's many-valued logic. Unfortunately, Gödel does not give an equation for implicational formulas, and certainly the 'consequence' $E(q) = \min[E(p), E(p \supset q)]$ is questionable as far as equality is concerned. Nevertheless, it should be obvious that the given theory can without any hesitation be transformed into a theory of the dogma, substituting grades of certainty for the notion of evidence.

Of course, the formal theory cannot give any indication as where the psychological roots of evidence may be found. Gödel does however include this question in his 'psychological program' and states (*MaxPhil II, 151*):

> Progr. (Psych.): Theorie der falschen Evidenz – wäre sehr wichtig als Hilfe, die Wahrheit zu finden, weil:
> 1. Falsche Evidenzen sind entweder eine Sünde oder eine Folge einer Sünde, wobei die Sünde wahrscheinlich meist in einer *iniquitas* des Urteils besteht. Also gibt das eine Theorie der *iniquitas* des Glaubens, daher auch eine der *aequitas*, d.h. der Wahrheit.

Concluding this section on evidence and grades of certainty, we present a truly philosophical (as well as theological) remark concerning religious belief and human reason, given by Gödel in the notebook *MaxPhil III* (25–26). It certainly leaves no

[43] The manuscript of Gödel's envisaged contribution to the joint work with Arend Heyting on the foundations of mathematics testifies that Łukasiewicz's system was indeed well known to Gödel. (The manuscript itself (Kurt Gödel Papers, Box 7a, Folder 10, item accession 040019) is briefly described by Dawson (1996, chapter V) and has now been entirely transcribed by Jan von Plato, Helsinki.)

[44] Kurt Gödel Papers, Box 6b, Folder 63, item accession 030086.

[45] Gödel's footnote: "Das sind in der Mathematik die verschiedenen 'Schichten' des Konstruktivismus."

doubt about Gödel's deep views about the close ties and interconnections between religion and science in general.

> Die Religion befiehlt, Annahmen zu machen (du sollst glauben), die natürliche Vernunft ebenfalls (z.B. weil eine Annahme *A* bei der Feststellung, was das Richtige zu tun ist, wieder hinausfallen kann). Vielleicht ist das richtige Prinzip zum Handeln und Glauben folgendermaßen: Es ist eine Anordnung sämtlicher Hypothesensysteme (wo jedes einzelne System alle Fragen entscheidet) gegeben. Man beginnt mit dem Initialsystem und hält jedes System solange fest, als ⟨es⟩ nicht widerlegt ist (dann das nächste). In praxi ist zuerst eine Anordnung der Sätze nach "Wichtigkeit" gegeben und das Initialsystem. Dann werden bei jeder Widerlegung eines Systems möglichst unwichtige Sätze geändert. Gewisse Sätze werden überhaupt nie geändert (die Glaubensartikel). Das sind die unwiderlegbaren Sätze.

9.8 The Dogma of Free Will

Up to this point, only the relation *between* dogmas has been considered without paying any attention to the actual contents of any specific dogma, thus following the typical operating principle in formal logic. Gödel's remark FT26 is a clear hint though that he did indeed consider a rational investigation of the individual dogma, thus following the scholastic method mentioned in Joseph Lehner's lecture (see Sect. 9.3, items 1–3). The text alluded to in remark FT26 is part of the dogmatic constitution '*Dei filius*' of the First Vatican Council. It reads: "[B]ut reason never becomes capable of apprehending mysteries as it does those truths which constitute its proper object. For the divine mysteries by their own nature so far transcend the created intelligence that, even when delivered by revelation and received by faith, they remain covered with the veil of faith itself, and shrouded in a certain degree of darkness, so long as we are pilgrims in this mortal life, not yet with God" (Manning 1871, 215). Note Gödel's addendum "(intellegi?)".

Whereas the ontological proof may certainly be regarded as a rational attempt to shed light upon the dogma of the existence of God as well as upon the dogma expressing monotheism, Gödel's notes reveal that he was considering other dogmas as well. One of the few remarks in his notebook "*Theologie I*", not directly concerning an attended lecture, addresses, amongst others, the dogmas of God's omnipotence and mercy, as well as the dogma of free will, already promulgated[46] by the council of Trent in 1547.

> 1.) Scheinbarer Widerspruch zwischen: Barmherzigkeit Gottes, Möglichkeit, den Willen zu kehren, Existenz der Hölle und der Bösen, tatsächliche Umkehr durch Gott bei den Guten.
> 2.) Scheinbarer Widerspruch zwischen: Gerechtigkeit Gottes, willkürlicher Gnadenwahl.

[46]Canon VI of the decree on justification reads: "If any one saith, that it is not in man's power to make his ways evil, but that the works that are evil God worketh as well as those that are good, not permissively only, but properly, and of Himself, in such wise that the treason of Judas is no less His own proper work than the vocation of Paul; let him be anathema."(Waterworth 1848, 45).

This passage is immediately followed by the attempt to eliminate the considered contradictions. In order to present Gödel's train of thought, his underlining is retained and cancelled passages are given in square brackets.

Mögliche Auflösung:

1. Gott hat keine Möglichkeit, den Willen zu kehren [ohne die Freiheit zu nehmen, und das will er nicht]. Widerspruch gegen die Allmacht. Und daher auch keine Umkehr bei den Guten durch Gott. [Aber erklärt durch: Er hat auf einen Teil seiner Allmacht ein für alle Male verzichtet. Darin besteht das Geschenk der Freiheit. D.h. Umkehr des Willens ist nur dadurch möglich, dass er die Freiheit entzieht, und das will er nicht.]

[2. Gott hat die Möglichkeit, den Willen zu kehren, tut es aber nur bei denen]

2. Gott hat die Möglichkeit, ihn zu kehren, tut es nicht, weil er sonst die Freiheit entziehen müsste. Daraus folgt: Er tut es nie. Also auch nicht bei den Guten.
3. Gott hat die Möglichkeit, den Willen zu kehren ohne Aufhebung der Freiheit, tut es aber nur bei denen, welche seiner Barmherzigkeit schon vorher würdig sind (also besser als die anderen).

Later writings show that Gödel tended towards the dissolution suggested in item (2.), strongly connecting the phenomenon of free will with the concept of 'possibility'. In *MaxPhil XIV*[47] (77–78) he writes:

> Der erste Gegensatz ⟨...⟩ drückt sich in folgenden [nur verschiedenen Seiten desselben Ausdrückenden] Begriffspaaren aus: Schöpfer–Geschöpf; Ursache–Folge; *Actus–Passio*; *Actus–Possibilitas*.[48] Das letzte Gegensatzpaar hat für unsere Erkenntnis nur deswegen Sinn, weil wir durch unseren freien Willen am Schöpfertum Gottes teilnehmen. [Gott schuf den Menschen sich zum Bild.] Und zwar sind wir nicht nur durch unser Handeln, sondern auch in unserer Erkenntnis aktiv. Was uns gegeben ist, sind die "Bausteine" der Welt (d.h. die Begriffe), und wir bauen die Welt daraus selbst auf. Für ein gänzlich passives Wesen würde die Erkenntnis einfach darin bestehen, dass ein Bild der wirklichen Welt in ihm wäre, und für Möglichkeit wäre kein Platz, oder doch nur in der Form der wirklichen Gedanken Gottes. (Ein solches Wesen könnte für den Ursachenbegriff Verständnis haben, ohne ihn in sich selbst realisiert zu haben.) D.h. also, es hätte kein wirkliches Verständnis für den Begriff des Möglichen (*casus irrealis*).

It is remarkable that the topic of free will already plays a central role in two notebooks on the foundations of quantum mechanics[49] written in 1935, just a short time before Gödel started to concentrate on theological considerations. Whereas these books contain a lot a detailed technical material, they also concentrate on philosophical aspects like Leibnizian monadology as well as vitalistic thoughts. The remarks on free will can be roughly divided into three categories, summarized as follows.

[47] Kurt Gödel Papers, Box 6b, Folder 72, item accession 030099.

[48] Gödel's footnote: "Wahrheit–Falschheit; zeitliche und nicht-zeitliche Existenz?; Wirklichkeit und Begriff; Ding–sich verhalten."

[49] Kurt Gödel Papers, Box 6b, Folder 78, item accessions 030106 and 030107. The contents of these books is described in (Lethen 2020a) and (Lethen 2020b), concentrating on monadology and vitalism, repectively.

- There is neither a causal nor a statistical description of free will. The relation between causality and free will may be considered as the archetype of complementarity.
- Free will and memory constitute the fifth spatio-temporal dimension. The relation between free will and the future corresponds to the relation between memory and the past: Whereas a Monad is able to shift it's viewpoint into the past along a path of viewpoints connected through a 'memory' relation, it is also able to choose between possible future viewpoints within a system of branching time. If one accepts that there is no branching into the past, the main difference between the two directions is a choice in depth on the one hand, and a choice in breadth on the other. Note the close connection to the above *MaxPhil* remark.
- Gödel is seeking for a general frame for a physical theory which, emerging after the theory of quantum mechanics, will include the notion of free will.

9.9 Conclusion

Although Gödel was baptized lutherian, and although he was able to give detailed reasons why he did not want to become catholic, he obviously developed a strong tendency towards a predominantly catholic theology as early as 1937 when he decided to attend theological lectures. His extensive theological notes written in 1938 and shortly thereafter clearly reveal a keen interest in the phenomenon of the catholic dogma, incorporating logical methods and stressing the analogy to theoretical physics. It is remarkable that it was also the subject of physics which, only a few years earlier, brought Gödel into contact with Leibnizian monadology, the philosophy of vitalism, and, last but not least, with the concepts of soul and free will. An analysis of these phenomena obviously led him to his theological worldview which he adopted when still in Vienna and about which further investigations of his *Nachlass* hopefully will be able to devolope a detailed picture.

9.10 Appendix: Literature on Dogmatic Theology

On the very first page of Gödel's notebook "*Theologie I*", one can read the following remark: "Probl.: Nach welchen Büchern studieren die Theologiestudenten?" Amongst other theological notes in his *Nachlass*, we have now discovered an extensive list of theological literatur,[50] problably first compiled in 1938 and later continued in Princeton, consisting of about 100 items. Retaining Gödel's numbering, we identify those books on the list which deal directly with dogmatic theology.

[50] Kurt Gödel Papers, Box 10b, Folder 47, item accession 050149.

6. Zobl, Johann. 1865. *Dogmengeschichte der katholischen Kirche*. Innsbruck: Wagner.
9. In this item, Gödel writes:

   Dogmatiken (kurze) nach:
   kurze: *Egger, Franc; Minges, Parth.; Pesch, Chr.; Hurter, H.; Diekamp, Fr.*
   lange: *Janssens, Lauz.* ($\overline{34}$); Jos. Bernhart ($\overline{34}$)

   The corresponding books are:

   (a) Egger, Franz. 1909. *Absolute oder relative Wahrheit der heiligen Schrift? Dogmatisch-kritische Untersuchung einer neuen Theorie*. Brixen: Weger.
   (b) Minges, Parthenius. 1908. *Das Verhältnis zwischen Glauben und Wissen, Theologie und Philosophie nach Duns Scotus (Forschungen zur christlichen Literatur- und Dogmengeschichte)*. Paderborn: Schöningh.
   (c) Pesch, Christian. 1908. *Glaube, Dogmen und geschichtliche Tatsachen. Eine Untersuchung über den Modernismus*. Freiburg i. Br.: Herder.
   (d) Hurter, Hugo. 1903. *Theologiae dogmaticae compendium in usum studiorum theologiae*. Innsbruck: Wagneriana.
   Or possibly:
   Hurter, Hugo. 1893. *Medulla theologiae dogmaticae*. Innsbruck: Wagneriana.
   (e) Diekamp, Franz. 1930. *Katholische Dogmatik nach den Grundsätzen des heiligen Thomas*. Münster: Uaschendorff.
   (f) Janssens, Laurentius. 1904. *Summa theologia ad modum commentarii in Aquinatis summam praesentis aevi studiis aptatam*. Freiburg i. Br.: Herder.
   (g) Bernhart, Joseph. 1930. *Der Vatikan als Thron der Welt*. München: Paul List.
   Or possibly:
   Bernhard, Joseph. 1935. *De profundis*. Leibzig: Hegner.

16. Scheeben, Mathias Joseph. 1873. *Handbuch der katholischen Dogmatik* (5 Bände). Freiburg i. Br.: Herder.
21. Braun, Joseph. 1926. *Handlexikon der katholischen Dogmatik*. Freiburg i. Br.: Herder.
72. This item reads:

   Revue on apologetics
Revue of faith and dogma } französisch (Father Bolger)

   These texts could not be identified.
73. Tanquerey,[51] Adolphe. 1959. *A Manual of Dogmatic Theology*. New York: Desclee.

   Here Gödel adds the comment "Notre-Dame empfohlen".

[51] Gödel's spelling is "Tenquery".

Acknowledgements The research for this article is a part of the GODELIANA project led by Jan von Plato in Helsinki, Finland. This project has received funding from the European Research Council (ERC) under the European Union's Horizon 2020 research and innovation programme (grant agreement No 787758) and from the Academy of Finland (Decision No. 318066). I would like to thank Jan von Plato, Maria Hämeen-Anttila, and Annika Kanckos for their encouraging support and many fruitful discussions about Gödel's work and his writings. I would also like to thank Eva-Maria Engelen who provided useful comments on my Gabelsberger transcriptions.

References

Adžić, Miloš, and Kosta Došen. 2017. Logic Lectures: Gödel's Basic Logic Course at Notre Dame. *arXiv preprint arXiv:1705.02601.*

Anderson, C. Anthony, and Michael Gettings. 1996. Gödel's ontological proof revisited. In *Gödel'96: Logical foundations of mathematics, computer science and physics—Kurt Gödel's legacy, Brno, Czech Republic, August 1996, proceedings*, 167–172. Association for Symbolic Logic.

Crocco, Gabriella, and Eva-Maria Engelen. 2016. *Kurt Gödel Philosopher Scientist.* Aix en Provence: Presses de L'Université de Provence.

Dawson, John W. 1996. *Logical Dilemmas: The Life and Work of Kurt Gödel.* Natik: A. K. Peters.

Gierer, Alfred. 1997. Gödel meets Carnap: A prototypical discourse on science and religion. *Zygon* 32(2): 207–217.

Gödel, Kurt. 1932. Zum intuitionistischen Aussagenkalkül. *Anzeiger der Akademie der Wissenschaften in Wien* 69: 65–66.

Gödel, Kurt. 1933. Eine Interpretation des intuitionistischen Aussagenkalküls. *Ergebnisse eines mathematischen Kolloquiums* 4: 39–40.

Gödel, Kurt. 1944. Russell's mathematical logic. In *The Philosophy of Bertrand Russell*, ed. Paul Arthur Schilpp, 123–153. Evanston and Chicago: Northwestern University.

Gödel, Kurt. 1951. *Some basic theorems of the foundations of mathematics and their implications.* In *Kurt Gödel, Collected Works. Vol. III. Unpublished Essays and Lectures*, eds. Warren Goldfarb, Charles Parsons, Solomon Fererman, et al., 304–323. New York: Oxford University Press.

Kanckos, Annika, and Tim Lethen. 2019. The Development of Gödel's Ontological Proof. *The Review of Symbolic Logic*, published online, 1–19.

Kanckos, Annika, and Tim Lethen. 2022. Kurt Gödel's reception of Charles Hartshorne's ontological proof. This volume.

Kaulen, Franz. 1868. *Geschichte der Vulgata.* Mainz: Verlag Franz Kirchheim

Lethen, Tim. 2020a. Monads, Types, and Branching Time – Kurt Gödel's approach towards a theory of the soul. In *Wider den Reduktionismus – Ausgewählte Beiträge zum Kurt Gödel Preis 2019*, eds. Oliver Passon and Christoph Benzmüller, 13–24. Berlin: Springer Spektrum.

Lethen, Tim. 2020b. Kurt Gödel's Anticipation of the Turing Machine: A Vitalistic Approach. *History and Philosophy of Logik* 41(3): 252–264.

Łukasiewicz, Jan. 1922–23. Interpretacja liczbowa teorii zdań. *Ruch Filozoficzny*: 92–93.

Manning, Henry Edward. 1871. *The Vatican Council and Its Definitions: A Pastoral Letter to the Clergy.* New York: D. & J. Sadlier.

McKinsey, John. C., and Alfred Tarski. 1948. Some theorems about the sentential calculi of Lewis and Heyting. *The Journal of Symbolic Logic* 13(1): 1–15.

Ott, Ludwig. 1957. *Fundamentals of Catholic Dogma.* Cork: Mercier.

Pius XII. 1950. Encyclical Humani Generis of the Holy Father Pius XII to our venerable brethren, patriarchs, primates, archbishops, bishiops, and other local ordinaries enjoying peace

and communion with the holy see concerning some false opinions threatening to undermine the foundations of catholic doctrine. http://w2.vatican.va/content/pius-xii/en/encyclicals/documents/hf_p-xii_enc_12081950_humani-generis.html
von Plato, Jan. 2013. *Elements of Logical Reasoning*. Cambridge: Cambridge University Press.
Simpson, Alex K. 1994. *The Proof Theory and Semantics of Intuitionistic Modal Logic*. Thesis, University of Edinburgh.
Sobel, Jordan Howard. 2003. *Logic and Theism: Arguments for and against Beliefs in God*. Cambridge: Cambridge University Press.
Sørensen, Morton Heine, and Pawel Urzyczyn. 2006. *Lectures on the Curry-Howard Isomorphism*. Amsterdam: Elsevier.
Spade, Paul Vincent, and Claude Panaccio. 2019. William of Ockham. In *The Stanford Encyclopedia of Philosophy* (Spring 2019 Edition), ed. Edward N. Zalta. https://plato.stanford.edu/archives/spr2019/entries/ockham/
Suttner, Ernst Christoph. 1984. *Die Katholisch–Theologische Fakultät der Universität Wien 1884–1984. Festschrift zum 600-Jahr Jubiläum*. Berlin: Duncker & Humblot.
Wang, Hao. 1987. *Reflections on Kurt Gödel*. Cambridge (Mass.): MIT Press.
Waterworth, James. 1848. *The canons and decrees of the sacred and oecumenical Council of Trent*. Chicago: Christian Symbolic Publication Society.

Chapter 10
Kurt Gödel's Reception of Charles Hartshorne's Ontological Proof

Annika Kanckos and Tim Lethen

10.1 Introduction

Anselm of Canterbury (1033–1109) is believed to be the first philosopher to have put forward an ontological proof or argument for the necessary existence of God. As will be outlined in the following section, two variants of this proof can be found both in his little book *Proslogion*, written in 1077/78, as well as in an answer to the monk Gaunilo who criticises the argument as not being valid. In 1962, Charles Hartshorne (1897–2000) published a formal modal logic proof (cf. Hartshorne 1962, 51) formalizing Anselm's second variant of the argument. Whereas it has been speculated[1] whether Kurt Gödel—when constructing his own ontological proof—was influenced by Hartshorne, he is reported to label the proof as "wrong" (cf. Wang 1987, 146 and Section 3), and no direct sign of such an influence could be traced until very recently.

Amongst theological papers in Gödel's *Nachlass*, we have now discovered (and transcribed from the Gabelsberger shorthand) a single sheet of paper which does not only show that he knew Hartshorne's proof very well, but also presents a remarkable simplification of the proof. The notes are accompanied by remarks closely related to Anselm's two variants of the Argument as presented in the *Proslogion*, which reappear in Gödel's 1970-version of the proof (Gödel 1970) in a kind of type-lifted manner.

[1] See for example (Adams 1995, 390–391).

A. Kanckos · T. Lethen (✉)
Department of Philosophy, History and Art Studies, University of Helsinki, Helsinki, Finland
e-mail: annika.kanckos@helsinki.fi; tim.lethen@helsinki.fi

E. Ramharter (ed.), *The Vienna Circle and Religion*, Vienna Circle Institute Yearbook 25, https://doi.org/10.1007/978-3-030-76151-6_10

In order to complete the picture, we show a close link between Anselm's and Gödel's conceptions of *positiveness* and *God*, thus underlining that Gödel's ontological proof may well have been influenced by other sources than Leibniz.

10.2 Anselm's Variants of the Ontological Argument

Before we turn to Hartshorne's proof and Gödel's notes on it, it should be well worth sketching Anselm's two variants of "the Argument,"[2] which are discussed in great detail in (Hartshorne 1962) and (Hartshorne 1965). Also, we will have a first glance at possible formalizations of certain elements of the Argument, which will serve as a basis for further discussions.

In his *Proslogion*, chapter II, Anselm defines God as "that than which nothing greater can be conceived," which can be taken as a main ingredient and a starting point for both of his variants. His main line of argumentation then runs as follows: As the "fool" is able to understand this very definition, God exists in his understanding. As we shall see later on, Hartshorne formalizes this circumstance as Mq (i.e. $\sim N \sim q$ in his terms), q representing the proposition "There is a perfect being, or perfection exists" (Hartshorne 1962, 50). But then, it is not possible that God exists in the understanding alone because it is greater to exist in the understanding *and* in reality at the same time. Although Hartshorne himself never formalized this part of the Argument,[3] it may well be captured as $\sim M(Mq \wedge \sim q)$ or equivalently as

$$N(Mq \supset q) \tag{10.1}$$

In chapter III, Anselm then changes his way of argumentation. Here the central claim is that it is impossible that God exists and at the same time can be thought not to exist, which might be captured as $\sim M(q \wedge M \sim q)$, which in turn is equivalent to

$$N(q \supset Nq) \tag{10.2}$$

While Hartshorne repeatedly insisted that this part of the argumentation in chapter III is by far stronger in a sense explained by him for example in (Hartshorne 1965, 11–12 and 89), Sobel (2003, 85) renders the two given formulas (10.1) and (10.2) as equivalent, stating that the equivalence "holds unconditionally." In actual fact, the

[2]When Charles Hartshorne refers to Anselm's ontological proof, he often simply uses the expressions "the Proof" or "the Argument" (cf. Hartshorne 1962, 30). Throughout this article, we will stick to the latter expression, while the term "proof" will always refer to the formal versions.

[3]The modal logic formalization of the Argument as given in *Proslogion* II can be found in (Sobel 2003, 84).

truth seems to lie somewhere in the middle, as it can be proved that the two modal principles are indeed equivalent in the modal logic **KB**. Without the axiom **B** neither of the two formulas is a logical consequence of the other. Details will be given in Sect. 10.5, where Gödel's notes on Hartshorne's proof are discussed.

10.3 Charles Hartshorne and the Ontological Proof

In 1971/1972, Hao Wang conducted extensive interviews with Gödel which later formed the basis for his book (Wang 1996). In these conversations, Gödel described Hartshorne as "an example of a contemporary metaphysician." (Wang 1996, 138) This quote is stated without any deeper context than the fact that "Gödel was in favour of metaphysics and opposed to positivism," which seems to indicate that Gödel had a generally favourable outlook on Hartshorne in 1971. Supporting this, but also bringing up a critisizm of Hartshorne's work, is a later quote from 1972: When asked to name some recent philosophers whom he found congenial, Gödel replied by naming William Henry Sheldon, Josiah Royce and Charles Hartshorne, also describing Hartshorne as in the tradition of the scholastics and that he had to "add what has been done by the followers of Leibniz." (Wang 1996, 141)

However, Gödel also expressed in 1972 more direct critizism of what he called "some great philosophers of twenty years ago," including Hartshorne, by attributing the lack of successors to the phenomenon that these philosophers "generalize things without any inhibition." (Wang 1996, 145) This may also be a specific reference to Hartshorne's ontological proof as using too general logical principles as its foundation.

At a later date (Nov 1972) Gödel again expresses strong criticism of Hartshorne and his ontological proof. Gödel then states that Hartshorne's lack of knowledge in mathematical logic leads to a negative effect resulting in an ontological argument which is wrong.

For easier reference we repeat Hartshorne's formal proof along with his original annotations in Fig. 10.1, as given in (Hartshorne 1962, 51). The two premises, based on Anselm's argument in chapter III of the *Proslogion*, can be found in lines 1 and 7, respectively. It should be mentioned that, although q abbreviates the first order formula $\exists x.P(x)$, a perfect being exists, the steps of the proof are purely propositional.[4] Also note that $\sigma \rightarrow \tau$ abbreviates strict implication or $N\sim(\sigma \mathbin{\&} \sim\tau)$ in Hartshorne's notation.

Underneath his proof, Hartshorne adds: "Those who challange the Argument should decide which of these 10 items or inferential steps to question. Of course one may reject one or more of the assumptions (1, 3, 7); but reject is one thing, refute or show to be a mere sophistry is another." Note that he does not include step 2, the law of the excluded middle, in his list of possible rejections.

[4]Harthorne himself uses the notation $(\exists x)P(x)$.

1. $q \rightarrow Nq$	"Anselm's principle"; perfection could not exist contingently
2. $Nq \vee \sim Nq$	Excluded Middle
3. $\sim Nq \rightarrow N \sim Nq$	Form of Becker's Postulate: modal status is always necessary
4. $Nq \vee N \sim Nq$	Inference from (2, 3)
5. $N \sim Nq \rightarrow N \sim q$	Inference from (1): the necessary falsity of the consequent implies that of the antecedent (Modal form of modus tollens)
6. $Nq \vee N \sim q$	Inference from (4, 5)
7. $\sim N \sim q$	Intuitive postulate (or conclusion from other theistic arguments): perfection is not impossible
8. Nq	Inference from (6, 7)
9. $Nq \rightarrow q$	Modal axiom
10. q	Inference from (8, 9)

Fig. 10.1 Charles Hartshorne's proof

10.4 Simplifications of the Proof

In (Schanno 2017), Alisha Schanno presents a simplified version of the above proof, constructed by a theorem prover consisting of no more than 10 lines of PROLOG code. As non-logical axioms, Mq (ip) and $N(q \rightarrow Nq)$ (ap) are used, which correspond to Hartshorne's lines 7 and 1, respectively. The logical axioms include the modal axioms $MNa \rightarrow Na$ (5), $Na \rightarrow a$ (t) as well as $N(a \rightarrow b) \rightarrow (Ma \rightarrow Mb)$ (km). The automatically constructed proof of q is then presented as

$$\texttt{mp(t, mp(5, mp(mp(km, ap), ip)))}$$

where mp indicates an application of *modus ponens* or, more correctly, of C. Meredith's *rule D* of *condensed detachment* which is implemented within the PROLOG system as *Robinson's unification algorithm*. It should be noted that the shortness of this proof does not merely depend on notational conventions. Even if we expand every single condensed detachment into two seperate steps (i.e. substitution and an application of modus ponens), the proof would only be half as long as Hartshorne's proof written in the same style.

Apart from being much shorter than Hartshorne's version, this proof shows two significant advantages.

- While Hartshorne heavily relies on the law of the excluded middle, Schanno's version is completely intuitionistic, even from a modal perspective. Here, the use of the rule km appears to be problematic at first sight, as it can only be deduced intuitionistically if the diamond operator M is used as an abbreviation for $\neg N \neg$. (Note that Hartshorne completely avoids the use of the operator M.) If both N and M are taken as primitive operators, the above rule km has to be added to the

Table 10.1 A tabular version of Anselm's Argument from the perspective of the "fool's understanding," represented by the column "some possible world"

	Actual world	Some poss. world	All poss. worlds	Remark
1.	Mq			`(ip)`
2.	$N(q \supset Nq)$			`(ap)`
3.		q		(1.)
4.	$q \supset Nq$	$q \supset Nq$	$q \supset Nq$	(2.)
5.		Nq		MP (3./4.)
6.	q	q	q	(5.)

underlying logic as an axiom, as was done by Schanno in her proof system. For further details see (Simpson 1994, chapter 3).

- It follows Eder and Ramharter's criteria (Eder and Ramharter 2015) for the formal reconstruction of Anselm's ontological argument very closely. Here, the third criterion runs as follows: "The structure of the formal reconstruction should represent the fundamental structure of the argument. It should be no more and no less detailed than is necessary to map the argument."

In order to clarify the second item we reproduce Table 10.1 which is taken from (Sobel 2003), using slightly different notation.

In this table, the column "some possible world" represents what Anselm calls the fool's "understanding," and his way of logical reasoning is presented in steps 3.-6. from exactly this (i.e. the fool's) point of view. Shifting the perspective into the actual world, Schanno's proof now simulates steps 3.-5. using axiom `km`, while the table's very last step corresponds to her application of the modal axiom **5**.

The use of the logic **S5**[5] in Gödel's and Hartshorne's proofs has often been criticized as too strong and Gödel himself is reported to have had "reservations about his ontological proof because of his doubt about using some principle in modal logic." (Adams 1995, 391, fn. *g*) Whereas modern theorem provers can show that the logic **KB** is sufficient for the proof (see Benzmüller and Woltzenlogel Paleo, 2014 and Benzmüller and Fuenmayor, 2018), they do not give much insight *why* this is so. The very last line in the above table now clearly reveals the underlying reason: If the necessity of q is established in some possible world, q holds in every possible world, including the actual one, $MNq \rightarrow Nq$.[6] But this fact actually carries far too much information for our special case if we are trying to establish God's existence in the *actual* world only. For this purpose it suffices to infer q from the fact that the necessity of q has been confirmed in some possible world, i.e. $MNq \rightarrow q$ (b), thus reducing the underlying logic to **KB** and the corresponding proof (using Schanno's notation) to

`mp(b, mp(mp(km, ap), ip)).`

[5]Or related theories involving axiom **5**.

[6]To make sure that the actual world is included, the axiom **T** expressing reflexivity is also needed.

One further reduction of the underlying logic can be based on Charles Hartshorne's own thoughts as presented in (Hartshorne 1962, 53). Here, he takes God's necessary existence as independent of any contingent facts: "[I]f God logically could be necessary He must be, since no contingent condition can be relevant." Due to God's nature, the proposition $MNq \supset Nq$ could therefore be taken as an additional axiom, thus avoiding the modal axioms **T**, **B** and **5** altogether.[7] From a philosophical point of view, this idea may be supported by the question why an underlying (theological) model should necessarily be symmetric or even euclidian as a whole, if only the specific property of being perfect (or God-like) is actually touched by these structural restrictions. As will be seen in Sect. 10.6, Gödel himself would probably have rejected the inclusion of this additional axiom which is purely based on the nature of God.

10.5 Gödel's Notes

Gödel's notes on Hartshorne's proof, which have been now discovered in his *Nachlass*,[8] consist of two pages written in logical notation and Gabelsberger shorthand. They are given in Figs. 10.2 and 10.3.[9]

The notes begin with a simplification of Hartshorne's proof, at the heart of which lies a version of Becker's postulate: modal state is necessary, $NNp \vee \sim MNp$, either Np holds everywhere or nowhere.[10] Gödel then deduces the modal axiom **5**, before he takes exactly those steps presented in (Schanno 2017). It is interesting to note that he had already taken these proof steps some ten years earlier, in a version of the ontological proof which has only very recently been discovered in his *Nachlass* and which has been presented in (Kanckos and Lethen 2019).

Besides the simplified proof, the notes reveal another interesting fact about Hartshorne's proof, given by Gödel in his note marked (°). Here he stresses the circumstance that God's non-existence must not be regarded as possible, because the assumption $M \sim p$ would lead to God's necessary non-existence. The underlying reason is given in the note marked ($^{\times}$). Here Gödel points out that Hartshorne's axiom $N(p \supset Np)$ implies the formula $N(\sim p \supset N \sim p)$, from which in

[7]Apparently, Hartshorne did not notice this simplification. This may be due to the fact that his proof uses an equivalent of the modal axiom **5**, instantiated by $\sim Nq \rightarrow N \sim Nq$ and annotated as "Form of Becker's Postulate: modal status is always necessary."

[8]Kurt Gödel Papers, Box 10b, Folder 49, item accessions 050156, on deposit with the Manuscripts Division, Department of Rare Books and Special Collections, Princeton University Library. Used with permission of Institute for Advanced Study.

[9]To improve readability, the overall layout has slightly been altered. Longhand is presented in *italics*.

[10]Note that Gödel uses the abbreviation p instead of Hartshorne's q.

$p \supset Np$ ^× (1)
Mp °
Np (also p)

Allg. richtiger *Mod. Schluss*

Bew.:
$NNp \vee \sim MNp$,
also $MNp \supset Np$.
Aber aus (1): $Mp \supset MNp$,
also $Mp \supset Np$

Hartshorne über das *ontol. Arg.*
Logic of Perfect., $\overline{62}$, *p.* 50–51
(Was er in erster Linie zeigt, ist, dass die Existenz Gottes keine *factual matter* ist.)

° Man muss annehmen, dass man das für die Existenz Gottes einsehen kann, aber nicht für die Nicht-Existenz (für seinen Beweis).

× Folgt aus dem Wesen Gottes, also gilt es auch für die *Negat.*, aber daraus folgt nicht $\sim p \supset N(\sim p)$. (Z.B. nicht, wenn p wahr und zufällig ist.)
Wohl aber, wenn gilt: $N(p \supset Np)$
Dann gilt auch: $N(\sim p \supset N(\sim p))$

Bew.: Es gilt beides, wenn keine mögliche Welt existiert, in der p (oder keine, in der $\sim p$). Wenn dagegen beide Arten möglicher Welten existieren, dann sind beide falsch.

Fig. 10.2 Page 1 of Gödel's notes. The reference is Hartshorne (1962)

Möglichkeit Gottes → Existenz Gottes (*Hartsh.* ^×)

sehr schlecht kritisiert von *John O. Nelson*
in *Rev. of Methaph. 17,* $\overline{63}$*, p. 235–42*

Antwort darauf: *Rev. of Methaph. 17,* $\overline{63}$*, p. 608*

× Er zeigt nämlich, dass die Existenz Gottes keine "*factual matter*" ist.

Fig. 10.3 Page 2 of Gödel's notes. The references are Nelson (1963) and Hartshorne (1964), respectively. (Note that Hartshorne's reply was published in 1964)

turn—in analogy to Hartshorne's proof—God's necessary non-existence can easily be deduced if $M \sim p$ is assumed.

Finally, Gödel's semantic proof at the bottom of the page establishes the logical equivalence of the two formulas in question. Whereas his argumentation seems to assume a universal accessibility relation between possible worlds, the following proof uses the logic **KB** only, thus showing that an underlying symmetric relation is sufficient. For brevity, we show only one direction of the equivalence, the opposite direction follows exactly the same pattern.

1.	$N(p \supset Np)$	assumption
2.	$Mp \supset MNp$	$N(\sigma \supset \tau) \supset (M\sigma \supset M\tau)$
3.	$\sim MNp \supset \sim Mp$	modus tollens
4.	$NM \sim p \supset N \sim p$	duality of M and N
5.	$\sim p \supset N \sim p$	modal axiom **B**: $\sigma \supset NM\sigma$
6.	$N(\sim p \supset N \sim p)$	necessitation

Note that the formula $N(\sim p \supset N \sim p)$ is actually equivalent to $N(Mp \supset p)$ which may be taken as a formalization of Anselm's Argument as given in *Proslogion* II, see Sect. 10.2.

10.6 *Der schlechte Weg* and Gödel's Type-lifting

In general, Gödel rejects both Anselm's original form of the Argument as well as Hartshorne's modal formalization as can be concluded from a note in his notebook *MaxPhil XIV* (107), published in (Gödel 1995, Appendix B). Here he states:[11]

> Wenn man annimmt $\varphi(x) \supset N\varphi(x)$ [weil aus dem Wesen von x folgend], dann ist es leicht beweisbar, dass es für jedes kompatible System von Eigenschaften ein Ding gibt, aber das ist der schlechte Weg. Vielmehr soll $\varphi(x) \supset N\varphi(x)$ erst aus der Existenz Gottes folgen.
>
> If $\varphi(x) \supset N\varphi(x)$ is assumed [as following from the nature of x], then it is easily provable that for every compatible system of properties there is a thing, but that is the bad way. Rather, $\varphi(x) \supset N\varphi(x)$ should only follow from the existence of God.

Instantiating φ with Gödel's predicate G (being God-like[12]) and arguing in the line of Anselm and Hartshorne ("weil aus dem Wesen von x folgend"), *the bad way* (*der schlechte Weg*) now proposes that anything that is God-like is so necessarily. The following "easy" (informal) proof starts with this very assumption and immediately leads to Gödel's conditional claim, where the antecedent expresses the compatibility of the system's (i.e. God's) properties.

(i)	$G(x) \supset NG(x)$	assumption
(ii)	$\forall x.[G(x) \supset NG(x)]$	generalization
(iii)	$\exists x.G(x) \supset \exists x.NG(x)$	$\forall x.(\sigma \supset \tau) \supset (\exists x.\sigma \supset \exists x.\tau)$
(iv)	$\exists x.G(x) \supset N\exists x.G(x)$	$\exists x.N\sigma \supset N\exists x.\sigma$
(v)	$M\exists x.G(x) \supset MN\exists x.G(x)$	$N(\sigma \supset \tau) \supset (M\sigma \supset M\tau)$
(vi)	$M\exists x.G(x) \supset \exists x.G(x)$	modal axiom **B**

[11] Transcription and translation by the second author.

[12] Note that "being God-like" is expressed by Hartshorne by the first-order predicate P, while Gödel uses the second-order predicate P for positiveness, see below.

In (Sobel 2006, 402), Jordan Howard Sobel takes Gödel's above quote as "strong evidence that Gödel thought his principles entailed [the modal] collapse,[13] and was not bothered by this, that he indeed considered it a welcome feature of the Leibnizian metaphysics he was devoting for himself." However, Sobel misinterprets Gödel's comment based on a misleading translation. The *Collected Works* (Gödel, 1995, Appendix B) translate "weil aus dem *Wesen* von x folgend" as "as following from the *essence* of x", and Sobel identifies the meaning of the word "essence" with the Leibnizian concept of the thing's "complete properties" and Gödel's formal definition of the notion of *essence*, thus missing the quote's intimate connection to Anselm's Argument.

While clearly rejecting Anselm's Argument on the object level, Gödel does adduce a kind of type-lifting of Anselm's argumentation from the object level onto the level of properties. Whereas part of Anselm's Argument may be formalized as $N(q \supset Nq)$, or simply as $q \supset Nq$ (see Sect. 10.2), in Gödel's proof, the axiom

$$P(\varphi) \supset NP(\varphi) \tag{10.3}$$

holds, because—following Gödel's own annotation[14]—"it follows from the nature of the property." If we transfer this statement into Anselm's language, a positive property *is so truly positive, that it cannot even be conceived not to be positive*, which parallels the Argument given in *Proslogion* III, where Anselm writes in his opening sentence: "And it assuredly exists so truly, that it cannot be conceived not to exist."

It is worth mentioning that Gödel's 1970 proof also contains a second variant of the axiom (10.3), namely $\sim P(\varphi) \supset N \sim P(\varphi)$. This variant, not being necessary for the proof, did not find its way into Dana Scott's version of Gödel's proof,[15] and thus appears to be by far less known. Its equivalent

$$MP(\varphi) \supset P(\varphi) \tag{10.4}$$

again may be regarded as a type-lifted version of the implication (10.1) taken from *Proslogion* II, *if it is possible that God exists, he exists in reality.* As with Anselm's two ways of argumentation, Gödel's two variants (10.3) and (10.4) can quite easily be shown to be equivalent in the logic **KB**. In the presence of his axiom $P(\varphi) \equiv$

---

[13] The expression "modal collapse" refers to the situation in which one can no longer distinguish between the modalities *possible*, *actual*, and *necessary*. For details in connection with Gödel's ontological proof, see (Sobel 2003, 132) and (Benzmüller and Woltzenlogel Paleo 2017).

[14] See his 1970 proof (Gödel 1970). Note the similarity to his "as following from the nature of x."

[15] In 1970, Gödel communicated his proof to Dana Scott. Scott's version has been published in (Sobel 1987) and in (Sobel 2003). A copy of this version, written in Dana Scott's hand, can also be found in Georg Kreisel's *Nachlass*, preserved in the Stanford University Archives (Georg Kreisel Papers, Box 45, Folder 5).

$\sim P(\sim \varphi)$, the logic **K** is already sufficient.[16] Returning to the aforementioned simplification in Sect. 10.3, it would be interesting to know if a type-lifted version of Hartshorne's axiom $MNq \supset Nq$, i.e. $MNP(\varphi) \supset NP(\varphi)$, could have been included into Gödel's concept of positiveness, "because it follows from the nature of the property."

While it is certainly hard to judge to which extent Gödel was actually influenced by Hartshorne's proof in this regard, it should be worth mentioning that the axiom $P(\varphi) \supset NP(\varphi)$ appeared in a different shape in exactly those earlier versions of the ontological proof which Gödel compiled before Harshorne's proof was published in 1962 (cf. Kanckos and Lethen 2019). In philosophical notes[17] written in 1954 or shortly after he states: "Daß die Notwendigkeit einer pos. Eigenschaft pos. ist, ist die wesentliche Voraussetzung für den ontol. Bew." [That the necessity of a positive property is positive, is the essential presupposition for the ontological proof.] which is reflected by the axiom $P(\varphi) \supset P(N\varphi)$. And in a footnote relating to the interpretation of positive properties as perfections he mentions the alternative "Oder: wenn $M\varphi$ eine Perfective, dann auch φ." [Or: if $M\varphi$ is a perfective, then φ is, too.] (Gödel 1995, 434), which in turn may be written as $P(M\varphi) \supset P(\varphi)$. In both cases, neither of the two variants is a logical consequence of the other.

10.7 Essence and Existence

While necessary existence already played a central role even in Gödel's earliest versions of the ontological proof, the notion of essence only enters the stage in an undated third version, first published in (Gödel 1995, 430) and interpreted as a complete axiomatization in (Kanckos and Lethen 2019). Here the symbol Ess_x plays the role of both a (unique) first order predicate (*the* essence of the object x) as well as a second order predicate qualifiying a first order predicate as an essence of x. While in the first case Ess seems to represent a function from objects to predicates, in his 1970-version of the proof ess finally had become a relation between objects and predicates, thus giving up the (formal) uniqueness of the concept of an essence.

In 1965 Charles Hartshorne expresses a very clear conception of the interplay between the notions of essence and existence, and the resemblence with Gödel's 1970-version, where he interweaves the two concepts, is striking. In (Hartshorne 1965), Hartshorne writes:

> An important point of the analysis is the inadequacy of the dichotomy *essence-existence*. A third term is needed, 'actuality'. An essence exists if there is some concrete reality exemplifying it; 'existence' is only *that* an essence is concretized, 'actuality' is *how*, or in what particular form, it is concretized. The particular form, the actuality, is always

[16] If the two variants are regarded isolated from the axiom $P(\varphi) \equiv \sim P(\sim \varphi)$, a simple countermodel can show that the logic **K** is not sufficient.

[17] *MaxPhil XIV*, 107, published in (Gödel 1995, 434).

> contingent [...] but it does not follow [...] that the existence is contingent. For existence only requires the nonemptiness of the appropriate class of actualities, and a class can be necessarily nonempty even though it has only contingent members. [...]
>
> Essence, existence, actuality—this trias is the minimum of complexity which must be considered if the famous Proof is to be correctly evaluated.

If existence requires the nonemptiness of the class of actualities, clearly necessary existence requires this very class to have every possible actuality as its member. Thus, it is only a small step that leads from necessary existence of an essence to Gödel's necessary existence of an object via its own essence.

It might be speculated whether Hartshorne's existence of an essence (independent of any instantiating object) lead to the much discussed[18] omitting of the conjunct $\varphi(x)$ in Gödel's definition of essence in 1970, which does appear in his preceding version and was later re-added by Dana Scott in (Scott 1970).

10.8 Anselm's Conception of God

As has been pointed out in (Kanckos and Lethen 2019), Gödel's interest in Anselm and his ontological argument is noted as early as 1925, attending a lecture given by Heinrich Gomperz. When Anselm's conception of God is discussed in the literature, the discussion is almost certainly based on the famous phrase "aliquid quo nihil maius cogitari possit" (*Proslogion*, chapter II). In connection with Gödel's treatment of the Argument it should be noted though, that the *Proslogion* also involves a very subtle conception based on grades of positivness. At the end of chapter XI, Anselm writes:[19]

> Sic ergo vere es sensibilis, omnipotens, misericors et impassibilis, quemadmodum vivens, sapiens, bonus, beatus, aeternus, et quidquid melius est esse quam non esse.
>
> So, then, thou art truly sensible, omnipotent, compassionate, and passionless, as thou art living, wise, good, blessed, eternal: and whatever it is better to be than not to be.

This is perfectly reflected by Gödel in his philosophical notes (*MaxPhil XIV*, 103, published in Gödel, 1995, 432) as "Der ontologische Beweis muß auf den Begriff des Wertes (p besser als $\sim p$) gegründet werden." [The ontological proof has to be grounded on the concept of value (p better than $\sim p$).] which is followed by several axioms concerning the "value" of properties.

Having captured the notion of positiveness in this way, Anselm goes on to describe the concept of God based on positiveness in chapter XXIV:

> Si enim singula bona delectabilia sunt, cogita intente quam delectabile sit illud bonum, quod continet iucunditatem omnium bonorum;

---

[18]See for instance (Benzmüller and Woltzenlogel Paleo 2016) or (Benzmüller and Fuenmayor 2018).

[19]Similar passages appear in chapter V as well as in the *responsio* to the monk Gaunilo.

> For, if individual goods are delectable, conceive in earnestness how delectable is that good which contains the pleasantness of all goods;

And in the following chapter he continues:

> Ama unum bonum, in quo sunt omnia bona, et sufficit. Desidera simplex bonum, quod est omne bonum, et satis est.
>
> Love the one good in which are all goods, and it sufficeth. Desire the simple good which is every good, and it is enough.

These passages reveal a clear correspondence to Gödel's definition of God, $G(x) \equiv (\varphi)\,[P(\varphi) \supset \varphi(x)]$ which appears in his proofs as early as 1941 (see (Gödel 1995, 429) and (Kanckos and Lethen 2019) for a corrected version). They even mirror the impredicativity explicitely surfacing with Dana Scott's axiom $P(G)$, being God-like is itself positive. This very impredicativity, which is based on Anselm's phrase "that good which contains the pleasantness of all goods," as well as the resulting antinomy are further investigated in (Lethen 2021).

The importance of accessing the notion of 'God' (*Gottesbegriff*) through his properties is further supported by a set of notes[20] mentioning for instance Descartes' ontological proof in a series of loose papers entitled "Max. u.ä." (Maximen und ähnliche) and dated "ca. 1941". Here Gödel writes:

31. Richtiger Beginn nach *Thomas Aqu.* = *First Truth* = Wahrheit über Gott, also:
    a.) Die einzelnen Sätze des Glaubenbekenntnisses in ihrem Sinn klären und Folgerungen ziehen (insbesondere der erste Satz).
    b.) Von den Eigenschaften Gottes ausgehen, insbesondere *ontolog.* Gottesbeweis (*Descartes*).
    c.) Insbesondere Abhandlungen lesen, die sich mit dem Gottesbegriff befassen (*August. & Hilarius, De Trinitate, Summa (De Deo u. Christologie), De divinis nominibus*).

31. Right beginning following Thomas Aqu. = First Truth = truth about God, i.e.:
    a.) To clarify the meaning of the sentences of the Creed and to draw conclusions (in particular the first sentence).
    b.) To start with God's properties, especially ontological proof of God's existence (Descartes).
    c.) In particular, to read treatises dealing with the notion of God (August. & Hilarius, De Trinitate, Summa (De Deo and Christologie), De divinis nominibus).

10.9 Conclusion

In his introduction to Gödel's ontological proof (Adams 1995), R.M. Adams writes:

> Among the historic sponsors of the ontological argument, it is not to Anselm or Descartes but to Leibniz that the parentage of Gödel's proof belongs, as scholars interested in the proof have long recognized.

---

[20]Kurt Gödel Papers, Box 6b, Folder 65, item accessions 030088, inserted into the notebook *Zeiteinteilung (Max.) II*.

Whereas it has already been pointed out in (Kanckos and Lethen 2019) that Gödel was well aware of Anselm's argument, the material now found in Gödel's *Nachlass* clearly underlines the whole breadth of the spectrum of possible influences on his own versions of the proof, including Anselm of Canterbury, Descartes, and Hartshorne. Nevertheless, it will at the same time be well worth considering the vast material on Leibniz' writings that Gödel left in his *Nachlass*.

Acknowledgements The research for this article is a part of the GODELIANA project led by Jan von Plato, to whom we remain grateful for his support. This project has received funding from the European Research Council (ERC) under the European Union's Horizon 2020 research and innovation programme (grant agreement No. 787758) and from the Academy of Finland (Decision No. 318066).

References

Adams, Robert Merrihew. 1995. Introductory note to *1970. In *Kurt Gödel, Collected Works. Vol. III. Unpublished Essays and Lectures*, ed. Warren Goldfarb, Charles Parsons, Solomon Feferman, et al., 388–402. New York: Oxford University Press.

Benzmüller, Christoph, and Bruno Woltzenlogel Paleo. 2014. Automating Gödel's ontological proof of God's existence with higher-order automated theorem provers. In *ECAI 2014, volume 263 of Frontiers in Artificial Intelligence and Applications*, ed. Torsten Schaub, Gerard Friedrich, and Barry O'Sullivan, 93–98. Amsterdam: IOS Press.

Benzmüller, Christoph, and Bruno Woltzenlogel Paleo. 2016. The inconsistency in Gödel's ontological argument: A success story for AI in metaphysics. In *Proceedings of the Twenty-Fifth International Conference on Artificial Intelligence*, ed. Subbarao Kambhampati, 936–942. Palo Alto, California: AAAI Press.

Benzmüller, Christoph, and Bruno Woltzenlogel Paleo. 2017. The ontological modal collapse as a collapse of the square of opposition. In *The Square of Opposition: A Cornerstone of Thought*, ed. Jean-Yves Béziau and Gianfranco Basti, 307–313. Basel: Birkhäuser.

Benzmüller, Christoph, and David Fuenmayor. 2018. Can Computers Help to Sharpen our Understanding of Ontological Arguments? In *Mathematics and Reality, Proceedings of the 11th All India Students' Conference on Science & Spiritual Quest, 6–7 October, 2018, IIT Bhubaneswar, Bhubaneswar, India*, ed. Sudipto Gosh, Ramgopal Uppalari, K. Vasudeva Rao, Varun Agarwal, Sushant Sharma, 195–226. Kolkata: The Bhaktivedanta Institute.

Eder, Günther, and Esther Ramharter. 2015. Formal reconstructions of St. Anselm's ontological argument. *Synthese* 192: 2795–2825.

Gödel, Kurt. 1970. Ontological Proof. In *Kurt Gödel, Collected Works. Vol. III. Unpublished Essays and Lectures*, ed. Warren Goldfarb, Charles Parsons, Solomon Feferman, et al., 403–404. New York: Oxford University Press.

Goldfarb, Warren, Charles Parsons, Solomon Feferman, et al. 1995. *Kurt Gödel, Collected Works. Vol. III. Unpublished Essays and Lectures*. New York: Oxford University Press.

Hartshorne, Charles. 1962. *The Logic of Perfection*. LaSalle, Il.: Open Court Publishing Company.

Hartshorne, Charles. 1964. What the Ontological Proof Does Not Do. *The Review of Metaphysics* 17(4): 608–609.

Hartshorne, Charles. 1965. *Anselm's Discovery – A Re-Examination of the Ontological Proof for God's Existence*. LaSalle, Il.: Open Court Publishing Company.

Kanckos, Annika, and Tim Lethen. 2019. The Development of Gödel's Ontological Proof. *The Review of Symbolic Logic*, 1–19.

Lethen, Tim. 2021. Kurt Gödel on logical, theological, and physical antinomies. *Bulletin of Symbolic Logic* 27(3): 267–297.
Nelson, John. O. 1963. Modal Logic and the Ontological Proof for God's Existence. *The Review of Metaphysics* 17(2): 235–242.
Scott, Dana. 1970. Notes in Dana Scott's hand. In *Logic and Theism: Arguments for and against Beliefs in God*, Jordan Howard Sobel. Cambridge: Cambridge University Press.
Schanno, Alisha. 2017. Untersuchungen zum Hartshorne'schen Gottesbeweis. Presented at the Science Fair *Jugend forscht 2017*.
Simpson, Alex K. 1994. *The Proof Theory and Semantics of Intuitionistic Modal Logic*. Thesis, University of Edinburgh.
Sobel, Jordan Howard. 1987. Gödel's ontological proof. In *On Being and Saying: Essays for Richard Cartwright*, ed. J. J. Thompson, 241–261. Cambridge, Mass.: The MIT Press.
Sobel, Jordan Howard. 2003. *Logic and Theism: Arguments for and against Beliefs in God*. Cambridge: Cambridge University Press.
Sobel, Jordan Howard. 2006. On Gödel's ontological proof. In *Modality matters: Twenty-five essays in honour of Krister Segerberg*, ed. Lagerund, Henrik, Sten Lindström, and Rysiek Śliwiński, 397–421. Uppsala: Uppsala universitet.
Wang, Hao. 1987. *Reflections on Kurt Gödel*. Cambridge, Mass.: MIT Press.
Wang, Hao. 1996. *A Logical Journey: From Gödel to Philosophy*. Cambridge, Mass.: MIT Press.

Part II
General Part

Chapter 11
The Vienna Circle in China: The Story of Tscha Hung

Yi Jiang

Preliminary Remarks

At the 13th International Congress of Logic, Methodology and Philosophy of Science of the IUHPS/DLMPS in Beijing, August 9–15, 2007, I had the honor and pleasure to organize an affiliated symposium on "East meets West in Philosophy of Science: Moritz Schlick and Hong Qian (Tscha Hung)". This small symposium took place covering the following papers: Friedrich Stadler, "Moritz Schlick and Hong Qian: Personal and Philosophical Encounters", Hans Joachim Dahms, "Remarks on Hong Qian's Life and Work", Christian Damböck, "Hong Qian, Schlick, Vienna and Causality", Richard Dawid, "Schlick, Hong Qian and the Causality Debate in Quantum Physics", Tobias Fox, "A Transcendental Consideration of Hong Qian's Causality", and Robert S. Cohen (Boston), "Personal Reminiscences of Hong Quian".

Already in 1992 Bob Cohen had co-organized the first Beijing International Conference on Philosophy of Science together with Chinese colleagues and departments. Its main papers were published in 1996 as *Realism and Anti-Realism in the Philosophy of Science* (Cohen et al. 1996). This book contains Bob Cohen's "Recollections of Tscha Hung" and Fan Dainian's "Hong Qian (Tscha Hung) and the Vienna Circle".

Two years later, in 1994, Bob Cohen again had initiated an "International Conference on the Vienna Circle and Contemporary Science and Philosophy: In Memory of Tscha Hung" in Beijing under the auspices of the IUHPS/DLMPS together with Chinese institutions and colleagues, which was not documented with proceedings, regrettably.

Y. Jiang (✉)
School of Philosophy and Sociology, Shanxi University, Taiyuan, China
e-mail: yijiang@sxu.edu.cn

E. Ramharter (ed.), *The Vienna Circle and Religion*, Vienna Circle Institute Yearbook 25, https://doi.org/10.1007/978-3-030-76151-6_11

In 2003 Bob Cohen (1923–2017), a good personal friend of mine and a long-term supporter of the Institute Vienna Circle, now Vienna Circle Society, donated his "Robert S. Cohen Collection and Archives" to the Institute Vienna Circle, which he had opened himself: https://www.univie.ac.at/vcs/biblio/rsca.htm#collection

Afterwards he proposed the abovementioned meeting in Beijing, also because he knew Tscha Hung personally and appreciated his life and work. In this sense, he had donated his valuable personal scientific library to the philosophy department of Tsinghua University, which hosted the congress in 2007. From the beginning we had planned the publication of this Beijing meeting as proceedings in honor of Tscha Hung = Hong Qian (Susieng, 1908 – Beijing, 1992) in the "Boston Studies in the Philosophy of Science". At the same time, we decided to include a missing contribution on Tscha Hung from a genuine Chinese point of view dealing with his comprehensive philosophical writings in Chinese language. In 1980 I had also the pleasure to meet Prof. Tscha Hung in person at the 5th Wittgenstein Symposium in Kirchberg/W. on ethics (Tscha Hung 1981), followed by an impressive exchange in Vienna on the occasion of the Neurath-Schlick-Centenary Symposium in 1982 (Tscha Hung 1982). Due to many circumstances and difficulties, the projected proceedings did never realize and Bob Cohen had passed away in 2017. I had met him for the last time in Boston in 2010. And the remaining speakers had used their papers for different occasions or were not prepared to deliver them for other reasons, regrettably. Only on the occasion of a meeting with Professor Yi Jiang (then from Beijing Normal University) in 2016 at the Wittgenstein Symposium in Kirchberg/W. I succeeded in getting this intended paper for our projected publication.

So, I decided to publish the promised article of Prof. Yi Jiang in one of our Vienna Circle Yearbooks, because he was willing to write it exclusively on my invitation, in remembrance of Tscha Hung as a completion of the former publication project. Now we are really grateful to Prof. Yi Jiang for his contribution and I am satisfied that the old project could be finished *pars pro toto* in this way. Thanks go to the author for his willingness to contribute this article, and to Dr. Helmut Spindler for valuable comments on this paper.

The subsequent article is published in commemoration of both Tscha Hung and Bob Cohen, close friends in the tradition of the Vienna Circle – the former a student of Moritz Schlick (1882–1936) at the University of Vienna, the latter a most important scholar and mediator of the Vienna Circle and Logical Empiricism in the English-speaking world after 1945. Special thanks go to Michael D. Cohen, son of Bob Cohen and personal representative for the estate of his father, for the permission to publish the subsequent photograph of Bob Cohen and Tscha Hung, ca. 1985.

Friedrich Stadler (Vienna Circle Society, Director – Institute Vienna Circle, Permanent Fellow, University of Vienna)

References

Tscha Hung (1981). "Wittgenstein und Schlick", in: *Ethik. Grundlagen, Probleme und Anwendungen*. Akten des 5. Internationalen Wittgenstein-Symposiums 1980. Hrsg. von Edgar Morscher und Rudolf Stranzinger. Wien: Hölder-Pichler-Tempsky, 437–442.

Tscha Hung (1982), Moritz Schlick und der Logische Empirismus", in: *Schlick und Neurath – Ein Symposium*. Hrsg. von Rudolf Haller. Amsterdam: Rodopi, 175–205.

Robert S Cohen, Risto Hilpinen, Qiu Renzong (Eds. 1996). *Realism and Anti-Realism in the Philosophy of Science. Beijing International Conference*. Dordrecht-Boston-London: Kluwer.

Photo: Bob Cohen and Tscha Hung, Beijing ca. 1985
Acknowledgement: Michael D. Cohen, MD, Saratoga Springs, NY.

Yi Jiang Abstract

2019 signaled the 110th anniversary of Tscha Hung's birth and the 90th year of the publication of the Vienna Circle's manifesto – *The Scientific World-conception: the Vienna Circle*. The two events are linked through the Vienna Circle's connection with Tscha Hung who studied with Moritz Schlick during the 1920s and 1930s. Tscha Hung is today known as a member of the Circle who participated in its meetings on Thursday evenings. He is important in Chinese philosophy not only as a representative figure of the Vienna Circle but also as a symbol of the arrival of scientific reason in China and as a witness of the encounter of Western and Chinese philosophy in China. It may be of interest to the philosophical community at large to know what form that encounter took after Tscha Hung's return from Vienna in 1937. In this paper, I will try to illuminate this encounter by telling the story of Tscha Hung's struggle with traditional Chinese philosophy as it is exemplified by his debate with Fung You-lan on metaphysics. I will also explore how Tscha Hung's philosophy led to a revaluation of some of the assumptions of the Vienna Circle. Finally, I will describe his relation to Chinese philosophy and metaphysics both in its historical and its philosophical dimension.

11.1 Tscha Hung and His Contemporaries in China in the 1930s and 1940s

Tscha Hung (Hong Qian, 洪谦 in Chinese) is today considered to have introduced the Vienna Circle to China. But there were, in fact, two others who had done so before him. They were the cousin philosophers, Zhang Shen-fu and Zhang Dai-nian. Zhang Dai-nian published a paper on the Vienna Circle's physicalism in the journal, *Da Lu* (*The Continental* in English) on March 1, 1933 when he was only 24. In 1934, his cousin Zhang Shen-fu, in turn, published a paper on the main currents in modern philosophy in which he expressed his appreciation of the philosophy of Vienna Circle. The two cousins had come to their understanding of the philosophy of the Circle by reading some its publications drawing on their own rich background in the natural sciences and their methodology. According to the historical documents, Zhang Dai-nian's ideas on the Circle were inspired by Zhang Shen-fu who was at the forefront of elucidating Chinese philosophy for the first generation of modern Chinese philosophers (Jiang and Bai 2010). Zhang Shen-fu's philosophy has been noted in China for its combination of dialectic materialism with logical analysis. However, his greatest contribution to modern Chinese philosophy was that he introduced Bertrand Russell's philosophy to China. Zhang had studied mathematics and philosophy at Peking University in his early years and so could well understand Russell's philosophy of mathematics and logic, including problems in the foundation of mathematics, and the relation of philosophy to mathematics. Because of his genius and his excellent training in mathematics, he could also fully appreciate the philosophy of the Vienna Circle in which he found something vivid, hopeful, promising and worth pursuing (Zhang Shen-fu 1985, 68). Zhang generalized the philosophy of the Circle, suggesting that the goal of analysis was to clarify both ideas and language in order to determine the reality of the external world (Zhang Shen-fu 1985, 66). Still, most of his work in philosophy concerned the study of Russell's philosophy rather than that that of the Circle. The reason of this change was probably the return of Tscha Hung from Vienna.

11.1.1 Tscha Hung's Philosophical Background

Tscha Hung wrote his PhD dissertation on the problem of causation in modern physics under the supervision of Moritz Schlick and received his Doctoral degree in 1934. For the following two years he went on to do research at the University of Vienna until Schlick was assassinated by an insane student in the summer of 1936. Tscha Hung returned to Peking at the beginning of 1937 and afterwards began to lecture mainly on the philosophy of the Vienna Circle at Tsinghua University. Zhang Shen-fu had left Tsinghua by then for political reasons. Zhang Dai-nian had also quit his position at Tsinghua and was concentrating on writing his outline of Chinese philosophy in the Peking University library. After the July 7 incident of 1937, Tscha Hung moved to Yunnan, a southwest province of China, with other faculties of Tsinghua University in order to evade the war with Japan and he then taught the philosophy of the Vienna Circle at the National Southwest Associated University (NSAU). There is a legend about his teaching at NSAU. Hung taught in the Department of Foreign Languages rather than in Department of Philosophy, for some of the professors of philosophy disliked the Vienna Circle. So, Hung regularly taught German in the Department of Foreign Languages and the philosophy of the Vienna Circle only occasionally in Department of Philosophy. After the Second World War Hung was invited to Oxford and went to do research at New College until 1947. He was then appointed professor and head of the Department of Philosophy at Wuhan University. From 1951 he served as professor and head of the Department of Philosophy at Yenching University, formerly Peking University, and in 1956 he became the director of the section of the history of foreign philosophy in the Department of Philosophy of Peking University. From 1965 he was also the director of the Institute of Foreign Philosophy, an independent research institution not affiliated with the Philosophy Department, till he resigned from that position in 1987. He remained, however, a professor at Peking University until his death in 1992.

While Zhang Shen-fu spent much of his life bringing Russell's philosophy to China, Tscha Hung made the study of the Vienna Circle his life's career. Some scholars have compared his introduction to the philosophy of the Circle to China with Ayer's introduction of the Circle to Britain, and they have concluded that Hung's introduction is more comprehensive and more accurate while Ayer's account is idiosyncratic (Fan et al. 1992). That comparison appears to be fair but it is, in fact, based only on Hung's work in the 1940s. Since the 1970s, Hung has advanced a variety of criticisms of some of the ideas of the Vienna Circle, expressing disagreement with Schlick and Carnap, in particular, on the problem of the foundations of knowledge and truth, and proposing an anti-foundationalism of his own. These criticisms have created a different image of Hung in contemporary Chinese philosophy, as being not only an advocate of the philosophy of the Vienna Circle (no matter whether his advocacy was accurate or comprehensive) but being also an independent thinker who reflected deeply on the problems the Circle posed. And it is in fact because he developed the philosophy of the Circle further in his own way that he has received global recognition. He was praised, for this reason, to have made a significant new contribution to the philosophy of the Vienna Circle and to have developed deep insights into logical positivism as well as into Wittgenstein's philosophy (Fan et al. 1992).

Tscha Hung had an extraordinary feeling for his tutor, Moritz Schlick, and the Vienna Circle, comparable with Schlick's feeling for Ludwig Wittgenstein and Zhang Shen-fu's for Bertrand Russell. He respected and admired Schlick personally. He treated Schlick, indeed, as his idol and insisted strongly on the philosophy of the Vienna Circle even when criticism and suspicions were advanced against it; and he did so by clarifying some ideas of the Circle (Hung 1990, 43–45, 69, 257). Though his ambition was to spread the philosophy of the Vienna Circle in China without any reservation, his spirited search for truth has influenced Chinese scholars for several generations. Ren Ji-yu, Hung's colleague in NSAU and himself famous in Chinese philosophy and a former director of the National Library of China, called Hung the first to introduce the philosophy of the Vienna Circle to China (Ren 2006). Wang Zi-song, remembering his own studies at NSAU, wrote that Tscha Hung was the only professor who had studied Western philosophy abroad at that time (Wang Zi-song 2007). Du Xiao-zhen, Hung's colleague and professor of Peking University, recalled Hung's support for her when she was translating Jean-Paul Sartre's writings into Chinese: "Hung was concerned with logic and verification, but he was also full of humanity in his heart. His persistent and proud humanism has influenced me for a long time, even after his death" (Du 2005). Jin Xi-ping, also Hung's colleague and professor of Peking University, praised Hung's silence in the Cultural Revolution which allowed him to keep his academic position (Jin 2000). Wang Wei, a student of Hung at Peking University, explained the climate at the Institute of Foreign Philosophy by saying that the senior professors, including Hung and Xiong Wei (who had studied with Martin Heidegger in the 1930s), were true scholars in the Chinese tradition, men like Confucius who teach but don't necessarily write. They recognized a commitment to the idea that it is difficult to express what can be expressed clearly and that we should keep silent on what cannot be expressed. They shied away from trying to express the ineffable. Only by keeping silent, they were convinced, can we show the meaning of language (Wang Wei 2006, 289). Taken together, these comments reflect Hung's attachment to the Socratic motto that I know I know nothing.

Tscha Hung published, in fact, only 31 essays in his entire life, ten of them in the 1940s, six in 1950–1979, and 15 after 1980. The essays published in the

1940s are collected now in his book on the philosophy of the Vienna Circle and the essays published after 1949 in his book on logical empiricism. For political reasons, Hung spent most of his time between 1949 and 1979 translating the works of Western philosophers. This undertaking is today recognized widely in the Chinese philosophical community as one of the great contributions of Hung to contemporary Chinese philosophy. His translation, compilation, and selectin of masterpieces of past philosophy, provided rare materials for academic research in China during that period. The two collections of masterpieces most respected by Chinese academics were *Selection of Original Writings of Western Classic Philosophy*, and *Selection of Original Writings of Western Modern Bourgeois Philosophy*. The former was published in five volumes between 1957 and 2013, including *Ancient Greek and Rome Philosophy* (1957), *Philosophies in the Western European Countries in the 16th–18th Centuries* (1958), *German Philosophy at the End of the 18th Century and the Beginning of the 19th Century* (1960), *French Philosophy in the 18th Century* (1963). The final volume on *Medieval Philosophy* was published only in 2013 having been translated and compiled by Hung's students and fellows twenty-one years after Hung's death. The *Selection of Original Writings of Western Modern Bourgeois Philosophy* was first published in 1964 and then expanded twice in two volumes re-entitled *Selection of Original Writings of Western Modern Philosophy* in 1993. Hung did the proofreading of the new edition shortly before his death. Both works have greatly advanced first-hand understanding of Western philosophy in China. They have set a standard for Chinese translation in being faithful, clear, and elegant. Hung upheld, in fact, very strict requirements for his translations. Even for a translation by a well-known scholar he asked a reviewer to undertake the proofreading, and sometimes even two reviewers. Meanwhile, he also encouraged young scholars to translate works under his careful guidance. He admired his collaborators and colleagues for their contributions but rarely mentioned his own work in them.

He fang-yu, a historian of modern Chinese history of ideas, has spoken of the unity of learning and academics in Hung's work, saying that what made him so highly worthy of respect in China was not only that he introduced the philosophy of the Vienna Circle as well as Western philosophy to China but that, in doing so, he expressed the true mentality of a Chinese philosopher in his much admired sense of mission and his willingness to dedicate himself to the times in which he lived (He 2007). Thus, Hung expressed his concern with academics in China even during the second Sino-Japanese War in the 1930–40s. He wrote, "at the start of the establishment of a new country and the final victory over the Japanese, I commit myself to doing academic research in order to grow our culture and our future. We must carry on our mission very carefully and deliberately so as not to disappoint our mission and the hope of generations." (Hung 1944) After 1949 he also gave expression to the need to study the history of Western philosophy. In 1957, he published a paper in the *People's Daily,* the top national official newspaper, in which he called for more attention to the study of history. This paper, however, gave him a political trouble in the political campaign prior to the Cultural Revolution. He was criticized politically because of his words were outdated in the official eyes and he was almost sent to labor camp in the countryside. Owing to some members of the Vienna Circle concerned with Hung's situation during the Cultural Revolution, Zhou En-lai became involved in protecting some of the senior professors at Peking University from political persecution, Hung was allowed to stay on the campus, instead, and permitted to continue translating but not to teach (Fan et al. 1992, 70). Hung did not complain of his unfair personal situation and, instead, during the Cultural Revolution devoted his time to the translation and compilation of writings of Western philosophers, which he considered to be the only thing worth doing academically and made him stay far away from political life. He was always seen as being aloof from political or material pursuits and self-contained in personality, but his behavior appears now as strong-hearted and full of feelings concerning the nation's destiny.

11.1.2 Fung You-lan and his Relation to the Vienna Circle

Fung You-lan is one of the philosophers we must mention in the relation to Tscha Hung. Their 1940s debate on the topic of metaphysics is one of the historically significant academic debates in modern China. It reflects the difference between a traditional Chinese conception of metaphysics and one derived from the Western tradition, especially from logical empiricism. Before turning to the content of this debate, I want to describe Fung's relation to the Vienna Circle and to logical empiricism.

There is no evidence that Fung had any personal contact with members of the Vienna Circle, except for Hung. We don't know whether he read Schlick's or Carnap's writings. What he learnt about the philosophy of the Vienna Circle may have come from Zhang's cousin first and from Tscha Hung later. Fung had studied for his PhD in the United States in the 1920s and had traveled in Europe in the 1930s and lectured in the States again in the 1940s. The experience would have made him familiar with the Vienna Circle. He may also have met some members of the Circle at the 8th World Congress of Philosophy in Prague in 1934. That Congress provided the first occasion for the Circle to present its philosophical ideas at an international event, in that the Circle proposed a panel organized by Frank on the frontiers of the philosophy of the natural sciences in which many members of the Circle made their presentations, including Schlick, Carnap, Lukasiewicz, Reichenbach, Nagel and Morris (Stadler 2001, 358–359). We don't know whether Fung attended that panel but he would have known of the philosophy of the Circle by then. Fung was invited once again to deliver lectures on Chinese philosophy in the United States in 1946, at a time when some members of the Circle were living there and had changed their previous formulations of the Circle's philosophy under criticism of American philosophers such as Quine. Fung may not have had contact with the members of the Vienna Circle in the US. But he had had a chance to meet Wittgenstein in Cambridge in 1933 when he traveled in Europe. The story is that he took tea with Wittgenstein one afternoon. They did not talk about philosophy but about cultural traditions in general. As we know, Wittgenstein had shifted his philosophy away from the *Tractatus Logico-Philoosphicus* by the time they met but that Fung also had no idea at all of Wittgenstein's new philosophy. Fung knew Wittgenstein only through his link to the Vienna Circle. Even though Fung had no personal contact with the Circle, his criticism of logical empiricism has been very influential in the Chinese philosophical community.

The public debate between Tscha Hung and Fung You-lan took place initially on November 11th 1944. Hung, as has been said, had returned to China in 1937. He had taught the philosophy of the Circle first at Tsinghua and had then moved to Yunnan to teach German in department of foreign languages and the philosophy of the Circle in department of philosophy. As Zhang Dai-nian recalled, Hung's ideas were much different from those of his colleagues in the department at Tsinghua including Fung You-lan and Jin Yue-lin. He went against any attempt to argue for ontology and disagreed with Fung You-lan's new metaphysics and Jin Yue-lin's realism in his writing *On Dao*. Hung took thus his distance from the department in academic respects (Zhang Dai-nian, 1996, 539). According to He Lin, Western philosophy had begun to settle in the Chinese philosophical community since the 1920s when the Chinese Philosophical Society was established in 1925 and the Society's Journal of *Philosophical Reviews* was first issued in 1927 (He Lin 2002, 25). The Society also organized annual meetings in which scholars in Western philosophy published their new research papers. It is worth noting that the study of Western philosophy involved not only the introduction of Western doctrines but their exploration in a distinctively Chinese way.

The debate between Tscha Hung and Fung You-lan took place at the annual meeting in Yunnan in 1944. The year before, Fung You-lan had published a paper entitled "New Metaphysics' Place in the History of Philosophy and Its Method" in vol. 8 of *Philosophical Reviews*, in which he had criticized the Vienna Circle and had argued that its rejection of metaphysics did not affect his own new metaphysics. Hung's critique of Fung aimed at precisely this point. He argued that Fung did not correctly understand the Circle's rejection of metaphysics and that his new metaphysics was also not immune to the rejection of metaphysics. According to a record of this debate made by He Lin, Fung immediately responded to Hung at the meeting. Jin Yue-lin and Shen You-ding, the famous logician, also argued for Fung and against Hung at that moment (Hu Jun 2002, 144–145). But there is no historical record of what Fung said to Hung and how Jin and Shen sought to justify Fung. Hu Jun, a professor at Peking University, comments that the debate derived, in fact, from a misunderstanding between Hung and Fung (Hu Jun 2002, 196–206). In my judgment there was, however, no such misunderstanding. Their disagreement derived rather from different understandings of the nature of philosophy and from the different methods the two used to deal with philosophical problems. I will discuss these differences in the next section of the paper.

11.1.3 *Jin Yue-lin as Tscha Hung's Contemporaries*

Among Tscha Hung's contemporaries were some other Chinese philosophers who concerned themselves with logical positivism. Professor Jin Yue-lin, mentioned above, is the most prominent among them. Jin studied political science at Pennsylvania State University and at Columbia University in 1914–1921 and he worked in Britain, Germany, and France for some years before returning to China in 1925 at the invitation of Tsinghua University. He, then, became the first professor of logic at Tsinghua and he was thus the first to systematically introduce modern logic and the frontiers of logical studies to China. We can reasonably compare his work as a logician with the contribution that Carnap has made to modern logic. Both declared modern logic to be central to modern philosophy. But Jin paid more attention to Bertrand Russell than to the Vienna Circle in his career. There are two reasons for this. Jin Yue-lin preferred Russell's philosophy, first of all, to logical positivism. The logical analysis Russell proposed was, for Jin, meant to achieve an understanding of the world, while the doctrine of the Vienna Circle concentrated more on the logic of language. In addition, Jin himself had strong ambitions in metaphysics which he combined with his research in logic and knowledge. Modern logic was for him just a more general form of logic which reveals the nature of things as far as their logical form is concerned. Jin Yue-lin interpreted his logic thus metaphysically (Wang Zhong-jiang 1993). He had for this reason little interest in logical positivism.

11.2 The Philosophical Debate Between Tscha Hung and Fung You-lan

Having described the historical context of the debate, I now want to look at the philosophical difference between its two protagonists. That difference, I have tried to show, was rooted in their different understandings of the nature of philosophy and the philosophical method they used. Let's look in details at the difference respectively in terms of Hung's criticism of Fung's understanding.

11.2.1 *On the Nature of Philosophy*

Tscha Hung insisted on the Vienna Circle's understanding of the nature of philosophy as a clarification of the meaning of our propositions. According to that understanding, the mission of philosophy was to guarantee that our formulation of human knowledge was clear. The theory of knowledge was, of course, the key to philosophy in order to know the external world exactly. Philosophy was thus to be understood as a practical way of thinking that could help people to learn more about the world and to act well in a moral sense. Tscha Hung viewed the theory of knowledge, not ethics, to be the center of his philosophy, even though Schlick had focused much on ethics in his philosophy. Hung had a good background in the natural sciences, which made him look at the world and deal with philosophical problems in a scientific way. He had studied physics, mathematics and logic under the guidance of Schlick in Vienna. By contrast, Fung You-lan, though he had studied philosophy in the United States and recognized the importance of logical analysis in philosophy, did not consider philosophy to be a theory of knowledge of the natural sciences. He argued, instead, for the tenet of Chinese philosophy that philosophy is a search for the meaning of life in which the nature and ground of human life are considered. He viewed the mission of philosophy to be a working for the highest ideal of being human and thought that we could get an ontological understanding of what we know only from the knowledge that we know nothing. It was from this traditional Chinese perspective that Fung criticized the Vienna Circle's rejection of metaphysics. It was obvious, however, that his understanding of metaphysics was not the same as that in the Western tradition. According to Carnap, traditional metaphysics is a study of what is beyond our experience and concerns the so-called nature of entities (Carnap 1996, 132). For Fung, on the other hand, metaphysics meant searching ontologically for a realm of imagination, a dream-like kingdom. He wrote: "The function of metaphysics is only the sublimation of the human mind without any particular knowledge of the world. It helps man raise his soul to the highest realm" (Fung 1986, vol.5, 167).

According to the Western tradition, there is no conflict between metaphysics and natural science in general; the two are understood to be complementary (Ribeiro 2015). But for Fung they are opposed in their methods. Fung argued that metaphysics is a study of the form of experience in a negative way, while the natural sciences deal with the content of experience in a positive fashion. In speaking of the form of experience Fung meant to refer to the emptiness of experience; this meant that metaphysical propositions cannot make any substantial claims about empirical matters. Metaphysical propositions are commensurable with any fact in the empirical world since they are empty of factual content. Fung thus proposed that approaches to metaphysics could be divided into two, the positive and the negative. "The positive approach is the studying of metaphysics by logical analysis, and the negative one is about what cannot be said. What cannot be said is a metaphysical approach to philosophy." (Fung 1986, vol.5, 173) The negative approach as he called it, is a searching for the realm of emptiness and the positive approach is a means for the searching. Fung's understanding of the nature of philosophy is obviously distinct from that entertained by Hung and the Vienna Circle. For them logical analysis is not a means for philosophy but the essential part of philosophy and research in the logical form of knowledge is not a means for philosophy either but the essence of all knowledge. As Schlick said, "every cognition is an expression or a representation. That is, it expresses a fact which is cognized in it. This can happen in any number of ways, in any language, by means of any arbitrary system of signs. All these possible modes of representation – if they otherwise actually express the same knowledge – must have something in common; and what is common to them is their logical form." (Schlick 1979, 156) Fung, however, had a radically different understanding of the nature of philosophy and logical analysis as thus upheld by Hung and the Vienna Circle.

11.2.2 On the Method of Philosophy

Tscha Hung agreed with the Vienna Circle that logical analysis aims at a clarification of the meaning of propositions, by which all the differences between various doctrines can be detected. He also rejected Kant's synthetic a priori judgements as a logical illusion. If a proposition is related to reality, he said, it must be justified by experience. If a proposition cannot be justified this way, it cannot be a proposition about empirical facts. Since a synthetic a priori judgement is supposedly not empirical, it cannot be justified (or falsified) by experience. The conclusion is that there cannot be such a proposition. Hung also rejected Fung's attempt to construct metaphysics by means of logical analysis. He argued that Fung's metaphysics was based on a category mistake which confused different concepts in the theory of knowledge. Thus, Fung's understanding of metaphysics was untenable (Hung 1989, 233, 236).

Fung thought, however, that he had justly criticized the logical positivism just by revealing the mistake in the Vienna Circle's rejection of metaphysics by means of logical analysis. He borrowed intuitions from common sense to analyze metaphysical propositions into two kinds and sought to reconcile the negative approach to metaphysics with our common judgements. But he failed to argue adequately for his new metaphysics in formulating his propositions. His four arguments for his new metaphysics, for example, are based on his understanding of commonsense intuition. The first is that everything is what it is. It is the law of identity in logic that A equals A. Fung explained this law by means of the proposition that what makes a thing to be itself is the reason that there must be something that makes a thing to be itself. It is evident that this proposition is about what causes a thing to be itself. And this, in turn, is said on the basis of the commonsense intuition that everything has a cause that makes it to be. He did not, however, consider the debates about this concept of causation in the history of philosophy. He simply proposed the proposition on the basis of commonsense. Fung's second argument was that everything must be there. He explained this by the propositions that existing things are existing and existing things exist based on what makes them exist. This explanation of existence by the existent is once again in accord with our commonsense. The third argument is the following inference. The primary thing is that existence is a movement and existence is about something. Thus, the conclusion is that existence of something is a movement of something or cause of existence. This reason of all the things is reduceable to the one. He claimed the one to be the Dao or Origin as the first cause of all movements. All these ideas came seemingly from Daoism which held that thing is derived from nothing and nothing is identical with Dao which is origin of everything in the world. Fung's fourth argument was the proposition that the existence of all things is called wholeness or completeness. It asserted thus the identity of wholeness with the existence of all things. But wholeness, if there is such a thing, is not the existence of all things according to Western metaphysics; it is, rather, the being beyond all things and the determination which determines all things. Fung's understanding of wholeness is, however, in accord with the common understanding of the concept in ordinary life. Thus, Fung's explanation of metaphysics is derived from our commonsense understanding of these terms, but covered in a philosophical clothing that is taken from the West. However, even from the point of view of a commonsense understanding, Fung's terms are actually not genuine concepts; they are but chaotic and fail to discriminate. This means, for Hung, that Fung's understanding of metaphysics is, in fact, unacceptable as a clarification of concepts and has nothing to do with our ordinary experience either.

11.2.3 On the Relation of Metaphysics and Knowledge

In his 1943 paper. Fung divided human knowledge into four kinds: 1, mathematics and logic; 2, metaphysics; 3, science; 4, history. But it is very odd that metaphysics, is according to his division, part of human knowledge. For Fung, metaphysics in the sense of traditional philosophy has been abolished by the Vienna Circle, but his own new metaphysics is not abolished as a part of knowledge; it is strengthened, rather, by the illumination of its very nature. Hung strongly disagreed with Fung's division of knowledge and he especially rejected the idea that metaphysics is included in human knowledge. He argued that Fung's metaphysical propositions like "mountain is mountain" and "water is water" and the like, are not empirical, for they have nothing to do with the actual world. They are mere statements of identity and as such are logical, but not metaphysical (Hung 1989, 184, 188). For Hung, traditional metaphysics is not a theory of knowledge, though it can be a guide to finding the meaning of life and help man to live well. But Fung's metaphysics has nothing to do with this. It is a total delusion that contributes nothing to our life and is less poetic than traditional metaphysics. It is thus reasonable to abolish it (Hung 1989, 191).

Some critics have argued that Hung misunderstood Fung's metaphysics and that his criticism to Fung is untenable. Chen Xiao-ping, a representative of these critics, claims in his recent paper that Fung's metaphysics can be analyzed in linguistic terms, which shows that a metaphysical proposition may be analytic only by its form rather than its contents (Chen 2019, 25–31). Chen reconstructs Fung's metaphysics in an analytic approach. According to him, Fung's metaphysical propositions, such as that what makes a thing to be itself is the reason why there must be something that makes a thing to be itself, is a logical inference based on the law of identity. In these propositions, Chen argues, Fung expressed his metaphysical insights in a logical way, for all the terms he used in them are logical, rather than empirical. Thus, Chen concludes that Fung's ambition was understandable. However, I don't think Chen's judgment is compelling, because it is impossible to put Fung's metaphysical propositions into a linguistic form. Fung's propositions are not logical or linguistic but senseless in metaphysics. For they just had a seemly logical form at first glance. But they could not be justified both in logic and experience, according to the Circle's principle of verifiability. In this sense, Fung's metaphysics should be in the list of metaphysics the Vienna Circle would reject. As the matter of fact, Fung had been keen on Chinese traditional philosophy which focuses on subliming people's horizon in spirituality, not on arguing for his ideas. He did not consider what language we should use to express our feeling of objects in the world, though he always criticized the failure of Chinese philosophy in logic.

Again. Chen comments that Hung misunderstood Fung's metaphysics because of a difference in their approaches to metaphysics. According to Chen, Hung did not understand Fung because Hung followed the Vienna Circle without advancing any novel idea, whereas Fung interpreted metaphysics creatively in his own way. Thus, according to Chen, Hung could not understand Fung correctly. Chen mentions the famous comparison Fung proposed between a carrying-on narration and a repeating narration. For Chen, Hung just followed the repeating narration while Fung practiced the carrying-on narration (Chen 2019, 28). I don't consider this criticism to be fair to Hung. Hung did not criticize Fung's misunderstanding of the Vienna Circle's rejection of traditional metaphysics on the basis of his own understanding of the Vienna Circle's philosophy; his criticism of Fung proceeded on a different, methodological level. Chen's accusation of Hung's work as a just repeating narration is unacceptable for two reasons. It is hard to argue, on the one hand, that the comparison of the repeating and the carrying-on narration is accessible to evaluation of the two philosophers' works. The most important one of criteria for judging the value of a philosophy is, I think, not the comparison but the novelty or creativity in philosophy comparing with other philosophical doctrines. Hung's work was absolutely novel to the Chinese philosophy at that time, while Fung's metaphysics was considered to be creative in understanding the traditional Chinese philosophy. According to the criterion, they were outstanding in their philosophical works. On the other hand, the comparison itself is unavailable in philosophical discussions. According to Joana Konova's entry on repetition in The Chicago School of Media Theory online,[1] the philosophical discussion of repetition ranges from Plato's understanding of repetition as reminiscence or recollection of an already existing knowledge (cf. *Phaedo*, *Meno*, *Philebus*) to Deleuze's "upgrading" of repetition as an "active force [...] producing difference" (Parr 2005, 223). In *Difference and Repetition* (1968), Deleuze discusses Nietzsche's concept of the "eternal return" as "the power of beginning and beginning again" (Deleuze 1994, 136). Deleuze counters Freud's understanding of repetition as a compulsive reiteration of the past which psychoanalysis attempts to stop by representing it, with the understanding that repetition allows for reinvention (Parr 2005, 224). In this sense it is hard, again, to argue that the carrying-on narration exceeds the repetition in philosophical discussions. On the contrary, the repetition is always more adequate in philosophy than the carrying-on narration. Tacit knowledge

[1] https://lucian.uchicago.edu/blogs/mediatheory/keywords/reptition/=08-05-2020

underlies the repetition, which provides us with the very authority for further knowledge or carrying-on narrations.

According to the judgement above on the comparison of repetitive and carrying-on narration, Fung's metaphysics is also taken account for in repetition. Derong Chen, a linguist in Chinese language and culture from Toronto University, in his book on Fung's new metaphysics interpreted that metaphysics as metaphorical. He compared Fung's metaphysics with Hegel and argued that Fung's ideas and metaphysical terminology were very close to Hegel, especially in his interpretation of Dao and Qi (Chen 2011, 153). This is surely an interesting approach to the interpretation of Fung's metaphysics from the perspective of methodology. Fung did provide a new way to understanding metaphysics metaphorically, different from the way the Vienna Circle understood metaphysics. That was, however, the methodological approach that Hung's used in his criticism of Fung's metaphysics. For Hung, it was unacceptable to account for a philosophical metaphysics in metaphorical terms, for a metaphysician could not be better in using metaphors in philosophy than a poet in poetry or a musician in music. As Carnap said, "metaphysicians are musicians without musical ability." (Carnap 1996, 30) On the methodological level, Carnap's criticism of metaphysics is directly applicable to Fung, when he said that "logical analysis, then, pronounces the verdict of meaninglessness on any alleged knowledge that pretends to reach above or behind experience. This verdict hits, in the first place, any speculative metaphysics, any alleged knowledge by *pure thinking* or by *pure intuition* that pretends to be able to do without experience." (Carnap 1996, 26, italic original) According to this verdict, Fung's is a speculative metaphysics of the kind that Carnap tried to eliminate by logical analysis.

11.3 Replies to Criticisms of the Philosophy of the Vienna Circle After 1979

As to the accusation, made by Chen Xiao-ping and others, that Hung's philosophy was just a repeating narration, we must reply that Hung was not simply concerned with introducing the philosophy of the Vienna Circle to China, he also set out to explain that philosophy in his own way. He provided his own understanding of this philosophy with deep insights, and especially so in his replies to criticisms of the Vienna Circle after 1979.

11.3.1 Tscha Hung's Philosophical Research After 1979

Historically speaking, the end of the Cultural Revolution was the beginning of Tscha Hung's new academic life. After 1979 he was extraordinarily active in international philosophical circles, as if he was recovering from a long winter. In 1980 he was invited to the fifth International Wittgenstein Symposium in Kirchberg am Wechsel, Austria, and visited the University of Vienna and Queens College in Oxford. He also attended the Schlick and Neurath Centenary in Vienna and visited Trinity College in Oxford in 1982. Two years later, he was rewarded with an honorary doctorate from the University of Vienna and he again visited Queens College, Oxford. In 1986 he was invited to deliver lectures at Tokyo University in Japan. He also attended the first conference on analytic philosophy and the philosophy of science in Hong Kong and visited the University of Hong Kong in 1988. Three years later, in 1991, some colleagues from Hong Kong proposed to published a volume in the honor of Tscha Hung's 85th birthday, and philosophers and scientists organized a conference on realism and anti-realism in the philosophy of science in his honor which was scheduled for 1992. Tscha Hung was delighted to accept theses invitations and prepared for the events when he unexpected died in February 1992. Robert Cohen, one of editors of the proceedings of the conference on philosophy of science in Beijing in 1992, recollected his communication with Hung during his time both in China and at conferences outside of China. He spoke of the comparison of Tscha Hung's work on the Vienna Circle in China with Ayer's in Britain and Hempel's in the United States, saying that "both the teacher and also the critic of his own mentors and colleagues, he was a scholar who quietly carried the roles of being the Ayer and the Hempel of China." (Cohen et al. 1996, xi) Fan Dai-nian, a well-known Chinese philosopher of science, also described Hung's academic activities in China in his paper on Tscha Hung and the Vienna Circle in the proceedings (Cohen et al. 1996, xvii–xxii). Fan together with two distinguished Chinese philosophers of science co-authored a summary of Hung's philosophical insights on the philosophy of the Vienna Circle in an influential paper that conceived of Hung, in the way that Cohen had done, as a critic of his own mentors and colleagues in Vienna (Fan et al. 1992). We can see from the Tscha Hung's last master pieces on members of the Vienna Circle, collected in the anthology, *Essays on Logical Empiricism*, published in Hong Kong in 1990, that he attempted to argue for their ideas with some criticisms.

11.3.2 Tscha Hung's Replies to Criticisms of the Philosophy of the Vienna Circle

These criticisms were based on a thorough understanding of the philosophy of the Vienna Circle. When Karl Popper claimed to have killed logical positivism with his own falsificationism, Hung replied in his lecture in Tokyo in 1988 that "it is a very rare thing in the history of philosophy that Popper, a world-famous philosopher, was proud of his murder of the logical positivism. But I believe that his murder failed." For his murder weapon, his falsificationism, was not sufficiently sharp. According to Hung, the general statements of natural laws are neither verified nor falsified by one or some protocol propositions. Verification and falsification are only exceptions when we defend scientific propositions by appealing to our senses (Hung 1990, 43–44). Popper had, of course, not been welcomed to the Vienna Circle when he was living in Austria and Hung did not know him personally until they met in Britain in 1980. According to Popper's autobiography (Popper 1974, 72–90) and according to Malachi H. Hacohen's account of Popper's life in Vienna in the 1920s–30s (Hacohen 1998), Popper admired the early writings of Schlick and attended seminars by Carnap and other members of the Circle. It is somewhat ironic that Popper was considered a member of the campaign against traditional philosophy, though he severely criticized logical positivism by means of his principle of falsification with regard to two central problems of epistemology: induction and the demarcation between science and metaphysics. Hung focused on the principle that no single factual assertion can be verified or falsified by observation. This, I believe, is indeed the correct way to understand the principle of verifiability, for Schlick and Carnap also changed their idea on the principle and spoke later of confirmation rather than verification or falsification.

Tscha Hung also replied to criticisms of logical positivism by Thomas Kuhn and Paul Feyerabend. As we know, Kuhn and Feyerabend proposed a social-historical approach to the study of science, which had a strong influence and was viewed as a development of logical positivism involving a second phase in the philosophy of science after the second world war. But Hung disagreed with their approach and accused them of being irrational. He claimed that the Kuhn-Feyerabend approach was a retrogression in the philosophy of science in that it was trying to analyze the structure of science in terms of society, history and culture with all their confusions, while logical positivism had tried to clarify that structure in logical terms. Hung argued that there is, in fact, a conflict and a mutual incompatibility between an analytic and a historical approach in the philosophy of science (Hung 1990, 257). Despite these disagreements, Hung collected some of Kuhn's and Feyerabend's writings in his edited anthology, *Selection of Original Writings of Western Modern Philosophy*, published in 1993.

As for Quine's criticism of logical positivism, Tscha Hung admired his rejection of the first dogma of reductionism but disagreed with Quine's rejection of the distinction between the analytic and the synthetic. As far as the first dogma was concerned, Hung said, "there is a conflict between the unification of science movement that Neurath and Carnap proposed based on the reductionism, and the development of the natural sciences up today. The point of conflict lies in difficulties in the reduction of biology to physics and chemistry, of psychology to neurophysiology and of the latter two to physics. Although scientists do not rule out the possibility of such a reduction in the future, they admit the very low of the probability. The big challenge is the theory of emergent evolution, according to which life or mind has an ad hoc form that one cannot infer or predict from natural law and any scientific theory. Thus, for a long time, most logical empiricists have seen this development as a historical event." (Hung 1990, 73) Hung disagreed with Quine's rejection of the distinction between the analytic and synthetic. It is unintelligible, however, that Hung's reason for disagreeing with Quine was that it would destroy the whole system of logical empiricism. Quine was, in fact, concerned only with rejecting the analytic-synthetic distinction. Hung thus failed to make his case against Quine. But Hung's disagreement with Quine was in line with the logical positivist attitude to Quine's rejection of the second dogma. Hung mentioned the serious responses of Carnap, Feigl and Waismann in his writings (cf. Hung 1990, 69). In the end, the dispute didn't, however, matter, since the logical positivists eventually changed their position on the two dogmas of empiricism that Quine had identified.

11.4 The Reevaluation of the Philosophy of Vienna Circle in the 1980s–90s

As a teacher of integrity and a witness to a stage of the history of philosophy – as Cohen put it in his recollections (Cohen et al. 1996, xv) – Tscha Hung did not only defend the original doctrines of the Vienna Circle, he also followed the subsequent development of that philosophy.

11.4.1 On the Development of Logical Empiricism

Tscha Hung denied, first of all, that logical empiricism was dead. In his 1986 Tokyo lectures Hung argued that no one would deny that logical positivism or empiricism has been one of the influential schools in contemporary philosophy, though it was not so fashionable today as it had been fifty years ago. He wrote, "there is such an epicenter any more today as that of logical empiricism in Schlick-centered Vienna fifty years ago, Carnap-Reichenbach-centered Los Angeles thirty-seven years ago, or Feigl-centered Minnesota later on. But the scientific conception of logical empiricism is still active in many countries in the West, for examples, with Ayer in London, B. F. McGuiness and A. Quinton in Oxford, Feigl in Minnesota, Hempel in Princeton, Haller in Graz, Weingarten in Salzburg, Stegmüller in Munich, Geymonat in Torino, Nash in Oslo, and H. L. Mulder in Amsterdam." (Hung 1990, 44).

Tscha Hung also noticed the development in logical empiricism in recent decades, such as the amendment of the concept of meaning, the transition from syntactic parsing to semantic analysis, the correction of the definition of verifiability, the concept of possibility and inductive logic, the containment of physicalism and the drift away from the unification of the sciences, and the adjustment and limitation of empiricism. All these are fully illustrated in publications by Feigl, Hempel, Kraft and Ayer who edited a series on the exact sciences in which the English translation of Schlick's early masterpiece, *General Theory of Knowledge*, was published in 1974. It was illustrated equally by H. L. Mulder, R. S. Cohen and B. McGuiness who edited a Vienna Circle Collection, by Haller and McGuiness who edited a series on the Return to Schlick, and by T. E. Uebel who edited an anthology, *Rediscovering the Forgotten Vienna Circle*. All of these publications highlighted, for Hung, the revival of the Vienna Circle and logical empiricism at the present time. But, as Hung added, it is needless to say that the study of ethics has been overlooked in analytic philosophy beginning with the Vienna Circle. It is arguable that, for him, a system of philosophy should include ethics as an essential part in the system as is the case, for instance, in Kantian and Marxist philosophy. Hung referred also to Bertrand Russell's criticism of logical positivism by complaining that this kind of philosophy had nothing to offer in philosophy but methodology (Hung 1990, 247–248).

It is somehow odd that Tscha Hung, one of the last living members of the Circle in the 1990s, would offer such a complaint of logical positivism. For Schlick had studied ethics and published his book on ethics at a moment of transition in his thinking in which he sought to apply the method of logical analysis to some of the traditional ethical problems and to interpret ethical statements as empirical claims about the means of maximizing happiness (Mihina 2016; Oberdan 2017). Neurath had also concerned himself with ethical problems and he had coined the term "felicitology" in his writings to characterize a version of ethics suitable for the scientific worldview of logical empiricism (Stuchlik 2011). Even Carnap had not entirely abandoned the study of ethics but had only restricted it in the appropriate way (Carnap 1996). Hung should have understood the Circle's attitude to ethics when he studied in Vienna. After half a century, however, he complained of the attitude of the Circle to ethics and accused it of a weakness in its philosophy. I believe that there is only one explanation for this shift: Hung realized the more important role of ethical study and moral philosophy in contemporary philosophy than he had previously done. But his comparison of the philosophy of the Circle with systematic philosophy is confusing. There is once again only one explanation for this. I assume that he was overloaded with the German sentiment for systematic philosophy, even though he was educated in the scientific outlook in Vienna.

Austrian philosophy has had a strong inclination toward the natural sciences and empiricism in contrast to German philosophy, as J. C. Nyiri and B. Smith have pointed out (Nyiri 1981, 1986; Smith 1994/1998). But Hung seems not to follow this Austrian trend even though he studied the natural sciences and mathematic logic with H. Reichenbach, Schlick, H. Hahn, F. Ehrenhaft and Carnap. German speculative philosophy had, in fact, dominated Chinese philosophical and intellectual circles in the 1920s when Hung was sent to Germany, initially to study the philosophy of mind with Rudolf C. Eucken rather than empiricist philosophy. Due to Eucken's death when Hung arrived in Jena in 1927, he went on to study physics, mathematics and philosophy at the University. The most attractive lectures, the ones that impressed him deeply, were delivered by B. Bauch and P. F. Linke, both Kantians and phenomenologists. Reichenbach, to whom Hung used to listen in Berlin, was the key figure to recommend him to go to Vienna to study philosophy with Schlick. It might be due to Hung's weakness in natural science that Schlick asked him to attend courses in mathematics and physics. But deep in his heart, Tscha Hung still greatly appreciated German philosophy, which he thought could provide insights into the meaning of the world and of human life. This may explain why in his later years he saw the Vienna Circle's view of ethics as a weakness in their philosophy and why he admired Schlick's early work on ethics. After 1949, Marxist philosophy occupied the whole field of philosophy in China, and that philosophy was, of course, intrinsically rooted in the speculative tradition of systematic German philosophy. Under the serious political control, Hung received this philosophy with some reluctance, but he realized absolutely that ethical problems needed to be probed in philosophy and that ethics was essential to any form of philosophical study.

11.4.2 On the Foundation of Knowledge and Theory of Truth

In his last years, Tscha Hung turned to the two fundamental problems in the theory of knowledge, the foundation of knowledge and the theory of truth. He published an essay on confirmation (*Konstatierungen* in Schlick's terminology) and a review on Schlick's writings on philosophical problems, in which he criticized Carnap's and Neurath's reductionism and their dependence on protocol statements. He also disagreed with Schlick's foundationalism which placed the concept of confirmation at the foundation of knowledge. He said, "Schlick's concept of confirmation or observation statements are much more empirical than his statements themselves. They fall in the field of psychology, rather than logic or grammar. Any claim of absolute certainty can be understood only in psychological terms, not in logical or grammatic terms, according to Schlick's concept." (Hung 1990, 105) And, "Schlick tried to find a stable foundation for scientific knowledge in indubitable experience and self-evident perception. But for me, this is another form of psychologism to which he was strongly atracted in his early years and I think it is fundamentally wrong. Indubitable experience and self-evident perception, as Schlick used to think about them, are subjective and psychological." (Hung 1990, 106) If protocol statements and confirmation cannot be the foundation of knowledge, what would it be for Hung? He did not give any answer to the question, but he was convinced that there was no such thing as a foundation at all. P. K. Chow correctly commented that Hung reminded empiricists that it is a waste of time to look for a foundation for knowledge, for it is not their mission to do so (Chow 1992). Hung proved to be thus an anti-foundationalist in the theory of knowledge (Fan et al. 1992, 76).

Tscha Hung also criticized Schlick's correspondence theory of truth in his essay. He said that Schlick was right in criticizing Neurath and Carnap, arguing that their coherence theory of truth deviated from empiricism since it made compatibility with protocol sentences the criterion of truth and thus essentially returned to conventionalism. Schlick proposed, instead, a correspondence theory; but Hung argued that this theory would lead to sophisms and an empiricism of sense-data, for it took the correspondence with sense-data to be the foundation of truth (Hung 1990, 161). For Hung, there is no final criterion of truth. It was unfortunate that Hung had no time to make his case for this understanding of the theory of truth due to his death in 1992. I assume that, with his anti-foundationalism and denial of final criterion of truth, Hung would have chosen a form of pragmatism with an analytic approach to protocol or confirmation sentences, but he would still adhere to the philosophy of the Vienna Circle with such amendments. This is, in fact, what he had been doing since the end of the 1970s.

It is precisely because of his insights that Hung has been praised internationally for his continuing defense of logical empiricism and his novel take on the theory of knowledge and the concept of truth. His reputation was highlighted at the International Conference on the Vienna Circle and Contemporary Science and Philosophy in Memory of Tscha Hung in Beijing in 1994. More than 40 philosophers from different countries participated in the conference and focused on the relation of Tscha Hung to the Vienna Circle, speaking not only of their personal relation with Hung, but also of his contribution to the philosophy of the Circle (Hu Nai 1995).

11.5 Tscha Hung and Chinese Philosophy

As a result of the debate with Fung You-lan, Tscha Hung has been seen as an opponent of the Chinese tradition in philosophy, as I have said above. Hung had, however, a strong feeling for Chinese philosophy. His philosophical ideal was to achieve a unity of spirit, to extend our minds, and to make intellectual progress, all of which are arguably characteristics of Chinese philosophy (Jiang Zhaoyang 2016). His faith in Chinese philosophy has both a historical and a philosophical dimension.

11.5.1 Historical Dimension

Historically speaking, Tscha Hung began his philosophical career with his first paper on Wang Yangming's philosophy in the *Journal of Xueheng* when he was studying in a preparatory class at the Southeast University in Nanjing. The paper was highly praised by Kang Youwei, a well-known Confucian and political reformer at that time, who recommended Hung to Liang Qichao, the famous historian, philosopher and politician. Liang, in turn, recommended Hung to Uno Tetsuto, a distinguished Japanese Confucian and scholar on Wang Yangming's philosophy, at Tokyo Imperial Universities. Due to sickness Hung was forced to return from Japan to Tsinghua University with Liang's support and he was recommended once again by Liang to go to Germany to study classical German philosophy of mind with Rudolf Eucken, the German idealist, who was influential in China in the 1920s. Though he moved on to study natural sciences and mathematics, because of Eucken's death before Hung's arrival in Jena, Hung was inclined to concern himself with the important problems of life when he studied in Jena. It is the reason why Bauch and Linke' lectures attracted much than Reichenbach, and he appreciated very much Schlick's early book on ethics. This shows that Hung had a strong feeling for the problems he saw essential to philosophical investigations. These problems are like the concerns of traditional Chinese philosophy, which posits that the mission of philosophical investigations is to pursue the highest level of spirituality and to accomplish the meaning of life. Meanwhile, Hung also recognized as his mission to make progress in intellectual life, to spread philosophical ideas to the public, and to strengthen the power of ideas. Under strict political control, he published papers in official newspapers and academic journals on such topics as Hume's *Essay on Human Understanding*, Kant's Nebular hypothesis, Mach's philosophy, and Carnap's physicalism. Before the Cultural Revolution, he also appealed publicly for the academic study of idealism and of contemporary Western philosophy. During the Cultural Revolution, as we have seen, he did much translating and editing of anthologies of the writings of Western philosophers for students, but abstained from academic research because that was not allowed until 1979. Hung did not, however, content himself with translation in this period of his academic life. He was eager to continue academic research during the Revolution, not only for a philosophical purpose but with the reconstruction of Chinese philosophy in mind. Hung's ambition and academic ideal burst out with energy after 1979. He attended many international conferences by invitation and published more than ten essays in top journals in China and in the world. In doing so, he was seen not just as a member of Vienna Circle but also as a distinguished Chinese philosopher. And he was proud of being Chinese philosopher. Though he had suffered much from political pressure during the Revolution, Hung did not complain of his pain at any later moment. This characteristic has also been seen as a feature of traditional Chinese intellectuals.

11.5.2 Philosophical Demission

In his philosophical thinking, Tscha Hung did not deny the importance of metaphysics at all. Instead, he admired the implications of metaphysics in Chinese philosophy. In his criticism to Fung You-lan's new metaphysics, Hung's target was not metaphysics itself but Fung's explanation of metaphysics.

For Tscha Hung, real metaphysics is necessary and indispensable to Chinese philosophy. But its significance is mostly practical in human life, rather than theoretical and doctrinal. Hung said, "metaphysics is a means to fulfil our inner life and the realm of experience, and to fill in the nothingness in our life with metaphysical propositions such as the immortality of the soul, free will and existence of God. It functions as poetry in culture to satisfy our desires for ideals and wishes in the mind, but it does not provide any scientific knowledge or theory." (Hung 1989, 116) In limiting the significance of metaphysics to practical life, Hung did not claim a priority of practice over theory or vice versa. He saw the two as equal, each functioning in a different field. "The meaning of life is infinite while the meaning of truth is finite. It is just as impossible to understand truth with the infinite meaning of life, as it is to experience life with the finite meaning of truth. The result will be a failure of both, neither being able to understand the meaning of truth nor to experience the meaning of life." (Hung 1989, 125) For Hung, Chinese philosophy gives priority in experiencing the meaning of life, an experience that can satisfy desires and fulfil our life, as artists working on artworks, Christians believing in God, and ordinary people experiencing and idealizing life metaphysically (Jiang 2016, 58). I am convinced that Hung appreciated Chinese philosophy as a metaphysical system, not only in the form of Confucianism, but also in that of Daoism and Chinese Buddhism. He studied Wang Yangming's Confucianism when he was young and retained an interest in Confucianist philosophy for his whole life. While he was not a Confucian scholar, he practiced Confucianism as a Confucian. He was a Junzi, 君子, who maintained a noble character and high standards as a man. He followed Zhang Zai's famous words, "to ordain conscience for Heaven and Earth, to secure life and fortune for the people, to continue the lost teachings of past sages and to establish peace for all future generations." (Zhang Zai 2009, 10) Hung had a strong sense of mission for improving academics and searching for truth in his time. He practiced the Socratic motto that I know I know nothing, as well as Confucius's saying in *The Analects*, "When you know a thing, to hold that you know it; and when you do not know a thing, to allow that you do not know it - that is knowledge" (Confucius 2009, 151).

11.5.3 Tscha Hung's Criticism of Fung You-lan's New Metaphysics

Tscha Hung's criticism of Fung's new metaphysics, as I have argued above, was not of metaphysics itself but of Fung's approach to metaphysics. Some commentators have argued that the debate between Hung and Fung might be described as one between a scientist who advocates the priority of science over metaphysics and a metaphysician who advocates the priority of metaphysics over science – a debate among Chinese intellectuals in the 1920s that was historically significant in the history of modern Chinese thought (Jiang 2016, 165–167; Li Wei-wu 2008, 2). This debate was highlighted in the so-called conflict of science and metaphysics (科玄之战). For a scientist like Ding Wen-jiang, philosophy should concern itself with knowledge of the empirical world and reject metaphysics in accord with the attainments of science, while a metaphysician like Zhang Jun-mai argued that philosophy should be studied in a metaphysical rather than a scientific manner (Ding Wen-jiang 1923, Zhang Jun-mai 1923). Some scholars have claimed that the debate reflects a shift of philosophical models in modern China and the difference between Chinese and Western philosophy (Li Wei-wu 2008, 2; Hu Jun 2013, 43–44). I consider that analysis too simple for providing an proper interpretation of the historical debate. Tscha Hung had, in fact, a clear conception of metaphysics in mind, as I have shown. His criticism to Fung's metaphysics was not to give up metaphysics, but to delimit metaphysics within philosophy. Hung and Fung were thus concerned with two different understandings of the nature of philosophy.

Strictly speaking, Hung did not think of Fung's metaphysics as part of philosophy at all whereas Fung took his new metaphysics to beat at the heart of Chinese philosophy as he understood it.

Let us look carefully at the crucial point of Hung's criticism of Fung's new metaphysics in their debate. Hung thought that Fung misunderstood the logical positivist rejection of metaphysics and that Fung's new metaphysics failed to reach its goal to replace the old metaphysics the Vienna Circle had rejected with his own new one. At the very center of Hung's criticism was Fung's misunderstanding of the method of logical analysis rather than of the nature of metaphysics. Fung argued that his new metaphysics was constructed by means of logical analysis. He correctly saw that traditional Chinese philosophy fell short of logical argumentation and inference. With an ambition to amend this, he tried to establish that metaphysical propositions are analytic and without any content concerning the world. He insisted, however, also that this kind of analytic proposition differed from logical propositions which are merely formal. He argued that the propositions of his new metaphysics were content-free, but affirmative of content concerning the world in an abstract or conceptual way. Hung counter-attacked that it was a contradiction to say that a formal or content-free proposition could have something to do with the world. In Hung's view, the formal proposition is logical and has to do only with the logical expression as syntactics does in logic or linguistics. But Fung's propositions tried to express something empirical about the world, though he believed at the same time that his propositions were also concerned with the highest realm of spirituality in the universe. Hung commented that Fung's failure to reconstruct the new metaphysics was due to his method in metaphysics. He said, "first, Fung tried to establish his metaphysics by appealing to the Vienna Circle's rejection of metaphysics in order to prove the base and place of his metaphysics in philosophy, but this metaphysics is just the opposite of the Circle's point of view. We could say that the Circle's rejection of metaphysics did not deny traditional metaphysics, but illuminated its nature. By contrast, Fung's new metaphysics took the risk to be rejected by the Circle's criticism. Second, Fung thought that it was possible to establish a metaphysics rising through the understanding to the highest realm of life by means of a formal metaphysics. But for the Vienna Circle, his contribution to metaphysics failed to go beyond and was indeed less great than traditional metaphysics. Third, with the method of metaphysics, it is possible to understand the significance of metaphysics for the philosophy of life through the intuitional approach of the Vienna Circle. But we would miss out on this significance if we applied Fung's formal approach." (Hung 1989, 192) Yang Guo-rong, a contemporary Chinese philosopher, agreed with Hung's criticism of Fung, saying that, "on the one hand, Fung tried to combine logical analysis with logical construction by learning from Zhu-Cheng Confucianism in order to reconstruct metaphysics, but he strayed from the positivist line. On the other, he tried to formalize the system of metaphysics in order to obtain the analytic propositions that are in accord with the positivist criterion of meaning. It seems that he has thus returned to positivism. The two intentions underline the direction to harmonize the tension and conflict of metaphysics and positivism" (Yang 1996, 96). Yang concluded that the new metaphysics was untenable in its interpretation of reality, even though Fung intended the mission of his philosophy to be an interpretation of the whole universe. Due to this, the new metaphysics failed. Although Fung tried to go beyond positivism and traditional metaphysics, his ambition to formalize his metaphysics without any empirical content made it inevitable that he eventually failed to fulfill his purpose to give up abstract logical deduction while also abolishing the speculative illusion (Yang 1996, 97). It makes sense to consider Fung's new metaphysics an attempt to reconstruct Chinese philosophy by means of a logical approach to metaphysics, but he failed, not because he misunderstood metaphysics but because he abused the logical approach to metaphysics. Hung was right when he dismissed Fung as having constructed a theoretical system of nothingness.

11.6 Tscha Hung and Metaphysics

Regarding with metaphysics, Tscha Hung did not just reject Fung's new metaphysics in a negative way, but discussed the significance of metaphysics in Chinese philosophy as well as in philosophy in general. As I read, Tscha Hung published an essay on metaphysics in which he analyzed metaphysical problems as involving a misuse of expressions of our experience of life with his distinction of three kinds of metaphysics and explained the relation of meaning, truth and metaphysics in detail. He adopted the model of propositions in Wittgenstein's *Tractatus Logico-philosophicus* in order to express his own understanding of metaphysics in the essay "Meaning, Truth and Metaphysics" (Hung 1989, 118–125).

11.6.1 The Effable and the Ineffable

To begin with, Tscha Hung analyzed the role of metaphysics in philosophy generally. He recognized that there is something ineffable in our knowledge but understandable, nonetheless, in immediate experience. He distinguished what can be ineffable from what can be effable in our knowledge. For effability is intrinsic to knowledge; the ineffable is not part of our knowledge at all. He said, "by being effable we mean what can be expressed in our language or in a system of signs, no matter what kind of knowledge it belongs to. And the knowledge is an expression of facts, which is the nature of knowledge in its own way." (Hung 1989, 104) What is ineffable in our knowledge, for Hung, is not senseless but due to our senses. For example, when I see a pitch of red, I experience it as red but cannot express the content of my experience of redness such that a blind man can know what an image of redness is like only from the knowledge of redness. Similarly, no one can express his experience of life if he has not enjoyed life at all. Hung argued in this way that we have an immediate experience of content in the string of our consciousness, but we cannot, in principle, express that meaning. At stake is the distinction between

Kennen and *Erkennen* as the Vienna Circle put it (Hung 1989, 105). The content and nature of our experience are ineffable, for we cannot find an effable way to express them in our language. And neither can we express the distinction between or the relation of two experiences. These are immediate experiences that are ineffable in any form of concepts.

Tscha Hung argued that all we can have knowledge of are relations of form, propositions of construction as Carnap called them. As Hilbert put it, a geometric concept can be defined only by being in accord with certain postulates, called implicit definitions. By an implicit definition Hilbert meant that the construction of concepts can be achieved only by propositions of construction which define the formal relation of concepts. Hung agreed with the Vienna Circle that all knowledge is about form, not about content and that knowledge is an expression of a calculation of formal relations, it is not an experience of things or objects. Hung excluded the experience of things or objects, in this way, from our knowledge of formal relation of concepts. He was thus demarcating the limits of knowledge, and by this means defended the position of metaphysics in philosophy. For Hung, metaphysics was about the ineffable in language, something which goes beyond the field of knowledge. It concerns personal experience that comes from our immediate relation with the world. It is an important part of our experience of the world and different from expressible scientific knowledge. Like poets and musicians, metaphysicians can communicate their feelings of the world in a poetical or musical way. Metaphysics thus expresses personal experiences, but not knowledge of the world. Hung distinguished natural sciences from the human sciences, *die Geisteswissenschaft*, just as the Vienna Circle did. He did not deny the importance of metaphysics for our experience of the world. Metaphysics can do what knowledge cannot, namely express the content of our experience in its own way. With the help of this understanding of metaphysics, Hung concluded that traditional Chinese philosophy was metaphysical in this sense, for Confucius and Mencius had already admitted the ineffable experience in their philosophy.

11.6.2 Criticisms of Three Kinds of Metaphysics

Tscha Hung criticized metaphysical approaches to knowledge from (1) a realist, (2) a reductive, and (3) intuitional perspectives.

Tscha Hung agreed with Kant's conception of the transcendent. Kant had described the transcendent as the problem of the reality of the external world under the assumption that this reality is independent of the phenomena. It is, on this view, a thing-in-itself that we are unable to know by means of the phenomena. The logical positivists had argued against this realism by means of their principle of verifiability. Realism is false, they had said, because it is unverifiable by empirical observation or the analytic propositions of logic. Hung sought to provide an alternative argument against realism. He borrowed Kant's distinction between the thing-in-itself and the phenomena and asserted their relation to be external rather than internal. He said, "it is necessary to postulate a transcendent object if the relation of phenomena and thing-in-itself has to be dealt with and is expressible. It is the case, we might say, that what must be expressed in our language is not only about the phenomena but about the thing-in-itself if a formal relation is expressible for reality. In principle, the relation must be available to objects on different levels. ...... Kant thought that the transcendent object is unknowable and that only phenomenal objects are knowable, because he did not understand the distinction between knowledge and experience. The transcendent object is not unknowable but it is unexperienced. Only phenomenal objects are experienced but they are unknowable." (Hung 1989, 110–111) Hung clarified in this way the importance of metaphysics for philosophy. Metaphysics is experienced but not perceived. We can understand our experience of the world but we cannot know the content of our experience. The content is experienced by our understanding of the world such as in the imagination or in the creation of experience; and this is the mission of metaphysics.

Tscha Hung further analyzed the reductive metaphysics which was proposed by some agnostics. They argued that there are unsolvable problems that science cannot deal with, such as the infinity of the universe and the nature of matter or force. But these are problems that metaphysics can deal with, according to these agnostics. Hung called this agnosticism an inductive metaphysics, for it claims that induction can resolve the metaphysical problems which are intrinsically unsolvable in science. But Hung argued that reductive metaphysics was due to a misunderstanding of the role that induction plays as a foundation of scientific knowledge. All inductive knowledge is empirical and falls thus into the domain of empirical science. It is absurd to claim the possibility of induction on a metaphysical level beyond empirical sciences. Hung said, "it is impossible to defend a metaphysics by means of induction, because, as a matter of fact, we cannot infer a metaphysical postulate from any inductive fact. And from a logical point of view, the reductive knowledge is transcendent just like the metaphysical and as such contradictory to the concept of induction." (Hung 1989, 112) For Hung, inductive metaphysics could not be a kind of metaphysics at all.

Tscha Hung, moreover, also thought that some philosophers in the humanistic tradition such as Schopenhauer and Bergson claimed an intuitional metaphysics for which knowledge could be intuitionally a priori. Schopenhauer argued that the natural sciences view objects only from the standpoint of the external phenomena, like a scientist observing and describing a house only from the perspective of its environment. But a metaphysician has been in the house and has looked around inside. Bergson also thought that scientists are limited to descriptions of objects by means only of temporary formulas or signs while philosophers concern themselves with objects in their intuition (cf. Schlick 1979, 99–135). But Hung argued that this metaphysics also appeals to experience in speaking of intuitional contents and that being transcendent is misleading in that it uses language when we attempt to express our experience. This kind of metaphysics should be considered self-defeating, for the intuitional content of experience is based on mental explanations intrinsic in our mind, and have nothing to do with transcendent things. Hung said, "the intuitional metaphysician tries to obtain transcendent knowledge from the intuition of experience alone, but the result is contradictory. If we had such a knowledge of the transcendent, we would be appealing to a formal relation rather than the content of the experienced object or consciousness, but this is the mission of science, not of philosophy" (Hung 1989, 115).

Tscha Hung criticized various kinds of metaphysics by means of his distinction between science and philosophy. For him, science deals with what is happening in the world while philosophy concerns the meaning of what happens in the world. In brief, science is a description and a report of facts in the world and philosophy is the analysis and assessment of such descriptions and stated reports. But there seems to be no room left for metaphysics both in science and philosophy according to this distinction, for metaphysics by nature tries to talk about the descriptions and reports of the world in the way science does but fails, and it does not work as an analysis and assessment of the descriptions and reports either. This is likely the Vienna Circle's position against any form of metaphysics. But the purpose of Hung's argument against metaphysics is not to give it up completely. In contrast to the Vienna Circle, he admitted the significance of metaphysics in traditional Chinese philosophy and culture.

11.6.3 Metaphysics in History

Tscha Hung analyzed the reason why metaphysics developed in history. He said that the reason was that metaphysicians were stubbornly holding on to the position that the truth or falsehood of a sentence was not determined independently but by its place in a system of metaphysics from which it could be inferred. And he argued that sentences of metaphysics could not be compared with reality, for the truth or falsity of a sentence in metaphysics could be determined only in and in accord with a system of metaphysics. In this sense truth in metaphysics is coherence among sentences and the criterion of truth is the principle of non-contradiction. It is evident that Hung implied that the coherence theory of truth is normally attributed to idealist metaphysics. It appears that, according to this theory, a metaphysical sentence can be true if and only if it is consistent with other sentences in the system of metaphysics it belongs to. Indeed, Hung thought of metaphysics as such a theoretical system in which a sentence is true due simply to its consistence with others. It looks in this way like a system of logic, for being true in logic is for propositions to be consistent with each other in a logical system. Fung might think of metaphysical propositions as analytic. But his understanding of analytic propositions was entirely wrong. The difference between the analytic and the synthetic is obviously not concerned with whether they are formal or not, but it is determined at the syntactical or a priori level. According to the logical positivists, a proposition is analytic if and only if its truth depends solely on the meaning of its terms, either because it is true (or false) by definition or by the conventions of language. While the logical positivists believed that the only necessarily true propositions were analytic, they did not define "analytic proposition" as "necessarily true proposition" or "proposition that is true in all possible worlds". Synthetic propositions were defined by them as propositions that are not analytic. These definitions applied to all propositions, regardless of whether they were of the subject–predicate form. Thus, under these definitions, the proposition "It is raining or it is not raining" was analytic, while for Kant it was analytic by virtue of its logical form. And the proposition "$7 + 5 = 12$" was analytic for the logical positivists, while under Kant's definitions it was synthetic. Obviously, Fung's metaphysical propositions are neither analytic nor synthetic. They are senseless both in logic and in the empirical world.

Tscha Hung also argued, as Schlick had already done, that metaphysical problems are unanswerable or unsolvable, because they are senseless due to their being inappropriately constructed. He said, "the metaphysical problems in the history of philosophy are those that no one knows what they are talking about; they are specious problems. Thus, the only way to resolve these problems is to explore their senselessness by showing that there is no such problem at all. That is the only answer to those problems. Truth and meaning in a system of metaphysics, if there are such, are outside logic: metaphysics is not a system of truth in the logical sense" (Hung 1989, 124). But Hung did not reject metaphysics completely; he confined it, rather, to its own proper field. Metaphysics is fundamental to the meaning of our life in the world; it has a place in the significance of culture. For Hung, our search for the meaning of life is the same as our search for the meaning of truth, though they differ in the way they search. Hung compared the meaning of life and the meaning of truth by saying that the former is infinite and the latter finite and that it is impossible to understand the latter by the former and vice versa. Some commenters have argued that Hung admitted thus a positive metaphysics and that this is to be found in the practical dimension (Jiang 2016, 134). But Hung actually did not make such a distinction between a negative and a positive sense of metaphysics. Although he admired the practical dimension in defending metaphysics in Chinese philosophy, he just wanted to limit metaphysics to its right field, rather than propose a general distinction within metaphysics. We have seen that Hung defended Chinese philosophy as a metaphysics. This kind of metaphysics is helpful for people to attain a sublime view of the spiritual realm by practicing principles a priori. Chinese philosophy is, for Hung, full of this kind of metaphysics and differs in this way from traditional metaphysics in the West, which studies abstract essences and seeks to transcend experience. By contract, metaphysics in Chinese philosophy is concerned with human experience in the world, especially so in Confucianism and partly in Daoism. Thus, metaphysics is central to Chinese philosophy. By defining Chinese philosophy as a metaphysics, Hung excluded it from the field of philosophy in the Western sense. This does, of course, not mean that Chinese philosophy is not philosophy at all, just because it seems contradictory to philosophy in the Western sense. It means that Chinese philosophy is different from Western philosophy in its understanding of the nature of philosophy and metaphysics. This has remained a controversial issue for Chinese philosophers up till now. Hung seems to defend the characteristic of Chinese philosophy as a kind of metaphysics, though different from the sense of philosophy in the West.

11.7 Conclusion

From the Chinese perspective, Tscha Hung is a pioneer in the modern encounter of Western and Chinse philosophy. From the international perspectives, he is a legend who witnessed the historical transformation of philosophy in the last century. I believe that he was also an extraordinary philosopher with deep and creative insights into the problems of philosophy. His story is a marvelous episode in modern Chinese philosophy and it tells us what happened in twentieth century Chinese philosophy. It is a unique chapter in the global history of modern philosophy in which Chinese philosophy emerged and engaged in the making of the world philosophy today. Tscha Hung played undoubtedly an irreplaceable role in this process through his involvement with the Vienna Circle and his reconstruction of Chinese philosophy.

Acknowledgements I am very grateful to Professor Hans Sluga for his careful reading of my draft and many helpful comments. I would like to thank Professor Fredrich Stadler very much for his great patience in waiting so long time for the completion of this work. Thanks are also due to Professor Li He, former editor of the *Journal of World Philosophy*, who allowed me to quote some materials from my Chinese paper on the reception of the Vienna Circle that was published in China in issue 6 of the journal in 2009.

References

Carnap, Rudolf. 1996. The Elimination of Metaphysics Through Logical Analysis of Language. In *Logical Empiricism at Its Peak: Schlick, Carnap, and Neurath*, ed. Sahotra Sarkar, 10–32. New York: Garland Publishing, Inc.

Chen, De-rong. 2011. *Metaphorical Metaphysics in Chinese Philosophy: Illustrated with Fung Youlan's New Metaphysics*. New York: Lexington Books.

Chen, Xiao-ping. 2019. Fung You-lan's Metaphysics and Logical Empiricism with Comments on Tscha Hung's Criticism to Fung's Metaphysics. *Studies in Philosophy of Science and Technology* 36 (2): 25–31.

Chow, P.K. 1992. Three Essays of Tscha Hung and His Philosophical Insights. *Philosophical Researches* 4: 78.

Confucius, 2009. *The Confucian Analects, The Great Learning & The Doctrine of the Mean*, Trans. James Legge. New York: Cosimo, Inc.

Cohen, R.S., R. Hilpinen, and Renzong Qiu, eds. 1996. *Realism and Anti-Realism in the Philosophy of Science*. Dordrecht: Kluwer Academic Publishers.

Deleuze, Gilles. 1994. *Difference and Repetition*. Trans. Paul Patton. London: Athlone Press.

Ding, Wen-jiang. 1923. Metaphysics and Science. *Nu Li Weekly*, April 15–17.

Du, Xiao-zhen. 2005. We Owe to Him. On the 100th Birth and 25th Death of J-P. Sartre. *Essays*. 3.

Fan, Dainian, Wengeng Hu, and Chunxiu Liang. 1992. The Propagator of Logical Empiricism in China – The Member of Vienna Circle: Tscha Hung. *Journal of Dialectics of Nature* 14 (3): 64–80.

Fung, You-lan. 1986. *Collection of San Song Tang*. Vol. 5. Zhengzhou: Henan People's Publishing House.

Hacohen, Malachi H. 1998. Karl Popper, the Vienna Circle, and Red Vienna. *Journal of the History of Ideas* 59 (4): 711–734.

He, Fang-yu. 2007. The Unity of Learning and Academics: Another Dimension of Tscha Hung's thought in the 1940s. *Reviews of Science and Culture* 4 (5): 15–26.
He, Lin. 2002. *Chinese Philosophy in Half a Century*. Beijing: Commercial Press.
Hu, Jun. 2002. *Analytic Philosophy in China*. Beijing: Capital Normal University.
———. 2013. *The Modern Shift of Chinese Philosophy*. Beijing: Peking University Press.
Hu, Nai. 1995. Minute of International Conference on Vienna Circle and Contemporary Science and Philosophy in Memory of Tscha Hung in Beijing. *Journal of Natural Dialectics* 1: 35.
Jiang, Yi, and Tongdong Bai. 2010. Studies in Analytic Philosophy in China. *Synthese* 175: 3–12.
Jiang, Zhaoyang. 2016. *The Aufheben of the Old Paradigm and the Establishment of the New Norms: Tscha Hung and the Modern Change of Chinese Philosophy*. Xiamen: Xiamen University Press.
Jin, Xi-ping. 2000. Two Puzzles on Heidegger's Case. *Open Times* 11: 29–35.
Li, Wei-wu. 2008. *The Model Shift of Modern Chinese Philosophy*. Beijing: Zhonghua Books.
Mihina, František. 2016. Moritz Schlick – Ethics from a meta-theoretical perspective. *Ethics & Bioethics* 6 (3–4): 109–118.
Nyiri, J.C., ed. 1981. *Austrian Philosophy: Studies and Texts*. Munich: Philosophia.
———., ed. 1986. *From Bolzano to Wittgenstein: The Tradition of Austrian Philosophy*. Vienna: Hölder-Pichler-Tempsky.
Oberdan, Thomas. 2017. Moritz Schlick. *The Stanford Encyclopedia of Philosophy* (Winter 2017 Edition), Edward N. Zalta (ed.), https://plato.stanford.edu/archives/win2017/entries/schlick/.
Parr, Adrian, ed. 2005. *The Deleuze Dictionary*. New York: Columbia University Press.
Popper, Karl. 1974. Autobiography of Karl Popper. In *The Philosophy of Karl Popper*, ed. Paul A. Schilpp, *The Library of Living Philosophers*, v. 1, 2–184. Chicago/New York: Open Court Publishing.
Ren, Ji-yu. 2006. Recollection of the Life in NSAU During the War. *Beijing Daily*, April 3, 20.
Ribeiro, Cláudia. 2015. The Complementarity of Science and Metaphysics. *Philosophica* 90: 61–92.
Schlick, Moritz. 1979. *Philosophical Papers* (Volume II), ed. H. L. Mulder and B. F. van de Velde-Schlick. Dordrecht: D. Reidel.
Smith, Barry. 1994/1998. *Austrian Philosophy, The Legacy of Franz Brentano*. Chicago: Open Court.
Stadler, Friedrich. 2001. *The Vienna Circle, Studies in the Origins, Development and Influence of Logical Empiricism*. Wien/New York: Springer.
Stuchlik, Joshua. 2011. Felicitology: Neurath's Naturalization of Ethics. *HOPOS: The Journal of the International Society for the History of Philosophy of Science* 1 (2): 183–208.
Tscha Hung. 1944. A Interpretation of Academics. *Thought and Times* 31: 35–47.
———. 1989. *The Philosophy of Vienna Circle*. Beijing: Commercial Press.
———. 1990. *Essays on Logical Empiricism*. Hong Kong: SDX Joint Publishing Company.
Wang, Wei. 2006. *Academic Essays of Wang Wei*. Shanghai: Shanghai Translation Publishing House.
Wang, Zhong-jiang. 1993. Jin Yue-lin and Positivism. *Philosophical Researches* 11: 47–50.
Wang, Zi-song. 2007. A Memory on Professors in Department of Philosophy of NSAU. In *Not Just for Memory*. Beijing: SDX Joint Publishing Company.
Yang, Guo-rong. 1996. *Positivism and China from Yan Fu to Jin Yue-lin*. Beijing: High Education Press.
Zhang, Dai-nian. 1996. *Collection of Zhang Dai-nian*. Vol. 8. Shijiazhuang: Hebei People's Publishing House.
Zhang, Jun-mai. 1923. The Conception of Life. *Tsinghua Weekly*., No. 272.
Zhang, Shen-fu. 1985. *Academic Essays of Zhang Shen-fu*. Jinan: Qi Lu Books.
Zhang, Zai. 2009. *Complete Works of Master Zhang, Supplements to Reflections on Things at Hand*. Trans. Chan Sin-wai. Hong Kong: Chung Hwa Book Co.

Chapter 12
Tolerance, Disagreement, and the Practical Dimension of Philosophy: Warren Hagstrom's Interview with Carnap

Adam Tamas Tuboly

12.1 Editorial Introduction

Rudolf Carnap's "Intellectual autobiography" from *The Philosophy of Rudolf Carnap* volume is perhaps one of the most often cited and quoted autobiography of the *Library of Living Philosophers* series. Carnap wrote an especially long piece; most of these autobiographies are no longer than 20–30 printed pages, while Carnap's essay was more than 80 pages. The autobiography is known for many reasons. First of all, the mature Carnap has either forgotten or intentionally masked many of his previous commitments and influences; among these, we find the socio-political lines of his thinking (see Damböck 2017; Tuboly 2017a) and many of his so-called continental philosophy sources like neo-Kantians and Husserl (see Friedman 1999; Richardson 1998; Carus 2016). But the autobiography is also known for its detailed discussions of Carnap's views (especially its second part, "Philosophical Problems", Carnap 1963a, 44–84). A variety of papers and books on Carnap's philosophy thus took his autobiography as a starting point to refute, to correct, or to support (recently, Anne Siegetsleitner (2019) had treated in detail Carnap's autobiography).

In that piece, Carnap often came back to an idea that was less discussed and noted while he was alive, but became the most significant guideline in his recent reevaluation; this is the *principle of tolerance* and the *metaphilosophy* that follows from it. The principle says that

> *In logic, there are no morals*. Everyone is at liberty to build up his own logic, i.e. his own form of language, as he wishes. All that is required of him is that, if he wishes to discuss

A. T. Tuboly (✉)
Institute of Philosophy, Research Centre for the Humanities Budapest, Hungary

Institute of Transdisciplinary Discoveries, Medical School, University of Pécs, Pécs, Hungary

E. Ramharter (ed.), *The Vienna Circle and Religion*, Vienna Circle Institute Yearbook 25, https://doi.org/10.1007/978-3-030-76151-6_12

> it, he must state his methods clearly, and give syntactical rules instead of philosophical arguments. (Carnap 1934b/1937, 52; original emphasis)

This liberating principle is the result of a long struggle within himself and the circle of his fellows in the Vienna Circle (Carus 2007). Carnap realized that many of his debates in the 1920s and early 1930s were conducted in a win-or-lose manner, that is, one party had to own privileged access to truth, i.e. only one of them was able to see the world as it is. At first, he had to transform these seemingly world-related debates into word-related ones, that is, he had to show that our philosophical debates concerned the very *languages* themselves we were about to use and not about fixed and independent furniture of a determinate world. But one might say that even if our debates are about language, there is one correct language that describes the world as it is; so we did not gain much after all. In fact, for a while, at the beginning of his linguistic turn, Carnap indeed accepted the legitimacy and existence of one fixed language that might capture everything needed. Languages are determined or constituted by their internal rules, captured by logic. So if one finds the right logic, the only valid forms of logical rules would mark the road to the right language.

But within a short time, and after many sleepless nights, Carnap changed his mind and went for the view that language is not a fixed issue, and especially not a once-for-all issue as there is no such thing as the right logic:

> The first attempts to cast the ship of logic off from the *terra firma* of the classical forms were certainly bold ones, considered from the historical point of view. But they were hampered by the striving after 'correctness'. Now, however, that impediment has been overcome, and before us lies the boundless ocean of unlimited possibilities. (Carnap 1934b/1937, xv)

Deviations from received norms in the past of logic were considered to be "bold" moves, detours from the right path. As time went on, philosophers and logicians find out that alternative logics are "alternative" only in their names, and they constitute entirely new domains, not worse, not better *in themselves* than others. They are alternative not in contrast to a right or normal logic, but alternative from the *conventional* domain and rules of logic. If you have to choose between them, you have to choose not on the base of which is the only legitimate and right one, but on other, more conventional grounds.

But this is just logic; as Carnap recaptured philosophy as a linguistic issue, as an activity of evaluating and choosing languages, he declared in his autobiography that philosophy in its entirety depends on practical considerations. Here is a longer quote:

> Since my student years, I have liked to talk with friends about general problems in science and in practical life, and these discussions often led to philosophical questions. [...] Only much later, when I was working on the *Logischer Aufbau*, did I become aware that in talks with my various friends I had used different philosophical languages, adapting myself to their ways of thinking and speaking. With one friend I might talk in a language that could be characterized as realistic or even as materialistic; [...]. Not that the friend maintained or even considered the thesis of materialism; we just used a way of speaking which might be called materialistic. In a talk with another friend, I might adapt myself to his idealistic kind of language. [...] With some I talked a language which might be labelled nominalistic, with others again Frege's language of abstract entities of various types, like properties,

> relations, propositions, etc., a language which some contemporary authors call Platonic. *I was surprised to find that this variety in my way of speaking appeared to some as objectionable and even inconsistent.* I had acquired insights valuable for my own thinking from philosophers and scientists of a great variety of philosophical creeds. When asked which philosophical positions I myself held, I was unable to answer. I could only say that in general my way of thinking was closer to that of physicists and of those philosophers who are in contact with scientific work. Only gradually, in the course of years, did I recognize clearly that my way of thinking was neutral with respect to the traditional controversies, e.g., realism vs. idealism, nominalism vs. Platonism (realism of universals), materialism vs. spiritualism, and so on. (Carnap 1963a, 17-18)

This might be considered as a biographical background of Carnap's tolerance and unconscious metaphilosophy: adapt yourself to your partner's requirement but always keep in mind the further goals of the discussion in order to arrive at balanced judgments of the chosen languages.

What Carnap hints in the above passage is a very practical attitude in choosing languages, or as he came to call them later, frameworks. As his tolerance led him to the acceptance of usually conceived-to-be problematic entities (such as abstract entities), Carnap became more and more explicit about his method and metaphilosophy. His famous paper from 1950, "Empiricism, Semantics, and Ontology", became a very definite statement on how one shall behave when confronted with the weight of decision and exploration of alternatives:

> The acceptance or rejection of abstract linguistic forms, just as the acceptance of rejection of any other linguistic forms in any branch of science, will finally be decided by their efficiency as instruments, the ratio of the results achieved to the amount and complexity of the efforts required. [...] The history of science shows examples of [...] prohibitions based on prejudices deriving from religious, mythological, metaphysical, or other irrational sources, which slowed up the developments for shorter or longer periods of time. Let us learn from the lessons of history. Let us grant to those who work in any special field of investigation the freedom to use any form of expression which seems useful to them; the work in the field will sooner or later lead to the elimination of those forms which have no useful function. *Let us be cautious in making assertions and critical in examining them, but tolerant in permitting linguistic forms.* (Carnap 1950/1956, 221, original emphasis)

Three things shall be noted here. One of them is the repeated emphasis on *history* and what history teaches us. For some reason, Carnap was confident in the idea that what history has taught, is that we shall take a tolerant and liberal position towards actions in science and philosophy, because dogmatism and uncritical prejudices tie one's hands. (Note that this is the very opposite view of Thomas Kuhn, who argued that what history has taught us is the inevitable, inherent, and innovative character of dogmatism in science; see Reisch 2019.) Secondly, Carnap laid down here a quite positive, somewhat idealistic, and utopian view on scientific and philosophical inquiry. Grant people freedom and some courage, and then they will weigh the merits and drawbacks of their positions and will choose what benefits them the most.[1]

[1]This view of Carnap about freedom, responsibility, and emphasizing the inner capability of making matured decisions about one's life and values goes back to Carnap's youth when he was a

Thirdly, "efficiency as instruments" are practical terms. Even though cognitive characteristics might be considered as relevant in choosing instruments, their practical dimensions dominate the situation. In choosing one framework, or language, or logic, one might be guided by certain theoretical considerations, but (1) all comes down to practical decisions at the end. To give Carnap's famous example from 1934, consider an apple lying on the table. Shall I eat it? Many arguments and considerations could be put forward (about its effect on health, about being possible poisoned, etc.), but all of them come down to implicationist suggestions: if you eat it, this and this will happen, if you do not eat, this and this will happen. While these could orient one if making a decision, the resolution of the situation could only be a definite *action* in this or that direction (Carnap 1934a).

Also, note that even those theoretical considerations that could guide you are the results of previous decisions and practical choices of frameworks and rules. Applying the rule of non-contradiction in a given situation presupposes that one has already *accepted* the rule of non-contradiction in a previous scenario. Thus one is getting closer to a dialectical circle (Carus 2007): your theoretical knowledge (which is utilized in making decisions) is the result of the acceptance of a linguistic framework, which is the result of a practical decision, influenced by previous theoretical knowledge.

These ideas are scattered through Carnap's writings from the mid-1930s to the late 1960s, but he became quite explicit in a less-discussed small booklet, *The Continuum of Inductive Methods* (1952). There he says,

> [A choice] is fundamentally not a theoretical question. A possible answer to a theoretical question is an assertion; as such it can be judged true or false, and if it is true, it demands the assent of all. Here, however, the answer consists in a practical decision to be made by *X*. A decision cannot be judged as true or false but only as more or less adequate, that is, *suitable for given purposes*. However, the adequacy of the choice depends, of course, on many theoretical results concerning the properties of the various inductive methods; and therefore the *theoretical results may influence the decision*. Nevertheless, the decision itself still remains a practical matter, a matter of *X* making up his mind, like choosing an instrument for a certain kind of work. (Carnap 1952, 53; emphasis added)

This passage shows quite clearly the dialectical process Carnap aimed at in describing the functioning of theory and concept formation. While he is often associated with a manly technical view of how theories are built up and how science works (within a complex web of algorithmic-like mechanistic processes), in fact, he had more subtle views, leaving room for further contextual and practical considerations. Here is another longer passage in support of this interpretation of Carnap's work.

> It is psychologically difficult to change a faith supported by strong emotional factors (e.g., a religious or political creed). The adoption of an inductive method is neither an expression of belief nor an act of faith, though either or both may come in as motivating factors. An inductive method is rather an instrument for the task of constructing a picture of the world

member of the so-called German Youth Movement. This idea is explored in great detail by Carus (2007) and in the chapters of Damböck et al. (2021).

> on the basis of observational data and especially of forming expectations of future events as a guidance for practical conduct. *X* may change this instrument just as he changes a saw or an automobile, and for similar reasons. If *X*, after using his car for some time, is no longer satisfied with it, he will consider taking another one, provided that he finds one that seems to him preferable. Relevant points of view for his preference might be: performance, economy, aesthetic satisfaction, and others. Similarly, after working with a particular inductive method for a time, he may not be quite satisfied and therefore look around for another method. [...] Here as anywhere else, *life is a process of never ending adjustment; there are no absolutes neither absolutely certain knowledge about the world nor absolutely perfect methods of working in the world.* (Carnap 1952, 55; emphasis added)

By denying any absoluteness and certainty in the world (and science is, of course, part of the world), Carnap opens up the possibility or even necessity of making practical decisions at every point of our practical and scholarly life. A theory is not better or more distinguished than any other tool that we make and use to find our way through the forest of life. Nothing proves to be an exception here; neither physics, nor mathematics, nor even logic. Everything comes down to decisions, to *feelings* as Carnap says in the interview.[2]

Although these points are summarized and systematized in various papers and books (most notably in the essays of Wagner (2009, 2012) and the 2007 book of A. W. Carus), the present interview below might be of particular importance because it is the longest discussion of Carnap himself about these issues. In fact, the interviewer of Carnap presumably did not know the richness of Carnap's metaphilosophy for a sociological study of science (as we will see, perhaps even Carnap did not know that).

In March 1961, the UCLA sociologist Warren O. Hagstrom was working on a project about the social organization of science. As a fellow of the Social Science Research Council, Hagstrom conducted interviews with scientists, mainly with mathematicians and physicists, to ask them about "patterns of communication and collaboration in the sciences". In the case of Carnap, Hagstrom noted topics in which Carnap was "unusually well informed", such as "relations between philosophers, logicians, and statisticians, attempts to introduce results from logic and statistics into philosophy, competing approaches to the interpretation of probability, and the social conduct of disputes about the foundations of mathematics."[3] Hagstrom

---

[2]There is another interesting line of thought in the interview, namely a finitist motive. While the logical space of reasons and possible paths might be infinite, our life and *practical possibilities* are finite, thus we have to make decisions. As Carnap says in the interview: "Here we are concerned with the fact that every philosopher and every scientist has to make up his mind which way to take because there is not enough time to try to all possible ways, and therefore he chooses his approach according to his mostly unconscious judgment or feeling or hunch of fruitfulness of this or that approach." This might even come close to the idea that Carnap often expressed (e.g. in the preface to the *Aufbau*), namely that philosophy is a collective enterprise; thus as a collective, philosophers might explore different paths, which might lead to different results, but as they contribute to the same enterprise, they do not fight or compete with each other but supplement each other's inquiries.

[3]One of the main examples of disagreement in the interview concerns debate over the foundations of mathematics; the other major example concerns, however, what is usually called the debate between ideal and ordinary language philosophy. By the time the interview was conducted Carnap

promised Carnap that he would quote the interview in his work confidentially without giving names, but later Carnap permitted him to use names and longer quotes as well.[4] After Hagstrom approached Carnap on the 5th of March 1961, he sent him back the transcript of the conversation in June 24, 1961, thus the whole interview happened quickly.

As Hagstrom was mainly a sociologist (of science), his questions concerned topics such as agreement and disagreement among scientists in the community, boundaries of science and transgressions of those boundaries, social reasons, and dimensions of acceptance and rejection of theories. What we can see from the interview is that Hagstrom tried to get from Carnap considerations of the above character. He starts, for example, by asking about differences – thematically and institutionally – between logic, mathematics, and philosophy. This was a well-known territory and social struggle for Carnap; as a philosopher of science, or philosopher of physics, he had troubles already back in 1920 with his doctoral dissertation. At first, he was sent back and forth between the philosophy and the physics department, and finally, Bruno Bauch took him at the department of philosophy. As Carnap (1963a, 11) recalled, "this experience with my thesis project, which seemed to fit neither into physics nor into philosophy, made clear to me for the first time what difficulties I would continually have to face in the future".

This might be a simple occasion of institutional difficulty in the highly hierarchical and authoritarian German system, but Carnap saw the sociological implications as well:

> If one is interested in the relation between fields which, according to customary academic divisions, belong to different departments, then he will not be welcomed as a builder of bridges, as he might have expected, but will rather be regarded by both sides as an outsider and troublesome intruder. (Carnap 1963a, 11)[5]

---

already read and replied to all of the paper in his forthcoming Festschrift, *The Philosophy of Rudolf Carnap*, which included numerous pieces that concerned the philosophy of language. But one of them was especially relevant in this context, namely Peter Strawson's "Carnap's View on Constructed Systems versus Natural Languages in Analytic Philosophy" (Strawson 1963). In his reply, Carnap emphasized his method of explication and its relevance to ordinary language and claimed that "language, whether natural or artificial, is an instrument that may be replaced or modified according to our needs, like any other instrument" (Carnap 1963b, 938). Regarding this debate (see, e.g., Franco 2018), this interview contains some nuances and further considerations as well; even about how Carnap viewed the history and ramification of analytic philosophy (or philosophies).

[4]See Hagstrom to Carnap, March 5, 1961 (ASP RC 089–50-03), Hagstrom to Carnap June 24, 1961 (ASP RC 089–50-02), and Carnap to Hagstrom September 23, 1961 (ASP RC 089–50-01). Nonetheless, Carnap's name does not appear in Hagstrom's book. See Hagstrom (1964). Nonetheless, Carnap's interview is often quoted by Hagstrom, referring to him as "an eminent logician".

[5]I argue elsewhere that the very same thing happened to Philipp Frank in the United States when he became an outsider both to sociology and philosophy with his integrated history and philosophy of science. See Tuboly (2017b) and Reisch and Tuboly (2021).

In fact, Carnap envisioned "in a few decades [...] a kind of department of pure and applied logic." This would have, as Carnap claimed, a cleaning effect as well; "I think this is the next field that will come free of this total melting pot that is called philosophy" (see the interview below). Though in Carnap's time and shortly after only Ph.D. programs and graduate schools were available in pure logic, later departments of logic were indeed established.

Hagstrom tried to get from Carnap as many social considerations as possible, though Carnap was reluctant to get involved in the sociologization of scientific issues. Typical questions of Hagstrom run like, "Do you suppose mathematicians and other scientists may resist logic (and philosophy) because of its basically critical functions?", "If a man's choice of an approach is a matter of individual feeling, what kind of experiences are most important in shaping this kind of preference?" The latter especially points toward the sociology of science as it relates to the question of which factors determine one's construction of scientific habit and method. That is, if you accept and make explicit the idea that in science it all comes down to practical issues of decisions, based on personal feelings and antecedent experiences, then a wide range of social determinants are opening up.

Interestingly, Carnap got the idea and the connotation of Hagstrom's question immediately. He started to talk about "the sociology of the origin of ideas", "the psychological origins of Einstein's ideas, [and the] historical origins". These are important and interesting topics according to Carnap; he even adds that some differences between philosophical schools and arguments "can be explained partly with sociological reasons". But he quickly continues as "after that is said we still have to solve the really philosophical problems, and they have to be solved with philosophical arguments not with sociological arguments". So, for Carnap, the sociology of science is a pre-study, or a parallel study of science and philosophy, not such a substantive field that might contribute to the *actual* set of scientific knowledge and cannot establish philosophical positions per se. To use some old distinctions, the sociology of science belongs to the context of discovery, and not to the context of justification.[6]

What we see here is also a crucial difference between actual sociology of science and logical empiricists' ideas about the field. While Carnap quickly falls back on some subjective factors, like feelings, and irrational influences, Hagstrom tries to force genuine issues like disagreement, capacity to cooperate, competition, institutional embededness, and so on. So one has to be very cautious about how close can Carnap's practically oriented metaphilosophy and voluntarist (Jeffrey 1994) conception of science be moved to sociological considerations about scientific knowledge; this warning for carefulness may not be novel or striking at all. What is more important, however, is that there was real tension in Carnap regarding the recognition of the subjective, voluntarist, practical, and even emotional basis of

[6]Note, however, that Carnap accepted a broader picture of philosophy of science as well, a so-called bipartite metatheory in which both the sociological and the logical inquiries have their own important place, contributing to the general enlightening program. See Uebel (2012).

knowledge and between the logically analyzable, reconstructable, and rationally comprehensible character of accepted intersubjective *knowledge*.[7] This interview, published here for the first time, contains the leitmotifs of Carnap's metaphilosophy and the subjective roots of all intersubjective knowledge that contribute to genuine scientific and philosophical disagreements.

12.2 The Interview

Interview of Professor Rudolph Carnap, Department of Philosophy, UCLA

By Warren O. Hagstrom

(This interview took place in Professor Carnap's home on March 15, 1961. This transcript has been slightly edited.)

Hagstrom: *Right now logic is being worked on by both mathematicians and philosophers. Is it possible to distinguish the logic done by each, or is it essentially the same?*

Carnap: Well, the new philosophy is different. By the new philosophy I mean the philosophy which to a large extent was developed in the present century, although the first systems originated in the middle of the last century. For example, there were George Boole and Frege ...

H: *Do you regard this as philosophical?*

C: As soon as they called their work logic they were conscious that it was in an intermediate field between philosophy and mathematics, because logic has always been cultivated by philosophers rather than mathematicians. In mathematics there was a critical situation in the nineteenth century. Mathematics had made very great progress in various fields, especially in analysis, the theory of real numbers and their functions, and this culminated in what we today call classical analysis, especially in France and Germany. This was a very fruitful development. As you know, in science there is often an alternation between periods in which there is much growth in size and content and new results, and periods in which there is a reconsideration of the foundations on which one has worked – whether they are solid, what are really primary concepts, are they entirely clear, does the whole edifice have a sound foundation on the ground. These re-examinations are usually not made when there is very much progress – they are so proud, so happy, they are finding new results, they have no time. Nobody is there to ask questions like, what about your basic concepts? Are they entirely clear from a philosophical point of view? They must be reconsidered; they are not[8] vital. So whether it is in physics, or chemistry, or mathematics, there are sometimes these different kinds of periods. So, after this period of rich development, a few people began to say, 'Are we quite

[7]The conception of rationality in Carnap was scrutinized recently by Carus (2017), while the intersubjective and political character of knowledge was detailed by Uebel (2020).

[8]"Not" is put into parentheses and underlined by pencil, seemingly by Carnap.

sure that all this wonderful growth is really on a sound basis?' And they began to study the bases again. This hadn't been done for quite some time; Newton and Leibniz had made some questions of that kind. They realized they needed a kind of logical analysis of their concepts. Where could they find a good theory of concept formation? Obviously in logic. Where were the people who cultivated logic? They were philosophers. So they looked in logic books and found them totally inadequate for what they needed. So instead of waiting for 800 years for the philosophers to develop a better logic to cope with their troubles, they sat down and made their own logic. Augustus Morgan[9] was one of them ... And Frege built the first really comprehensive system in the 80's of the last century, and his big work appeared in 2 volumes, in 1893 and 1903.[10] It included very strange formulas; if you look at them you are frightened, you think you cannot understand it, and none of the other mathematicians looked at what he did. It was so hard to understand; he had entirely new ideas.[11] But he was sure that it was a really important thing. First, that mathematics needed a better foundation. He said it was scandalous that mathematicians talk about numbers as their main objects, and no mathematician was able to give a clear definition of what number is. He wrote one or two satirical things so very ironic – he made fun of the various definitions given of number. His big work is called (Basic Laws of Arithmetic).

I think the reason logic was now made by mathematicians was probably the need for usable logic, not merely that logic was an interesting field. He built it in a very comprehensive way, so that the old dream of Leibniz was fulfilled, that what we need is a logic of relations, not only of attributes. He used this logic to show that every mathematical concept, beginning with 1, 2, 3 ... the general concept of a number, to integers, to rational numbers, to complex numbers – every concept occurring in pure mathematics can be defined on the basis of purely logical concepts. That was his great achievement. Part of that had already been done by mathematicians before; they had shown that in speaking about real numbers, it can be translated into speaking of rationals, integers, and finally natural numbers. But they thought you cannot analyse the concept of natural number; these things are so natural, so basic, so elementary and simple, there is no point in asking for analysis into anything simpler – they are the simplest possible concepts in that field. Kronecker[12] expressed this in his half-jocular statement that God made the natural numbers, everything else in mathematics is man's handiwork. Frege said it is not

[9]Augustus De Morgan (1806–1871) was a British mathematician and logician; "De Morgan's laws" are named after him. The laws concern the expression of conjunction and disjunction in terms of each other with negation.

[10]Carnap has in mind Frege's *Grundgesetze der Arithmetik* (Jena: Verlag Hermann Pohle, Band I/II), translated recently completely by P. Ebert and M. Rossberg (with C. Wright) as *Basic Laws of Arithmetic*: *Derived using concept-script*, Oxford: Oxford University Press, 2013.

[11]About the detailed reception of Frege's work, and how his reception influenced him, see Dale Jacquette, *Frege: A Philosophical Biography*, Cambridge: Cambridge University Press, 2019.

[12]Leopold Kronecker (1823–1891) was a German mathematician who criticized heavily Cantor's set theory. The mentioned saying of Kronecker comes presumably from a 1886 lecture of his.

the case that natural numbers cannot be analyzed; the mathematicians forget the main function that numbers have in human life, which is not in mathematics but in everyday life when we apply them to factual questions – how many books are on the table? When I say the number of books is 5 what do I mean by it? If I can answer that question, I have found the meaning of number. And so he gave a definition, using only the concepts of existence and the ordinary connectives like 'and' and 'not' and 'or' and 'identity.' 'There is an x, and there is a y, and their is a z,' (let's say 3 instead of 5) 'such that x is not identical with y, y is not identical with z, and x is not identical with z, such that x is a book on this table, y is a book on this table, z is a book on this table, and every book on this table is ether identical with x, or identical with y, or identical with z' – This whole statement is a translation of 'There are 3 books on this table,' or 'The number of books on this table is 3.' Here we have an analysis of '3' into existence, negation, conjunction, and so on.

That was really a great achievement. I happen to have been a student of Frege; I studied with him.[13] He was entirely unknown. Russell later[14] read his books and said that some things, such as this definition of number, had been found long before {him} by Frege and were discovered independently by him {Russell} later; Russell didn't know of these books – they were not even reviewed in the mathematical periodicals.[15] Frege was an old and rather bitter and disappointed man, because no mathematician paid any attention to him – nor any philosopher before Russell and Whitehead. I remember what a great satisfaction it was for me, when the Association for Symbolic Logic was founded,[16] – I was at Harvard for one year,[17] and Whitehead gave a talk on how fortunate it was that the Association was founded. And then he made some remarks about the development of logic in these 100 years and said that Frege was the greatest logician in the nineteenth century. Most of the people there, I believe, didn't even now the name. And now everybody knows Frege's name.

---

[13]Carnap has spent the winter semester 1910/11, the summer semester 1913 and the summer semester 1914 with Frege in Jena. His lecture notes of the courses "Begriffschrift I, II" and "Login in Mathematics", have been published in *Frege's Lectures on Logic: Carnap's Student Notes 1910–1914*, trans. and ed. by Erich H. Reck and Steve Awodey. Chicago and LaSalle, Illinois: Open Court, 2004.

[14]Although Russell has read some of Frege's work already in 1895, he did not understand it that time due to Frege's special symbolism. He famously notified Frege, however, that he has found a contradiction in his system in 1902, already after he [Russell] was working on his own system.

[15]In fact, though most of them were dismissive, critical and deeply misunderstood his endeavor, Frege's first book, the *Begriffschrift* earned at least six reviews by such figures as Reinhold Hoppe, Kurd Lasswitz, Paul Tannery, Karl Michaëlis, Ernst Schröder, and John Venn.

[16]The Association for Symbolic Logic was founded in late 1935 (after a year of discussions at Harvard); its first president was C.J. Ducasse and vice-president H.B. Curry. On its early history see C.J. Ducasse and H.B. Curry, "Early history of the Association for Symbolic Logic," *The Journal of Symbolic Logic* 27 (3) (Sep. 1962): 255–258.

[17]Carnap has spent the academic year 1940/41 at Harvard. See Greg Frost-Arnold, *Carnap, Tarski, and Quine at Harvard: Conversations on Logic, Mathematics, and Science*, Chicago: Open Court, 2013.

So the development of this new logic was a strange thing, in between philosophy and mathematics. Most mathematicians didn't pay much attention to what was happening, but now interest is very much increasing, especially here. When I came to this country in 1936[18] there was a little beginning; a very few philosophers had begun to study modern logic. I said everywhere that this is a very important instrument for philosophical thinking.

There are different views about the nature of logic. Strictly speaking, I would not say, as most philosophers have said since Aristotle and Plato, that logic is a part of philosophy, a philosophical theory. I would say that as we see it now – we see it now clearer than Frege and Russell could see it – philosophy can use and should logic in order to come to a better analysis of the concepts with which philosophers work and in which they are interested and a clarification of the problems of philosophy, but it is a method, it is not really part of logic [philosophy]. A problem in logic, such as, for example, whether a sentence of such and such a form is logically developed or not, may take a long time to solve; when you have solved it you may say, this is a result in the field of logic. This is not a philosophical result. Why should there be a philosophical question whether or not this formula is valid? It is a question of a pure, abstract, purely formal nature. So if we understand mathematics in a very very wide sense as a science or system of knowledge of pure form without content, then this is basic mathematics: logic is the basics of mathematics. And it serves, then, in all fields, since every scientist is interested in whether something is derivable or not, whether this is compatible with that, and so on, and so he must use the methods of logic. I try sometimes to explain this to philosophers by saying that the relation is similar to the relation between mathematics and physics. Mathematics is not part of physics. Physics is concerned with nature, mathematics is not about nature. Physicists have to use mathematics as a method. I think that is the same with logic. I think that the fact that for 2000 years logic was part of the work of the philosopher was merely for the historical reason that any new field is first studied by men who just need it, and then it takes a long while until the field is independent. Remember psychology: first the philosophers had it. There was no one else around – the zoologists don't do it, the astronomers don't, so philosophers had to, since any new interesting thins is done by philosophers. Then a few decades ago the psychologists said that 'we have our own science, we don't want to be part of philosophy,' so they made themselves independent.

H: *That was about the same time as logic emerged...*

C: Yes. I talked with Boring[19] about it once when I was at Harvard. His story of the history of the Harvard psychology department was interesting, how they had to struggle with the philosophers. Instead of being happy that psychologists were ready

---

[18] Carnap arrived to the States in December 1935 for the tercentennial of Harvard, celebrated in late 1936, where he was an invited speaker and took his honorary doctoral degree.

[19] Edwin Boring (1886–1968) was an American experimental psychologist at Harvard. He wrote a lot about the history of psychology, e.g. *A history of Experimental Psychology* (New York: Century, 1929), and *Sensation and Perception in the History of Experimental Psychology* (New York: Appleton-Century-Crofts, 1942).

to form their own department, they said 'No, you are philosophers and you stay here in the philosophy department.' They had to struggle for years to convince the philosophers and the administration that not everybody knows psychology, that it's a science. It cannot be part of philosophy, although there are philosophical questions about psychology just as about physics and about history.

H: *With logic it was different, wasn't it? The philosophers resisted the development of logic, especially as applied to philosophy.*

C: Yes. Historically it is interesting how these resistances and inhibitions go. I remember I said to Boring that there should be a new department of the logic and methodology of science, as the basis of the whole scientific structure. Would your university be willing to make such a thing? Harvard was leading in many fields. He said 'If you had gone through the years of struggle to make a new department of psychology, you wouldn't do it for a second time. I'm all in sympathy with your ideas, but make your own struggles wherever you are.' I said, 'I'm not a man of struggle. I write books and I think about problems, and I'll leave it to other people to do it.'

H: *Do you think such a form of organization will occur? Berkeley has a graduate curriculum in logic and methodology.*[20]

C: Yes, they have a Committee. I think this is just a beginning; in a few decades, I don't know how long it will be, there will be a kind of department of pure and applied logic. What they call methodology there is simply applied logic, in some sense. The scientists are more and more aware that they need some study of logic and methodology, so it might be that they will press the university to make a department, because they want to send their students there, just as the physicists and astronomers send their students to the mathematics department. Just as you cannot do astronomy without mathematics, and students learn basic mathematics in a mathematics department; so they will also say to their students, go and learn their logic and methodology, they you will work better in your field, especially if your field is just growing and needs new concepts, and new kinds of concepts, then you must learn something about concept formation. This is very important in many fields. I am often asked by people in economics about the nature of their concepts, or by psychologists – about operational definitions, for example. They ask if all definitions must be operational, and I say no. And they say, you positivists are responsible for this restriction; they say that psychologists who make this restriction say they have learned it from Bridgman[21] and the Vienna Circle positivists. I say yes, that was 20 years ago, but we have learned something, that

[20]The University of California, Berkeley has initiated in 1957 an interdisciplinary graduate program about logic and methodology of science, following the ideas of Alfred Tarski. The program still exists today; about its history, see Paolo Mancosu, "The Origin of the Group in Logic and the Methodology of Science," *Journal of Humanistic Mathematics* 8 (1) (January 2018): 371–413.

[21]Percy W. Bridgman (1882–1961) was a Nobel Prize winner American physicist; among philosophers he is mainly known for his work on philosophy of science, especially on operationalism. About the early history of Bridgman and the Vienna Circle, see Sander Verhaegh, "The American

we must be more liberal with respect to forms of concept formation, and that is especially important for psychology. So they would like to have a place where they send their students. Of course today they send them to the philosophy department. But it is only certain things in philosophy; they have no time to study all of history of philosophy, aesthetics, metaphysics, and so forth. If there were a department of logic and methodology, students from various fields would come over and learn their methods. I think this is the next field that will come free of this total melting pot that is called philosophy.

H: *Where would you find yourself then? In philosophy or in this new area?*

C: I am interested in both. I am a logician, that is to say, I like to work in purely formal, entirely abstract problems, just because they fascinate me... But then again I am also a philosopher and interested in some of the basic problems of philosophy. In recent years, for example, I have been interested in inductive reasoning, how it is to be justified[22]; and so I combine them. I try to develop something out of deductive logic, forms that you can transfer to inductive logic and to confirmation of hypotheses in science. So I am interested in both pure logic and in philosophical problems, and in the philosophy of logic and the use of logic in philosophical problems. So if this split in departments were to occur today I would find it hard to say; first I would ask to teach in both departments. If some strange administration were to say that professors could teach in only one department, I would find it hard to say. Some physicists have the same problem. Von Neumann[23] was a mathematician who wrote on quantum mechanics, Weyl[24] was a physicist but wrote purely mathematical books; so it would be with me.

H: *Some of the scientists and mathematicians I have talked to express some kind of hostility to philosophy. When they call certain problems 'philosophical' it seems to mean that they are not interested in them and feel them none of their business. Have you had this kind of reaction?*

C: Very much. One must distinguish there – in Chicago, and partly also here – it may be different in Chicago now – the professors of mathematics were not really interested in foundations problems, with one exception. A few of their students somehow became interested, by reading this or that, and so when I gave from time to time a course in the logical foundations of mathematics a few of them came over; sometimes I had from 1/3 to 1/2 students from mathematics in such a seminar. When they took this course I told them they also needed some logic,

---

Reception of Logical Positivism: First Encounters, 1929–1932," *HOPOS: The Journal of the International Society for the History of Philosophy of Science* 10 (1) (Spring 2020): 106–142.

[22]See, e.g., Carnap, "Inductive Logic and Inductive Intuition," in Imre Lakatos (ed.), *The Problem of Inductive Logic*, Amsterdam: North Holland, 1968, pp. 258–314.

[23]John von Neumann (1903–1957) was a Hungarian mathematician, physicist, and computer scientist.

[24]Hermann Weyl (1885–1955) was a German mathematician, physicists and philosopher. He worked quite early on the theory of relativity; in his philosophy of physics, Weyl followed the path of Edmund Husserl's phenomenology and Brouwer's intuitionism at the same time until the late 1920s.

and many took some, also sometimes a semantics course. Then the problem arose when they asked to write a PhD thesis with me, and to my regret I had to tell them it was practically impossible. Of course they could write with me, but it must be in philosophy, which meant they had to learn the whole of history of philosophy, the whole of metaphysics, the whole of value theory, ethics, and aesthetics; but the students had studied mathematics and were interested in it. In Berkeley, as you said, this situation is now different. A student can write a dissertation in the boundary field. It was not possible there, and likewise for physicists it was not possible there. So we must distinguish among people who say 'Oh this philosophy.' There are two kinds of people who say that. Mathematics students would say that about those kinds of philosophy in which they were not interested; the other one was the resistance the students found in the mathematics department against the kind of philosophy in which they were interested. The mathematics department would count only a small number of non-mathematics courses toward completion of the degree in mathematics. Some professors discouraged students from taking philosophy courses as being of little value for work in mathematics. There is today an increasing number of young mathematicians who are interested in foundations; the growth is slow, as usual for such things. But after seeing how it has grown in the last 25 years, I am confident that the growth will go on, that mathematicians will see that logic is important for mathematics. Once they study it some come to the conclusion that this is not philosophy; logic itself, the propositional calculus, the predicate calculus... these formalized systems of logic are abstract systems which are not essentially different from the systems of group theory, abstract algebra, abstract probability, topological spaces, metrical spaces.... It is just another such system, so why not say that logic is part of mathematics? So some do it this way. The reason I now have few students in mathematics is that now the mathematics department has a professor interested primarily in mathematical logic; he is a pupil of Tarski,[25] the leading mathematical logician in the country. This is professor Chang.[26] Another young professor with us, Montague,[27] also a pupil of Tarski, teaches mathematical logic in the philosophy department. They are close friends and have created a colloquium[28] where mathematicians and philosophers can talk. So now, at long last, a close collaboration between some of the mathematicians and some of the philosophers exists. We have to see what will become of this field.

As I said, although this is just a subjective view, the purely forma part will either become a department itself or part of mathematics.

[25]Alfred Tarski (1901–1983) was a Polish logician and mathematician who emigrated to the United States in 1939. Previously he was a leading member of the prominent Lwów-Warsaw School of logic and mathematics; in the US, he thought logic and mathematics at the University of California, Berkeley from 1942 till his death in 1983.

[26]Chen Chung Chang (1927–2014) was a Chinese logician who obtained his PhD from Berkeley in 1955 under Tarski's supervision. Later he became a professor of mathematics at UCLA.

[27]Richard Montague (1930–1971) was an American mathematician, philosopher, and linguist.

[28]Chang, Montague and Tarski initiated a bi-weekly Logic Colloquium in the late 1950s, and organized it through the 1960s.

H: *Do you suppose mathematicians and other scientists may resist logic (and philosophy) because of its basically critical functions?*

C: This is right. When we write about the work done by a mathematician or some other scientist, the writing is basically critical. We say, 'Where are his definitions? These are not definitions as they should be. We have learned in modern logic more exact ways of doing some of the things he does.' There is a feeling that an outsider is criticizing their work. But I believe that was the same when the physicists began with quantitative method and measurement and so forth, since Galileo and in the last 300 years. They applied some mathematics, not especially new, and then probably there came some mathematical friends to them and said 'You should study this mathematics more carefully; in the differential calculus, for example you define acceleration as rate of change of velocity in time, and you assume the truth of some theorems which have not been proved, such as that every function must have a derivative at every point. You should look into the theory which tells you under what conditions there is such a thing instead of relying on intuition.' So perhaps the physicists listened to this and were grateful that the work had been done and they could use it.

H: *Maybe scientists like to hang on to intuitive ideas and the logician is the enemy of such.*

C: Well, of course, there were intuitive ideas in physics, and the mathematicians took them and made them precise and took them out of the realm of intuitive guesswork. Logic has served in a similar way. In mathematics the logician wanted to dig one level deeper in the groundwork rather than adding another level at the top.

H: *Does this have clear consequences for mathematics?*

C: Well, at the beginning they said 'you just talk about mathematics and give mere definitions, which may be more exact, but can you show that there is any point at which your investigations become relevant to mathematics?' This has been answered, I think. Today a great deal of mathematics and the whole of analysis is based on set theory, and set theory is nothing but a developed mathematical logic; today mathematicians proceed in a quite different way from what they did even 50 years ago. They are now on a sounder basis; but they say, 'set theory is really mathematics.' Well, you can call it mathematics or call it logic. It was developed by a mathematician who, like Frege, worked quite alone and was unappreciated by other mathematicians, Georg Cantor[29] of Germany. His work had some difficulties because some contradictions and antinomies were found in his work. So the mathematicians call set theory mathematics and are fairly comfortable with it, but if it were called logic they would feel it was entering mathematics from outside. So it is all a terminological distinction, and came about by historical accident. It is conceivable that in Greek times mathematicians would have felt the need for an exact logic to use in geometry and would have cooperated with philosophers to

[29]Georg Cantor (1845–1918) was a German mathematician, known mainly for his early works on set theory.

develop it or developed it themselves. One can imagine that logic, developed in fact by Aristotle, could have been developed by mathematicians. Then it would have been called mathematics all the time, and that would have removed some of the terminological inhibitions by mathematicians against using it. I think logic has been part of philosophy nor for any intrinsic reason but because of the general hospitality of philosophy, which has included besides logic, astronomy and psychology and other fields.

H: *Could we shift the discussion to cases of disagreement now? I believe some of them more or less arose out of set theory. I'm interested in disagreements as more of less limiting cases in science, because science is generally a conspicuously successful way of settling disagreements about beliefs.*

C: There are, of course, disagreements in science; without them there would be no progress.

H: *As I understand it, you played a significant role in some of the disputes about the foundations of mathematics. For example, didn't you take part in a famous debate on the subject in the 1920's?*

C: Probably you have in mind the symposium at the Königsberg meeting, Versammlung Deutscher Naturforscher und Ärzte, which corresponds to the AAAS, in 1930.[30] I had agreed to participate in a discussion of the three main directions in the foundations of mathematics which were developed at that time. I was an instructor in Vienna and talked about the logicist foundations of mathematics, meaning by that the conception of Frege and Russell, that mathematics can be developed on the basis of logic, or that concepts can be defined and theorems can be proved from logic. Then there was the school of Hilbert and other people in Göttingen, which was represented by von Neumann, who later went to the Institute of Advanced Studies at Princeton; this view was called Formalism. And then there was the young Dutch mathematician, Heyting,[31] a pupil of Brouwer,[32] the founder and leader of the intuitionist movement. I thought it would be good for these three outlooks to come together and compare notes, so to say – to see to what extent we would agree and to clarify the whole problem situation. So we had these three talks and then a nice discussion[33]; I think it was quite fruitful. I have always kept contact with these other schools; we must learn from each other and see where the differences exist. Misunderstanding is easy because each point

[30]Carnap talks about "The Second Conference on the Epistemology of the Exact Sciences in Königsberg" that took place on September 5–7, 1930. For some details see Friedrich Stadler, *The Vienna Circle: Studies in the Origins, Development, and Influence of Logical Empiricism*, Dordrecht: Springer, 2015, 161–166.

[31]Arend Heyting (1898–1980) was a Dutch mathematician and logician, mainly known for his work intuitionism.

[32]L.E.J. Brouwer (1881–1966) was a Dutch mathematician, logician, and philosophers, mainly known as the founder of intuitionism.

[33]Papers of the symposium were published later in Paul Benacerraf and Hilary Putnam (ed.), *Philosophy of Mathematics: Selected Readings, Second Edition*, Cambridge: Cambridge University Press, 1983 (originally 1964), pp. 41–65.

of view has developed its own terminology; sometimes the same terms are used, sometimes terms which those of different points of view are liable to misinterpret. Communication is sometimes quite difficult.

H: *But it is possible?*

C: I think so. Brouwer was especially difficult in that respect, and still is in some respects. He is old but still alive and quite lively. He came to Princeton when I was there in 1952–53, and he gave a few lectures; I also talked privately with him. The first contacts were when Brouwer came to Vienna and gave some talks there, and we then discussed them.[34] It was so hard to communicate; he was rather firm on his views and firm on the use of his terms; when we pressed him to say clearly how he meant to use these concepts he became impatient... I could much more easily talk with Heyting. I welcomed it as a great approach to better mutual understanding when Heyting published some treatises in the Berlin Academy Proceedings of calculi, of formal systems, first of logic and then of the basic fields of mathematics on the basis of his research in them.[35]

H: *This came much later didn't it?*

C: Yes, but while I was still in Vienna:[36] it must have been between 1930 and 1932. There Heyting did not only talk about it and outline some systems, as Brouwer had done, but constructed the whole system with all the rules, so that one could see which deductive processes are allowed and which are not, which forms of definition are accepted. We corresponded and I told him that this was the first time we really could understand intuitionism. General statements are not enough, we want to see how you work, and the most exact form of that is when you give the whole thing, all the rules, so we can see what you do. Then we can compare these rules with the rules of classical mathematics and see which are rejected. That was big progress. I insisted that when we get together we speak about these different language forms instead of saying that one has a wrong view, the other a right one. Let us say that you have language and I have mine, and let us also attempt to understand the motivations behind it: why do you reject my language? Do you think it is wrong? Do you think it is doubtful whether these procedures can be used? For what reasons? By speaking about languages and language rules instead of making it so philosophical, it is more concrete in the sense that we have something to grasp and talk about. I have often done this, also with von Neumann and others who were followers of Hilbert. I like to discuss with those who have views different from mine, and I always learn very

[34]Brouwer was invited to Vienna by the physicist Felix Ehrenhaft in 1928. Some members of the Vienna Circle attended his lectures; famously, Ludwig Wittgenstein was there as well, and many people note this event as his return to philosophy.

[35]See, e.g., Heyting, "Die formalen Regeln der intuitionistischen Logik" (1930) and "Die formalen Regeln der intuitionistischen Mathematik I, II" (1931), both published by the Deutsche Akadmie der Wissenschaften zu Berlin. In his *Logical Syntax of Language* (published originally in German in 1934, translation in 1937, Kegan Paul, Trench, Trubner & Co Ltd.), Carnap noted already that Heyting's work gave a much better approximation of what intuitionism might be consisted in, and thus made a solid ground for further discussions (see §16, "On Intuitionism").

[36]Carnap left Vienna before the winter semester of 1931 to take his new position in Prague.

much from it. I like to find forms of questions and discussions by which that is possible. In the case of many philosophical divergences I have found that this can be done by talking about the different forms of language people have. For instance, when I talk with a nominalist and he says 'There are no abstract entities, this is just mere talk,' then I say 'Let's compare two languages, call the one the nominalist language and the other the language with abstract concepts in it. Let's first compare them with respect to things upon which we both agree, then with respect to practical questions, not philosophy but, say physics – which form is suitable there? Why?' Then I can understand my opponent much better. We may not agree with each other, but I know the reasons for the disagreement. He might say, as Heyting, that he refuses certain forms of sentences and deductions, on the grounds that these are not defined by the basic intuitions that I have. I then try to put myself in the place of the classical mathematician, in the place of the intuitionist, and so forth; I think the classical mathematician would say 'my intuitions are clear intuitions that there are such things as real numbers and that they follow such-and-such laws; why is that to be rejected?' I think this is psychologically sound and not to be rejected. The classical mathematician sees this in the way that Brouwer accepts the whole of classical theory for the natural numbers because our intuitions show us such a thing as a series. ... Brouwer says that all of mathematics is based on this intuition. Then there could be another man whose intuition will tell him that much more complicated structures are there. Does he not have the same rights? It seems to be a difference in degree; the other is bolder, the other more cautious. The bolder man would perhaps agree – I agree at least, when I try to take a neutral stand – that he cannot have quite the same confidence that there is not perhaps a contradiction in his system. No mathematician, even if he were to talk only about natural numbers, can be absolutely sure that tomorrow there will not be a contradiction. It would be enormously surprising, no one would believe it today. But there have been so many surprises in the development of mathematics that no one would dare to say that such a thing would never happen.

Now let's add something else. This is not quite as safe as this first field, because the forms are more complex, so we would have to add that this feeling of safety, which is a subjective matter, would have a lesser degree there. So even in the most restricted fields there can be no absolute safety, and in the richest fields, the most complex ones that are still alive,[37] no one has yet seen a contradiction. Sometimes they may reveal[38] contradictions, as Cantor's theory did. And again authors will make qualifications and restrict the rules, which will have been too general. The same is true as of a law in physics; first it is stated in too general {form}, later we see that it has exceptions. Thus in mathematics we must admit that we cannot be sure. This is psychologically hard; mathematicians often have become mathematicians because they didn't like the uncertainties of natural science, where new experiments or new theories can change many things. Mathematicians are probably men who

[37] In the original transcript, "alive" is underlined and marked with a question mark.

[38] In the original transcript, "reveal" is underlined and a pencil note says "contain?".

like certainty, clarity, great exactness and precision. I have great liking for that too, but we must make the admission that there is also uncertainty, that it is a matter of degree, and therefore there is no point in making dogmatic statements. Statements such as Brouwer would make, that 'This is the only sound and admissible part of mathematics.' Let's say 'This is the safest part;' and practically all mathematicians know this by now, the great majority have learned this from Brouwer; the great majority have not accepted Brouwer's restrictions, but practically all of them, when they make a new proof or a new theorem, in any field, whether analysis, abstract algebra, and so forth, then when they use some of these controversial axioms, the axiom of choice in set theory, indirect proof in mathematics in general, they call attention to it. They say 'I have tried hard, so far as I tell, we cannot prove this theorem without it.' Sometimes later someone finds a proof without it, and then there is great relief because we have a proof in a safer field. But they would not say, 'Since you have used that axiom, the proof is not valid', they would say 'We cannot be quite as safe, you are skating on thin ice, but all ice is thing, there is no absolutely safe ice.' So foundations are made in different ways. I expressed this in the beginning with Brouwer as 'Let's make distinctions and not prohibitions.' Let us not speak of inadmissible proofs but relatively unsafe ones.

H: *Has it taken a long time for this relatively tolerant and skeptical point of view become prevalent?*

C: It is partly a matter of temperament. Some people still talk in relatively thick[39] terms, some intuitionists talk on more tolerant terms. So it is a personal matter and you cannot say that 'this school takes this attitude' ...

H: *Is there any tendency, in arguments on this topic, for other topics to become involved in addition to those which distinguish the schools? So that, for example, a position on intuitionism comes to represent much more than a position on a few points? ... I mean, so that it becomes hard to pin down the disagreement, rather than being able to state that it depends on a few propositions?*

C: Oh yes. In the beginning it is often not quite clear where the differences really are, although both opponents pretend they know exactly what it is. Brouwer was very polemical against Hilbert. Hilbert had great prestige, was the leading mathematician in Germany, and was at Gottingen. Brouwer was at Amsterdam; he would say 'Our differences is this: I assert this and you assert that and I am against this' and so on. Then Hilbert would say 'No, this is not at all the difference;' so not only did they differ on mathematics, they differed on the historical situation, what Hilbert's view was, what was possible, what the difference between the two was. The main polemics occurred in the Berlin Academy; Brouwer got very temperamental, and said it was scandalous how they proceeded, how they pretended to know things which they had no good reason to assert.[40] He became quite

[39]Carnap crossed out this word and wrote another with a question mark with pencil, but it is unreadable.

[40]It is unknown to exactly what Carnap refers here. In the literature, the Hilbert-Brouwer controversy is a well-discussed issue, concerning, on the one hand, the foundational debate

emotional and indignant; one wouldn't expect such a thing in such a field as mathematics, so remote from human action.

H: *Has this vanished now?*

C: Yes, today polemics would be – a few man are more temperamental than others, but in general formalists and people of my view are not polemical. It is human in the beginning, things are so emotional. You remember in psychology also, behaviorism, gestalt psychology, and their opponents; every school has to fight, and has to fight hard even for their right to exist as respectable. They existed but were called crackpots by the old psychologists. When they become respectable they talk in more calm terms.

H: *In a way there are competing programs, and those interested in a particular program like to persuade others to adopt it.*

C: Yes, that is quite true. Thus in physics, psychology, philosophy, and even mathematics, to a limited extent, it is really the differences are with respect to their assertions, or rather their feelings, about where are the most promising methods. For example, from a higher point of view, there is not much difference between the analytic philosophy in England and the analytic philosophy in this country; from any other point of view, if you talk with an existentialist, or a neo-Thomist, or a dialectical materialist, or any outsider, they would say these are brothers. But brothers fight especially hard with each other, so the English analytical philosophers say the Americans are wrong in constructing artificial systems, etc., while the philosopher's task is to analyze the natural language.[41] While we say just analyzing the natural language is trivial, of course you have to construct systems. I would say the difference is really with respect to the feeling each has about his . . .[42]

Let's not formulate the differences as philosophical research, for this is not really the basis. The basis is really a feeling concerning the fruitfulness of different methods. – That is not an exact statement at all, I could not make an exact statement. I have a feeling that they way that people in England proceed, I would not say that this is wrong, that the methods and results are wrong, I would rather say I would proceed in a different way. I doubt very much that much of great interest will come out of their approach. Surely they will obtain many results, which will be right if the method is followed correctly.

Here we are concerned with the fact that every philosopher and every scientist has to make up his mind which way to take because there is not enough time to try

regarding mathematics, and on the other hand, Hilbert's act of removing Brouwer from the editorial board of "Mathematische Annalen". See, e.g., Per Martin-Löf, "The Hilbert-Brouwer controversy resolved?", in Mark van Atten et al. (eds.), *One Hundred Years of Intuitionism (1907–2007): The Cerisy Conference*, Basel/Boston/Berlin: Birkhäuser, 243–256. See also J. Posy Carl, "Brouwer versus Hilbert: 1907–1928", *Science in Context* 11 (2) (Summer 1998): 291–325.

[41]One of the most important documents and official initiators of the debate is Peter Strawson's paper in the Carnap volume of *The Living Philosophers* series, see "Carnap's Views on Constructed Systems versus Natural Languages in Analytic Philosophy," in P.A. Schilpp (ed.), *The Philosophy of Rudolf Carnap*, La Salle, Illinois: Open Court, 1963, 503–518.

[42]The original transcript has "end of side one", and the second half of the sentence is missing.

to all possible ways, and therefore he chooses his approach according to his mostly unconscious judgment or feeling or hunch of fruitfulness of this or that approach.

Therefore the difference is that the British have a feeling that their approach will be fruitful and we have a feeling that our approach will be fruitful. Let's not make a polemic out of this, because we cannot know it and in 20 years everybody will know all this. Then the results on both sides are there, and maybe they will say 'they have found very fruitful results on the other side, so let's be grateful both were done,' or maybe they will say, 'On one side they found much more fruitful results; while on the others worked and worked hard, and found some results, but not so much.' And then there will be no polemics either.

H: *Is it really possible to settle this kind of argument rationally?*

C: You can talk about it and ask for the reasons why one approach will more easily lead to fruitful results than the other. This is very hard to talk about. When I say to my students why I take one approach I give practical reasons such as that 'We have found such-and-such results this way, and this is encouraging; since I made these discoveries many years ago I have modified the approach somewhat but not changed its basic direction, because I believe it to be a fruitful approach; but if you think otherwise you must try another way.' We must not waste too much time in talking about the relative promise of this or that approach. We can provide some reasons for taking one approach or provide some results which we think are evidence of the fruitfulness of an approach. But there is no use for a long polemic, because it's not the same as if there were two theories, such as theories about the historical origins of something like the French Revolution. In such an area, people might make different assertions, but in five years we might find out that one or the other is wrong. Similarly with different hypotheses about things like the expansion of the universe; these are really incompatible, and after some years we will know who is right. But here the difference between philosophical schools is not of this nature. They pretend that it is, but only because they describe their basic attitude as if it were an hypothesis. They say, 'We believe the essence of language is so-and-so, the essence of communication is so-and-so, and that the natural language is the paradigm of communication.' And we think – not I, but some who take such views – present a counter-thesis, 'The essence of communication is so-and-so, therefore language constructed artificially, according to certain requirements will be the best means of communication and that we should study constructed languages.' I would say that these two theses are too wobbly to be good as a basis of discussion. I would say rather that 'You like to work with those natural languages, we like to work with constructed languages. Why do we do this? Because we have the feeling that it will lead to good solutions for philosophical problems and therefore will be fruitful for philosophy – and therefore for science because certain philosophical problems, if they are solved, will help science. And finally it will be clear which philosophical approach is more fruitful when the results are in – but the results will not look this way, 'The essence of communication is thus and so, and school A was right.' The results will be that 'We have learned that language works in such and such a way, and these forms have certain advantages and that others have certain disadvantages, and therefore we advise the physicists, for their purposes, to use one form, and we advise the pots, for other purposes,' – we need not advise him, he chooses himself – 'to use a natural language modified by himself,' and so on. And so the form of the polemic is wrong I think.

H: *If a man's choice of an approach is a matter of individual feeling, what kind of experiences are most important in shaping this kind of preference?*

C: What has shaped it is not so important because it might come out of historical accidents, that he grew up in that country, moreover in that part of that country, where there was this ideology or this tradition, or that he has happened to come from a study of physics or a study of history or has been a physician, or that his father was a physician or so-and-so. These are accidental, personal things and they are not so important – even if a whole group shares them. For example, we might see that in Germany such a kind of philosophy arises, and in America a different kind, because the whole social atmosphere in Germany differs in certain respects from that in America. The sociology of the origin of ideas is also an interesting problem, but they are not really important for solving the problem in the substantive fields. Some people, for instance the Marxists, think so. When they want to polemicize against Einstein, they say Einstein's ideas came out of a Bourgeois atmosphere, therefore they must be wrong. Of course it is a good question, what were the psychological origins of Einstein's ideas, the sociological and historical origins. But this is interesting for other reason – for sociology, etc. – but you cannot use it as an argument against the theory of relativity. So where it comes from is an interesting question, but the arguments on theories must be in the field of the theory. The Marxists do that quite wrong.

Among philosophical schools likewise, in some respects it is good to study the historical origins, because then we get a better psychological understanding why it is this way or that way. So it might be good to understand why the Marxists think this way, and Dewey and the pragmatists think this way, perhaps this can be explained partly with sociological reasons. But after that is said we still have to solve the really philosophical problems, and they have to be solved with philosophical arguments not with sociological arguments. So if you find out that perhaps the formalist way of thinking, of Hilbert but partly also mine in stressing the formal character of logic, comes psychologically out of such-and-such a type – probably an introverted type, probably what Sheldon[43] calls a cerebrotonic type – perhaps they find that, an they may find that the intuitionists are another type. While this is nice psychologically, but this does not solve the problems of the foundations of mathematics, for those problems we either must think in that field, and, therefore, when I talk with Heyting, – while I am interested in what kind of man he is – but this is only in order to understand his ways of thinking, but I really want to find out something about a problem in the foundations of mathematics, is it this way or is it that way? Then I have to think about mathematics, not about the person. Then

[43]William Herbert Sheldon (1898–1977) was an American psychologist and worked on so-called "somatotype psychology:" he tried to find correlations between body types, behavior and social settings.

I might take a different approach for historical or psychological reasons, but we finally want to solve the same problems. So the genuine problems in distinction to these problems which are pushed in the foreground by polemics which are not the genuine problems – the genuine problems are solved by working in the field, and after 10 or 20 years, I believe the formalists, and my direction, and the intuitionists, will have done their work and all three will have been fruitful. We see that already today. Therefore the judgment of history will not be that one was right, the others were wrong, but only, 'What did they contribute?' And the total result will be a certain view of mathematics which is much clearer and more satisfactory than any of us in these three schools or any other school can give today. And if you study that historically you will probably see – it seems to me, it need not be so – that some of them came from intuitionism, and some came from logicism, and some from the Hilbert school. So we are then grateful to all three of them.

H: *So you would accept some sort of synthesis?*

C: Yes. In this situation; that is not a general view of mine. But in this situation we see already the great merit that Hilbert had because we see clear today on so-and-so[44]; and I think Frege had a great merit, and we see today that certain accepted things in mathematics came from Frege, he saw it first and expressed it first; and now there is also the case of intuitionism. Although it is accepted only by a tiny minority – the others are either entirely uninterested or interested but opposed to it – we should consider, although many mathematicians do not see that, that certain distinctions are made in mathematics now which came from Brouwer. I mentioned some, such as whether a certain proof can or cannot be made in a certain region. The basic ideas of these were first emphasized by Brouwer, for his philosophical reasons, but now they are accepted on a neutral basis as important distinctions. These play a very great role in what is called the theory of recursive functions, what may be called the theory of effective logical methods. By 'effective' we mean that the answer will be found in a finite number of steps. Many mathematicians are now interested in the theory, but I believe, although such historical statements are precarious, that at least one of the sources was the great emphasis of Brouwer. He made it in a way that is not accepted by saying that some means are admitted and the others prohibited, but the emphasis on the distinction we owe to him. For reasons such as this I believe that a synthesis will be made containing elements contributed from all three. In other situations, involving two schools, one might be completely vindicated.

H: *Could we go back to the problem of the origins of an individual's position? If, for example, a person's associates, such as a student's teachers, play a decisive role in steering him toward one approach rather than another, it would also follow that if the person were to change his mind and select another approach there would be social pressures against him. It would be harder to effect a redirection of interest in the sciences then.*

[44] In the original transcript, "today on so-and-so" is underlined by Carnap and marked with a question mark.

C: But that is hard to know, because these things operate mostly unconsciously, I think.

H: *But, in general, are a man's associates likely to agree with him on basic approaches, the men he sees every day?*

C: I would certainly say that the fact that a man is in such-and-such a surroundings, first as a student, that this has a great influence. Then when he continues to do research, his department and his peers and superiors, may influence him. But it would be hard, at the end of the life of a thinker, to look at the ways he has chosen, find that he did not select the customary ways of thinking – here he proposed a new theory, there he proposed new things, things only supported by a small minority – why did he make such choices? Why did he have to fight to defend his minority position? Perhaps his promotion was delayed, perhaps he couldn't find a job at all. Frege, for instance, was rather old when I studied with him, and nobody thought of promoting him to full professor. This is an interesting question, but it is not easy to answer. How much was the direction of his thinking determined by his colleagues or his teachers? Or by his own thinking and reading? It is very hard to say. All these things are effective, but in an individual case it is hard to say.

H: *What effect do you think the various unsolvability theorems of Gödel have had on the relations between the different schools, up to now, especially on the relations between the position you hold and the formalist position?*

C: Why do you pick out this theorem?

H: *It has been suggested to me that this very strongly weakened the formalist position.*

C: Yes, but that is sometimes exaggerated. Gödel himself never said that this really destroys the formalist point of view. Bernays,[45] when he wrote the second volume of the Hilbert-Bernays book[46] on formalist mathematics, in the preface he says that Gödel's results are very important but that it should not be taken to mean that his aim is impossible; all he has to do is to admit more liberal means in the proof of the consistency of mathematics. He recognized that they could not do it in the way they thought, but the means were not very exactly specified; they were called finitist means but nobody said what exactly that included – they may not[47] have felt that a full list was not possible. Brouwer denied explicitly that any mathematician could ever give the full list of the deductive rules he could regard as valid; it was always conceivable that the next day he would invent an entirely new type of deductive procedure, not covered by the old rules, and he would see clearly that it is valid. Brouwer had said that before Gödel. He could say after Gödel's results that they were not a great surprise to them, that they did not damage their program, because

[45]Paul Bernays (1888–1977) was a Swiss mathematician, known for his work on set theory and the philosophy of mathematics.

[46]David Hilbert and Paul Bernays, *Grundlagen der Mathematik II*, New York/Berlin: Springer, 1939. The first volume, with the same title, also at Springer, was published in 1934.

[47]"Not" is underlined with pencil in the transcript, and a question mark is put by pencil on the margin by Carnap.

they had always said that everything is undecideable, that you can never give the total list of rules for the whole of classical mathematics.

Some people said that Gödel's results would destroy the whole of the Hilbert position, but I think again that these are matters of degree. From Hilbert we have no very clear-cut utterances. In the joint book he supplied the basic ideas but most of the writing, especially of the second volume, was by Bernays. So we have clear utterances by Bernays and Ackerman[48] and other students of Hilbert; I think they have learned a good deal from Brouwer and the result of Gödel, and they taken an attitude now that there is no fixed line. Before they thought there was a certain realm of deductive means which we call the finitist means, and metamathematics should be restricted to that. Later they said that the means to be used in metamathematics was more or less a matter of degree. They might not have said it in quite this way, but what they said goes in this direction. Gentzen,[49] who has constructed some partial proofs of consistency, has used means which Hilbert was not thinking of at all originally. He used, for example, a rule called the rule of transfinite induction in set theory, a rule which by its very name indicates that it goes beyond Hilbert's finitist methods. But both Bernays and Gentzen have said that this rule should be considered finitist.[50]

So the situation is that the differences between the schools are no longer quite as sharp. At most the intuitionists are still on their old position, although Heyting is much more willing to talk to others than Brouwer was. But the other schools look at the matter as looking for the means which are necessary for reaching the goals we want to reach, in metamathematics, among them proofs of consistency for stronger and stronger theories; and we will use all means we find necessary as long as it seems to us to be valid. We should never claim they are finitist in a very strict sense, never claim that they are absolutely safe against self-contradiction. As we have learned this in mathematics we have learned it in a milder way in metamathematics, although here we try to be somewhat more restrictive than in mathematics. But it is like mathematics in that we try to find methods which seem to be valid and intuitively acceptable but still which enable us to enlarge the scope of our metamathematical methods and perhaps then to discover, by this enlargement, proof for the consistency of a stronger theory than we had before. That seems to be the tendency and the feelings of the people. The formalists are not as dogmatic as it seemed at the beginning. Mathematicians have the feeling that we must discover things and see how much we can justify. This is the natural development.

---

[48]Wilhelm Ackermann (1896–1962) was a German mathematician who obtined his PhD under Hilbert in Göttingen in 1925. Ackermann is known for his work on first-order logic and for his joint textbook with Hilbert that was published in 1928 (*Grundzüge der theoretischen Logik*, Springer).

[49]Gerhard Gentzen (1909–1945) was a German mathematician and logician, known for his wok on natural deductions.

[50]Carnap's views on finitism, with the historical context in mind, is discussed in detail by Greg Frost-Arnold, *Carnap, Tarski, and Quine at Harvard: Conversations on Logic, Mathematics, and Science*, Chicago: Open Court, 2013.

H: *Could we spend just a few minutes on the subject of the interpretation of probability? I believe you have some distinctive views on the subject. Are the opposing points of view here as clear-cut and have they attracted as much interest as the foundations of mathematics?*

C: No. This is only now beginning. For a long time the frequentist interpretation of probability was the only one accepted by the great majority of mathematicians who were working in probability. Mathematical statisticians, who applied the concept of probability, are not the same people, but they also took the frequency interpretation as the only one. Mises[51] said explicitly that this is the only scientific interpretation of probability, and Reichenbach took the same view. There existed another interpretation since Keynes,[52] who was perhaps the first, which he called probability as a logical concept, and mine is essentially in that vein – call that the logical interpretation of probability. Keynes was the first. Harold Jeffreys,[53] an English geophysicist, was the second; his Theory of Probability was on this basis. But their views were not accepted at all by the great majority. So there was a small minority of people who were willing to think along other lines. A few philosophers, perhaps; Wittgenstein, although he did not develop his views. And now I have tried to develop this concept in greater exactness, and it has been done by Keynes and Jeffreys – although it is hard to tell if it is the same concept as Keynes' and Jeffreys'. It is in the same direction, and there are minor differences which don't matter. The main point is, Do some statisticians and mathematicians think there is something to it? I think Keynes and Jeffreys were too polemical on the following point: they said the frequency interpretation of probability must be entirely rejected, that it is not a good concept at all. I think their concept and the frequency concept are both good ones, for different purposes, and quite different definitions, and they have different roles in science. The frequency concept, or what we can call statistical probability, – which is not the same as frequency although it is based on frequency – there can be no doubt that it is of the greatest importance because it occurs in more and more scientific fields. The physicists, the man in publish health, use this kind of interpretation; their probability statements are factual statements which have been subjected to a lot of study. For example, in quantum theory and the insurance business. The probability interpretation of quantum theory is now generally accepted by physicists. When you say such-and-such a phenomenon, say a radioactivity phenomenon, e.g., that a particle of a certain kind has a particular probability of being emitted in a certain period of time, you are making a physical

[51]Richard von Mises (1883–1953) was an Austrian mathematician, known for his work on probability, statistics and aeronautics. Von Mises was involved both in the Vienna Circle around Moritz Schlick and in the Berlin Circle around Hans Reichenbach. He emigrated to the United States (after a six year detour to Turkey) in 1939 and taught there at Harvard until his death.

[52]John Maynard Keynes (1883–1946) was a British economist who published his famous and influential *A Treatise on Probability* in 1921.

[53]Sir Harold Jeffreys (1891–1989) was an English mathematician and geophysicist. He published his book, *Theory of Probability* in 1939, and thus initiated the revival of the Bayesian view of probability.

statement of fact. So this is the first concept, a frequency concept. There can be no doubt that it is a very important and indispensable concept.

So the question can only be, can there be and should there be a second concept of probability not based on frequency?...

H: *You say, then, that your theory would not lead to a revision or reformulation of mathematical statistics?*

C: I would say that their concept of probability is a good one, and if my concept is also a good one, then it should be added and also dealt with in mathematical statistics. I think mathematical statistics really needs a non-frequency concept as well. There are now a few statisticians, there seems to be a slight beginning, some interest in a non-frequency concept of probability. I would say as a second concept, but it is not yet quite clear. Among statisticians it is especially the Italian, de Finetti,[54] and the American Savage,[55] that work in this direction. Savage has written the Foundations of Statistics in 1955; he calls his concept the personal concept of probability because by it is meant the probability for the evidence that this person has at a certain time, e.g., for the prediction that tomorrow it will rain so much. It is also called 'subjective probability' by the critics of Keynes and Jeffreys, although Keynes and Jeffreys did not accept this word. I accept it only with hesitation, because it has the connotation that these are as subjective as taste in art, etc. It is not subjective in that sense; it is only subjective in this sense, that if you have made different observations than I, then on the basis of the evidence that you have, the probability of rain tomorrow will have a different numerical value. But you calculate from your evidence on the same rules that I do on my evidence, so these rules are objective. If we find valid rules, which is the big question; I think we can find such rules. I would rather say that the probability is 'relative to evidence' than that it is subjective or personal', which is what Keynes himself already said. So it is a logical relation between a body of evidence and an unknown statement which I call the hypothesis or sometimes the prediction. This is calculated not by making observations in nature but by a logical analysis of the evidence of the prediction or any complex hypothesis it might be and of their logical relations to each other, logical relations which may be defined in terms of deductive relations. On the basis of these relations we express a relation between the prediction and the evidence by saying the probability is 2/3, or whatever it is. So a numerical value of probability is a logical relation between two statements. This is the nature of the logical concept of probability. In my view this is the concept which is the basic concept for inductive reasoning, so I also use the term 'inductive probability' for it, although that is not generally accepted.[56]

[54]Bruno de Finetti (1906–1985) was an Italian statistician and actuary.

[55]Leonard J. Savage (1917–1971) was an American mathematician; his well-known book, *The Foundations of Statistics* was published in 1954.

[56]Carnap's final views on probability are to be found in "A Basic System of Inductive Logic, Part 1", in *Studies in Inductive Logic and Probability*, volume 1, edited with Richard C. Jeffrey (Berkeley: University of California Press, 1971), and in "A Basic System of Inductive Logic,

H: *Is this mostly a matter for philosophers then? Would you expect it to be of as much value for mathematicians?*

C: Oh yes, very much. I think their methods – Savage, who is a statistician, says more on this, although I have studied mathematical statistics because I think this is the area in which the concept will first be of importance in applications. Savage has said that the present methods in statistics are not uniform, that they have come about through historical methods and do not fit well with one another. If you give a statistician a problem of this kind he uses that method, a problem of that kind, that method, from another author, since the first author was not interested in the problem, say in quality control. There are various kinds of problems, such as decision problems in quality control, point estimation, interval estimation, the problem of whether two samples come from the same population, and so forth. The methods for attacking different types of problems are not on the same basis. We want to have one basic conceptual framework, one basic system of axioms for statistics, such as those in mathematics and physics, so that when we go from one problem to another we don't take an entirely different outlook. And Savage says that his concept will lead to greater uniformity and would be of great advantage, and that on this basis new methods could be developed which would be fruitful. I believe this too, and that my concept is closely related to his; he didn't believe this at first, but I believe I have now convinced him, or half convinced him, that we are going in the same direction. I accept the same axioms as he does and go further; I say we must ask new questions and the answer are given by certain axioms, and that we can only do this if we make certain changes in the conceptual[57] of what he does. I did this from the beginning and saw only later this is the relation. I also have additional axioms and therefore have a stronger theory, which will do an even better job in the same direction.

H: *Are these largely the efforts of a few individuals so far?*

C: Yes, chiefly I alone. Now I have some mathematicians, some friends of mine, who work with me in developing it. I am grateful for this, since they are more familiar with some fields like measure theory in mathematics.

H: *Los Angeles mathematicians?*

C: Yes. They are philosopher-mathematicians, you might say. One is Kemeny[58] at Dartmouth, who was a philosopher at Princeton when I was there. The other is Richard Jeffrey[59] (not to be confused with the English Jeffreys), who is Assistant Professor of Philosophy at Stanford and was a student of mine at Chicago. They work with me, along with a few younger students and mathematicians.. .

I would be interested in seeing copies of your published results.

End.

Part 2", in Richard Jeffrey (ed.), *Studies in Inductive Logic and Probability*, volume 2, Berkeley: University of California Press, 1980, 17–155.

[57]This is marked by Carnap with question marks.

[58]John G. Kemeny (1926–1992) was a Hungarian-born American mathematician and computer scientist. He exchanged many ideas with Carnap on semantics and model theory in the 1950s.

[59]Richard Jeffrey (1926–2002) was an American philosopher and logician, known for his work on probability theory. He worked with Carnap on probability from the late 1950s until the death of Carnap in 1970.

Acknowledgments The interview is contained in the Carnap Archive at Pittsburgh: Rudolf Carnap Papers, 1905–1970, ASP.1974.01, Special Collections Department, Archives of Scientific Philosophy, Hillman Library University of Pittsburgh. All rights reserved; I take the chance here to express my gratitude for their permission to publish the interview here.

I was supported by the MTA Lendulet Morals and Science Research Group and by the MTA Premium Postdoctoral Scholarship.

References

Carnap, Rudolf. 1934a. Theoretische Fragen und praktische Entscheidungen. *Natur und Geist* 2 (9): 257–260.

———. 1934b/1937. *The Logical Syntax of Language*. London: Kegan Paul, Trench, Trubner & Co.

———. 1950/1956. Empiricism, Semantics, and Ontology. In *Meaning and Necessity*, 2nd ed., 205–221. Chicago: University of Chicago Press.

———. 1952. *The Continuum of Inductive Methods*. Chicago: University of Chicago Press.

———. 1963a. Intellectual Autobiography. In *The Philosophy of Rudolf Carnap*, ed. P.A. Schilpp, 3–84. La Salle: Open Court.

———. 1963b. Replies and Systematic Expositions. In *The Philosophy of Rudolf Carnap*, ed. P.A. Schilpp, 859–1013. La Salle: Open Court.

Carus, André W. 2007. *Carnap in Twentieth-Century Thought: Explication as Enlightenment*. Cambridge: Cambridge University Press.

———. 2016. Carnap and Phenomenology: What happened in 1924? In *Influences on the Aufbau*, ed. Christian Damböck, 137–162. Dordrecht: Springer.

———. 2017. Carnapian Rationality. *Synthese* 194: 163–184.

Damböck, Christian. 2017. *"Deutscher Empirismus". Studien zur Philosophie im deutschsprachigen Raum 1830–1930*. Dordrecht: Springer.

Damböck, Christian, Günther Sandner, and Meike Werner, eds. 2021. *Logical Empiricism, Life Reform, and the German Youth Movement/ Logischer Empirismus, Lebensreform und die deutsche Jugendbewegung*. Cham: Springer. Forthcoming.

Franco, Paul L. 2018. Ordinary Language Criticisms of Logical Positivism. *HOPOS: The Journal of the International Society for the History of Philosophy of Science* 8 (1): 157–190.

Friedman, Michael. 1999. *Reconsidering Logical Positivism*. Cambridge: Cambridge University Press.

Hagstrom, Warren. 1964. *The Scientific Community*. New York/London: Basic Books.

Jeffrey, Richard. 1994. Carnap's Voluntarism. In *Logic, Methodology, and Philosophy of Science IX*, ed. Dag Prawitz, Brian Skyrms, and Dag Westerståhl, 847–866. Amsterdam: Elsevier.

Reisch, George. 2019. *The Politics of Paradigm: Thomas S. Kuhn, James B. Conant, and the Cold War "Struggle for Men's Minds"*. New York: SUNY Press.

Reisch, George, and Adam Tamas Tuboly. 2021. Philipp Frank: A Crusader for Scientific Philosophy. In Philipp Frank, *The Humanistic Background of Science*, ed. George Reisch and Adam Tamas Tuboly. New York: SUNY Press, 1–67.

Richardson, Alan. 1998. *Carnap's Construction of the World – The Aufbau and the Emergence of Logical Empiricism*. Cambridge: Cambridge University Press.

Siegetsleitner, Anne. 2019. Carnaps Autobiographie als Autobiographie. *Deutsche Zeitschrift für Philosophie* 67 (2): 236–250.

Strawson, Peter. 1963. Carnap's Views on Constructed Systems versus Natural Languages in Analytic Philosophy. In *The Philosophy of Rudolf Carnap*, ed. P.A. Schilpp, 503–518. La Salle, Illinois: Open Court.

Tuboly, Adam Tamas. 2017a. Carnap's *Weltanschauung* and the *Jugendbewegung*: The Story of an Omitted Chapter. In *Integrated History and Philosophy of Science*, ed. F. Stadler, 129–144. Cham: Springer.
———. 2017b. Philipp Frank's decline and the crisis of logical empiricism. *Studies in East European Thought* 69 (3): 257–276.
Uebel, Thomas. 2012. The Bipartite Conception of Metatheory and the Dialectical Conception of Explication. *In Wagner* 2012: 117–130.
———. 2020. Intersubjective Accountability: Politics and Philosophy in the Left Vienna Circle. *Perspectives on Science* 28 (1): 35–62.
Wagner, Pierre, ed. 2009. *Carnap's Logical Syntax of Language*. Basingstoke: Palgrave Macmillan.
———, ed. 2012. *Carnap's Ideal of Explication and Naturalism*. Basingstoke: Palgrave Macmillan.

Part III
Reviews

Chapter 13
Review: Johannes Feichtinger/Franz L. Fillafer/Jan Surman (Eds.), *The Worlds of Positivism. A Global Intellectual History, 1770–1930*. Cham: Palgrave Macmillan 2018, 367 pp., ISBN: 978-3-319-65761-5

Robert Frühstückl

Throughout its history, positivism has been highly controversial. Quite apart from the positivism dispute in twentieth-century German sociology, even in today's philosophy of science the adjective "positivist" is sometimes used to signify an apparently narrow-minded, overly scientistic philosophy, which either relies on an outdated concept of observability, or otherwise is blind to the importance of epistemic and especially non-epistemic values in scientific practice. That these uses of the term often echo nineteenth-century debates, particularly in Germany, and are frequently used to demarcate the intellectual field, is just one aspect of positivism and its history that gets highlighted in *The Worlds of Positivism. A Global Intellectual History, 1770–1930*, edited by Johannes Feichtinger, Franz L. Fillafer and Jan Surman. This is a collection of historical essays which follows the formation and development of positivist philosophical positions as well as the political agendas they sometimes supported through a global array of political, social and cultural contexts.

Interestingly, its publication in 2018 was complemented by the appearance of two further collections of essays on the history of positivism, though with a different and narrower focus on the founding figure Auguste Comte. Just a year before *The Worlds of Positivism* came out, *The Anthem Companion to Auguste Comte*, edited by Andrew Wernick, was published, followed a year later by *Love, order, & progress. The science, philosophy, and politics of Auguste Comte*, edited by Michel Bourdeau, Mary Pickering and Warren Schmaus. Whether the close timing of the publication of these volumes indicates a resurgence of interest in the history of positivism is uncertain. It seems to suggest, however, that it is high time to take a closer look at the

R. Frühstückl (✉)
University of Bielefeld, Bielefeld, Germany
e-mail: robert.fruehstueckl@uni-bielefeld.de

E. Ramharter (ed.), *The Vienna Circle and Religion*, Vienna Circle Institute Yearbook 25, https://doi.org/10.1007/978-3-030-76151-6_13

historical formation and development of one of the central intellectual movements and influential philosophical positions of the nineteenth and twentieth centuries.

The Worlds of Positivism offers a unique and rich perspective that covers a remarkably wide time scope with an equally wide geographical range, which takes the reader from Colonial India to Brazil, the Ottoman Empire, Spain, Germany, France, Austria, Poland, and Russia, as well as Norway, Denmark, Sweden and Finland. The volume consists of thirteen individual chapters and is structured in four parts, opening with a general introduction from the editors and concluding with a resumé by Feichtinger and Fillafer which critically summarizes and analyzes the individual contributions along four main themes emerging from the book; that is, (i) the relationship between the natural sciences and the humanities; (ii) society and social engineering; (iii) the relationship between positivism and religion; and (iv) positivism in different geopolitical contexts of the nineteenth and twentieth centuries.

What makes the volume special, particularly for intellectual historians, is the editorial approach to the subject, aptly captured in the title of the introduction: *Particularizing Positivism*. One cannot but agree with the editors when they identify the question of the universality of claims to knowledge, in both a philosophical and a political sense, as one – perhaps *the* – crucial issue of nineteenth-century positivism. However, instead of evaluating these claims to universal objective knowledge about nature and society philosophically, the editors propose to follow them through their individual historical instantiations. The individual contributions to this volume manage to bring out the various conditions and configurations underlying claims to universalism in the specific contexts in which they emerged. The significance of this viewpoint becomes even clearer in the context of the editors' explicit avoidance of a diffusionist model of intellectual history. Rather than operating with the conception of a ready-made doctrine of positivism which, after its manufacturing in Europe, is merely transferred to other parts of the globe, the authors focus on the individual historical figures who were key in the processes of appropriating positivist ideas within their respective national, cultural, and social contexts. The outcome is an insight into what the editors fittingly call "the 'brokered worlds' of positivism" (p. 4).

This allows for a multitude of historical insights. For example, in the contributions on positivism in Bengal India and the Ottoman Empire, the authors pick up on the familiar appeal of positivism to different national elites and their quests for modernization, which lies not only in its scientific optimism but also in Comte's criticism of European colonial and imperialistic politics. But they also highlight the preconditions of this appeal and explain how different historical agents managed to incorporate these ideas in their respective agendas; for example, how positivism allowed them to reconcile this scientific worldview with traditional religious ideas and their associated social order.

The volume also sheds light on the processes and controversies by which positivism emerged as a distinct school of philosophical thought. Particularly interesting are the juxtaposition of the contributions focusing on the beginning to mid-nineteenth century Germany and France respectively and the comparative perspective it facilitates. It is not only the relationship between its French and British instantiations that is frequently highlighted throughout the book, but also the complex relation between Comte and his disciples, most notably Emile Littré. Readers can learn how Comte's brand of positivism was instrumental in shaping emerging social sciences, but also how his critical views on psychology and introspection came to be seen as problematic by some who argued for the necessity to ground the social sciences in a subjective method.

Those readers with a special interest in the history of the Vienna Circle and of logical empiricism will find many illuminating aspects of nineteenth-century Austrian intellectual history that connect the positivism of the nineteenth century to logical empiricism and also shed a critical light on Otto Neurath's doctrine of a specifically anti-idealist strand in the history of Austrian philosophy, sometimes also referred to as the Neurath-Haller thesis. The volume also connects Hans Kelsen's brand of legal positivism and even follows Alf Ross, a student of Kelsen's, who later established the school of Scandinavian legal realism.

Finally, the volume provides insights into the relationship between the positivism of the nineteenth century and the emergence of analytic philosophy at the turn of the twentieth century, specifically through focusing on the significance of Wilhelm Dilthey, and particularly on the early Rudolf Carnap. At this point, the book connects well with familiar efforts in the history of the philosophy of science that concentrate on the neo-Kantian historical roots of logical empiricism, and it provides further insight into the eventual split between continental and analytic philosophy.

These are only spotlights directed at particular sites of a much larger stage that the book occupies. Intellectual historians and philosophers with a broader interest in history will find many more fascinating insights in this book.

Chapter 14
Review: Donata Romizi, *Dem wissenschaftlichen Determinismus auf der Spur*. Freiburg/München: Verlag Karl Alber 2019, 503 pp., ISBN: 978-3-495-49103-4

Marij van Strien

It is still commonly thought that when quantum mechanics was introduced, its shock lied in the fact that it went against a doctrine of determinism which was until then firmly established in science, and which had been almost universally accepted among scientists of the eighteenth and nineteenth century. Romizi's book *Dem wissenschaftlichen Determinismus auf der Spur* demonstrates the problems with this view by giving a rich account of the varieties of determinism and indeterminism in physics before the introduction of quantum mechanics in the 1920s. Her history of the concept of scientific determinism reveals that in the nineteenth century, there were various ways in which scientists could doubt or reject the absolute validity of determinism in science. What makes a history of determinism especially interesting is that it connects various broad issues in the history and philosophy of science: the question of the proper domain and limits of science, the truth claim of scientific theories, the ideals of unification and reduction in the sciences, the issue of free will and its religious contexts, as well as science popularization and the expansion of science into a general world view. By exploring these different aspects of the history of determinism, this book gives a rich picture of the history of philosophy of science in the eighteenth, nineteenth and early twentieth century.

In the introduction, Romizi criticizes a view which she describes as the 'received opinion': physics was homogeneously deterministic until the introduction of quantum mechanics. According to Romizi, this received view stems from "an extremely generalizing, unified image of the physics which preceded quantum mechanics" ("einem extrem pauschalisierenden, einheitlichen Bild der der Quantenmechanik vorangehenden Physik"). She argues that part of the problem lies in the use of the concepts of 'classical physics' and 'Newtonian physics', to describe the entire domain of physics from Newton until the revolutions of relativity theory and

---

M. van Strien (✉)
University of Wuppertal, Wuppertal, Germany
e-mail: vanstrien@uni-wuppertal.de

E. Ramharter (ed.), *The Vienna Circle and Religion*, Vienna Circle Institute Yearbook 25, https://doi.org/10.1007/978-3-030-76151-6_14

quantum physics: these concepts create an image of a unified and stable framework in physics, even though this period was actually rich in developments as well as in debates and disagreements about foundational issues in physics.

Romizi points out that the term 'classical physics' is in particular problematic because the domain of physics as we know it was only established during the nineteenth century: in the early eighteenth century, mechanics was still a part of mathematics, while the study of e.g. thermal and electric phenomena belonged to natural philosophy. However, Romizi does use the term 'classical' to describe a shared paradigm during a certain period: following Pulte (2005), she uses the word 'classical' to denote an account of scientific theories which was generally accepted in the eighteenth century, but gradually disappeared during the nineteenth century. According to Pulte, this "classical mathematical philosophy of nature" was characterized by the ideal of building up theories deductively from axioms of which the validity is beyond doubt, so that theories can be established as absolutely true and universal. With this conception of scientific theories, theories which are formulated mathematically can be taken to correspond with the world as it really is. Romizi argues that with this conception, the world itself is ultimately conceived of as mathematical, and that determinism, being inherently mathematical, follows automatically.

By the late nineteenth century, this classical understanding of science had disappeared: the ideal of absolute truth was abandoned, and scientific theories were now variously thought to be based on conventions, hypotheses, or empirical generalizations. The relation between theory and reality was problematized, and theories were increasingly conceived of as models, rather than as direct representations of reality. This general development, which has also been described by Helmut Pulte and Gregor Schiemann,[1] is described in detail by Romizi, who argues that this development brought along a weakening of determinism in physics.

This development indeed shows that the period between Newton and the revolutions of the early twentieth century was far from being a stable period of classical, Newtonian physics. This point could be pushed even further, however, by noting that already the period which Romizi describes as being dominated by a classical philosophy of nature was far from homogeneous. Recently, Marius Stan has drawn attention to the heterogeneity of mechanics in the eighteenth century:[2] during this period, there were different conceptions of matter and different accounts of what

[1]Helmut Pulte (2005). *Axiomatik und Empirie. Eine wissenschaftstheoriegeschichtliche Untersuchung zur Mathematischen Naturphilosophie von Newton bis Neumann*. Darmstadt: Wissenschaftliche Buchgesellschaft; and Pulte (2009). "From axioms to conventions and hypotheses: The foundation of mechanics and the roots of Carl Neumann's 'Principles of the Galilean-Newtonian Theory'". In: *The significance of the hypothetical in the natural sciences*, ed. Michael Heidelberger; Gregor Schiemann, 71–92. Berlin/New York: de Gruyter. Gregor Schiemann (2008). *Hermann von Helmholtz's mechanism: The Loss of Certainty: A Study on the Transition from Classical to Modern Philosophy of Nature*. Dordrecht: Springer.

[2]Marius Stan (2017). "Euler, Newton, and Foundations for Mechanics." In: *The Oxford Handbook of Newton*, ed. Christopher Smeenk; Eric Schliesser, 1–22. Oxford University Press.

the basic principles of mechanics were. By the early nineteenth century, there were two dominant frameworks in mechanics: the 'analytical mechanics' of Lagrange, which was highly abstract and axiomatic, and the 'physical mechanics' of Laplace, characterized by atomism.

Romizi's account of how exactly determinism was weakened by the late nineteenth century requires a more detailed look at what 'determinism' means. In the introduction to her book, Romizi calls attention to the problem that giving a strict definition of a certain concept can get in the way of describing the history of this concept: one has to allow for a certain flexibility in the way concepts are used, and for the possibility of ambiguities and shifts in meaning. For this reason, she avoids giving a strict definition of determinism. Instead, she describes three different aspects of determinism, which may be called dimensions, allowing for variations in the strength of conceptions of determinism along these dimensions. These distinctions work well to describe the main shifts in conceptions of determinism that took place during the nineteenth century.

The first of these dimensions, which she names 'solidity' ('Festigkeit'), has to do with the structure of scientific theories: in a theory which is maximally deterministic in this sense, predictions are derived deductively from the theory, which is built up deductively from first principles, corresponding with the 'classical' scientific ideal described above. Romizi shows how in the course of the nineteenth century, determinism got weakened along this dimension, as the idea of absolute certainty in natural science was given up, and statistical laws based on empirical data earned a place in science.

The second dimension, which Romizi names 'depth' ('Tiefe'), concerns the status of determinism: should determinism be interpreted ontologically, describing a feature of nature, or should it rather be understood epistemologically, as a feature of our theories or as a heuristic principle in science? Romizi argues that also in this sense, determinism was weakened during the nineteenth century, as it became more common to conceive of determinism in a non-ontological way. She argues that the deterioration of determinism in this sense already started with Kant, according to whom causality is not a feature of nature itself, but rather but rather a logical and conceptual presupposition of our (scientific) cognition of nature. The development towards less ontological conceptions of determinism corresponds with a shift in the status of scientific theories: especially towards the end of the nineteenth century, it became more common to think of theories as models with which one may be able to make accurate predictions, but which do not have to correspond with nature in every respect. When the idea that scientific theories should exactly correspond with nature is given up, whether or not a scientific theory is deterministic does not necessarily have to correspond with whether the reality it describes is deterministic. Romizi points out that the possibility of a gap between theory and reality could be used to question determinism as well as to defend it.

The third dimension is the 'width' ('Breite') of determinism, which has to do with the domain of mathematical science: determinism is strongest in this sense when it is argued that not just a specific theory, but nature as a whole is deterministic. Specifically, the issue here is whether determinism is restricted to mechanics or physics, or whether also e.g. organic and mental processes are taken to be deterministic (and approached mathematically). Romizi argues that in contrast to the other two dimensions of determinism, determinism was not weakened along this dimension during the nineteenth century, but in fact only reached its peak during the second half of the nineteenth century. Here, Romizi builds on Ernst Cassirer and Ian Hacking,[3] who have both argued that the concept of scientific determinism only became established through a famous lecture by the physiologist Emil Du Bois-Reymond in 1872, *On the limits of our knowledge of nature*. In this lecture, Du Bois-Reymond argued that all occurrences in nature are strictly determined through the dynamics of atoms. Romizi gives an interesting account of the context of public lectures, science popularization and the demand for a scientific worldview which gave rise to Du Bois-Reymond's grand statement of universal determinism. Hacking has argued that, despite some sporadic earlier occurrences, the term 'determinism' only became established through this lecture. Notably, he notes that Laplace, who is most commonly seen as the originator of scientific determinism, did not use the term. To account for the fact that during the period in which the 'solidity' and 'depth' of determinism were at their peak, the term determinism itself had not yet been established, Romizi introduces the concept of 'implicit scientific determinism'. She argues that until around the mid-nineteenth century, determinism was simply taken for granted, and points out that it makes sense that an idea is only formulated explicitly once it starts to be questioned and alternative conceptions become possible. However, it has to be remarked here that although Laplace did not use the term 'determinism', he described the corresponding idea pretty exactly, as did a number of earlier authors in eighteenth century France.[4]

With these three aspects of determinism, Romizi draws important distinctions, which help to clarify shifts in the conceptions of determinism as well as corresponding developments in philosophy of science during the nineteenth century. Nevertheless, at some points the analysis could have been sharper by drawing sharper distinctions between determinism and related concepts such as causation, predictability and reversibility. Notably, the account of the criticisms of determinism of Maxwell, Boussinesq and Saint-Venant could have been sharper in this regard. During the 1870s, all three of these authors argued against scientific determinism, and for the possibility of free will. Maxwell and Saint-Venant both emphasized the de facto unpredictability of many natural processes, because of our inability to have exact knowledge of initial conditions and the possibility of unstable systems in which a very small difference in initial conditions can have large effects. The fact that in practice, our capacities for prediction are highly limited, for them also motivated doubts about whether determinism holds in principle. But what sets

[3]See Ernst Cassirer (1956). *Determinism and indeterminism in modern physics*. Trans. O. T. Benfey. New Haven: Yale University Press. Originally published as *Determinismus und Indeterminismus in der modernen Physik* (Göteborg, 1937); and Ian Hacking (1983). "Nineteenth century cracks in the concept of determinism." In: *Journal of the history of ideas* 44 (3), 455–475.

[4]See Marij van Strien (2014). "On the Origins and Foundations of Laplacian Determinism." In: *Studies in History and Philosophy of Science* 45(1), 24–31.

Boussinesq apart is that he argued that the laws of mechanics *themselves* allow for genuine indeterminism: one can construct systems within mechanics for which, for given exact values of initial conditions, the equations of mechanics fail to determine what will happen. This argument, based on the laws of mechanics, is fundamentally different from arguments based on our observation of unpredictable behavior.

All in all, the book shows a rich spectrum of ways in which the absolute validity of scientific determinism could be doubted or rejected during the nineteenth and early twentieth century. To this end, Romizi discusses the views of a wide range of physicists and philosophers, including Fechner, Boutroux, Maxwell, Renouvier, Peirce, James, Mach, Boltzmann, and Zilsel. She rightfully points out that a rejection of determinism does not always amount to an acceptance of indeterminism: the question of determinism vs. indeterminism can also be regarded as undecidable or rejected as meaningless. She further points out that, contra Paul Forman and as argued before by Michael Stöltzner and Deborah Coen,[5] a rejection of determinism or even an explicit acceptance of indeterminism does not have to indicate a step towards irrationalism. Above all, Romizi's analysis of the history of determinism shows that in the nineteenth and early twentieth century, many scientists carefully considered the question of the proper domain and limits of science.

[5]See Paul Forman (1971). "Weimar culture, causality, and quantum theory: adaptation by German physicists and mathematicians to a hostile environment." In: *Historical Studies in the Physical Sciences* 3, 1–115; Michael Stöltzner (2003). *Vienna Indeterminism. Causality, Realism and the Two Strands of Boltzmann's Legacy*. Ph.D. Dissertation, Universität Bielefeld; Deborah Coen (2008). *Vienna in the age of uncertainty: science, liberalism, and private life*. University of Chicago Press.

Chapter 15
Review: Bernt Österman (Ed.), *"Skriv så ofta du kan": Brevväxlingen mellan Georg Henrik von Wright och Eino Kaila 1937–1958* ["Write as often as You Can": The Correspondence between Georg Henrik von Wright and Eino Kaila 1937–1958.]. Helsingfors: Svenska Litteratursällskapet 2020, 284 pp., ISBN 978-951-583-522-2 (Also available as an e-publication with extensive critical editorial material at: https://vonwright.sls.fi/#/home.)

Sami Pihlström

As the Vienna Circle transformed twentieth-century philosophy in the 1920–30s, Finland was a poor small country in the margins of Europe with a legacy of a bloody Civil War. One of the relatively few Finnish academic philosophers active in those days, Eino Kaila (1890–1958), established connections with Moritz Schlick, Rudolf Carnap, Hans Reichenbach, among others, already in the 1920s, hoping to bring "scientific philosophy" and the revolutionary ideas of logical empiricism to his remote home country where he felt intellectually isolated. (Kaila's relations to the Vienna Circle have been explored, e.g., by Ilkka Niiniluoto in this *Yearbook*, vol. 19, 2017.) Having worked at the University of Turku in the 1920s, Kaila became Professor of Theoretical Philosophy at the University of Helsinki in 1930 and was arguably the leading intellectual of the young nation when a bright student came to visit him in 1934. This student was Georg Henrik von Wright (1916–2003), Kaila's distant family member.

The philosophical friendship between Kaila and von Wright, now documented in their published correspondence spanning more than two decades, was crucial

S. Pihlström (✉)
University of Helsinki, Helsinki, Finland
e-mail: sami.pihlstrom@helsinki.fi

E. Ramharter (ed.), *The Vienna Circle and Religion*, Vienna Circle Institute Yearbook 25, https://doi.org/10.1007/978-3-030-76151-6_15

for the development of Finnish philosophy but important for the broader picture of twentieth-century philosophy, too. von Wright got his first professorship in Helsinki soon after WWII, and in 1948 he became Ludwig Wittgenstein's successor in Cambridge. He chose to return to Finland in 1951, however, and somewhat later (1962) succeeded Kaila as a highly distinguished member of the Academy of Finland. As one of Wittgenstein's literary heirs and one of the editors of his posthumous work, von Wright made a lasting contribution to the reception of Wittgenstein. Due to his foundational research on modal and deontic logic, the logic of norms, theory of action, and the structure of explanation and understanding, among many other topics, he was one of the giants of analytic philosophy during the second half of the twentieth century.

In comparison, Kaila remains internationally less widely appreciated. He wrote his main academic work in German and became somewhat eclipsed as the center of analytic philosophy moved to English-speaking countries after WWII. (Kaila's work has later been made available in English in the collection *Reality and Experience* [1979] and the translation of one of his main Finnish works, *Human Knowledge* [2014, originally 1939].) Yet, Kaila was a highly original thinker who subscribed to logical empiricism but never identified himself as an "analytic philosopher". He was, rather, a "synthetic" philosopher of nature seeking to integrate the foundations of the scientific worldview into a coherent philosophical system, restlessly pursuing such a holistic theory until the end of his life. Kaila was also enormously influential as a teacher and public intellectual in Finland. He trained a generation of future professors, and his most creative and independent student was undoubtedly von Wright.

The new book and website collecting Kaila's and von Wright's correspondence in 1937–1958 are based on the meticulous research conducted at the von Wright – Wittgenstein Archives at the University of Helsinki. Bernt Österman has with his collaborators put together an impressive volume. The traditional book contains a preface by Thomas Wallgren (the Director of the Archives), a 100-page scholarly introduction by Österman, and all the available correspondence consisting of 85 preserved letters (41 by von Wright and 44 by Kaila), some of them short notices but others longer explorations of both philosophical and "academic-political" as well as general cultural matters. In addition to all this, there is, in the electronic version, further extremely useful material for future scholarship: a comprehensive critical apparatus, extensive notes (with, e.g., short notices on the people, works, and places mentioned in the letters), as well as facsimile reproductions of the original handwritten or typed letters. There are also two appendices (items originally appended to the letters) only available in the online edition: von Wright's 1939 talk on induction at the Moral Sciences Club in Cambridge and Kaila's 1940 speech at the 300th anniversary jubileum of the University of Helsinki.

As Österman explains (6), because Kaila's correspondence was mostly destroyed – in Finland, only 14 letters by him are preserved in the archives of the Finnish Literature Society – the collection is uniquely important. In contrast, as many as ca. 23,000 letters by von Wright are preserved at the Finnish National Library and the University of Helsinki (6). It must be kept in mind, however, that von Wright and Kaila mainly exchanged ideas in personal meetings. There are occasionally gaps of several years in the correspondence simply because in those years both were based in Helsinki. In any event, some very important periods in the two philosophers' careers are manifested in the letters, including von Wright's student years in the late 1930s and his Cambridge period in 1948–51, as well as Kaila's only visit to the United States in 1948.

Notably, no wartime letters exist, though. This is because Kaila and von Wright were both based in Helsinki during WWII, as von Wright was liberated from front service due to a medical condition. The question concerning the two philosophers' relations to the Third Reich is of some historical interest, and the letters are an invaluable source for anyone examining this issue. It seems that both were horrified of the Nazis but failed to understand the scope of what was going on, viewing (like so many others) the Soviet Union as an even worse enemy for culture and civilization in general. Kaila was naïve in his belief in Germany's "historical mission" – something that Österman aptly describes as a "gigantic misjudgment" (*"en gigantisk felbedömning"*, 95). There are letters from 1940–41, but they are all from the "interim peace" period (as it is called in Finland) between the Finnish 1939–40 "winter war" against the Soviet Union and the "continuation war" that started in June 1941.

von Wright had planned to continue his studies in Vienna, but the developments in the 1930s made that impossible. However, some early letters do describe his trip to Vienna in 1937, when he was touring Europe with his close friend Göran Schildt. On April 15, 1937, he tells Kaila about a meeting with Kurt Gödel in Vienna but also reflects on the demise of the Vienna Circle: *"Det verkar som om någonting hölle på at dö ut här i Wien."* ("It seems as if something were in the process of dying here in Vienna.") A couple of years later, von Wright has already arrived in Cambridge, only then realizing that Wittgenstein was actually there and had returned to philosophy (von Wright to Kaila, March 5, 1939).

The correspondence also exhibits a certain kind of delicacy in the process of von Wright's development from Kaila's student into a disciple of Wittgenstein and eventually a professional philosopher in his own right, obviously shaped by but independent of both great teachers. Focusing on problems of reality rather than language, Kaila was never a fan of Wittgenstein; von Wright becomes gradually careful in discussing Wittgenstein in the letters. In one of his first characterizations of Wittgenstein to Kaila on April 5, 1939, he notes that Wittgenstein is rather far from the lines of thought he had learned in Helsinki but that there may be a "common core" and a minor change of orientation might be necessary for him. On June 29, 1939, when von Wright was still visiting Cambridge as a student and had become better acquainted with Wittgenstein, he tells Kaila that his meetings with Wittgenstein have revealed "entirely new worlds" that he had never imagined before (*"helt nya världar, som jag aldrig anat"*). On July 6, 1939, Kaila perceptively notes that von Wright seems to have more "congenial sympathy" to Wittgenstein's way of thinking than he himself does (cf. Österman, 83).

After Wittgenstein's death in 1951, von Wright decides to leave his professorship in Cambridge and to return to Helsinki, writing to Kaila that, "[a]long with you", the now deceased Wittgenstein was "the person who meant the most to my spiritual development" (June 9, 1951). Kaila's response is enthusiastic, expressing a new confidence in the future of philosophy "in this poor country" (June 17, 1951). In addition to Wittgenstein, the letters testify about the two philosophers' many international connections, with multiple occurrences of names like Moore, Broad, Tarski, Popper, and Gödel, among many others. Kaila also mentions a visit to Victor Kraft's summer house in Austria – at a time after the *Anschluss* and close to the outbreak of the war (July 6, 1939).

Österman explains, with reference to von Wright's 2001 autobiography, that on the one hand Wittgenstein liberated von Wright from Kaila's logical empiricism but on the other hand von Wright maintained from his student years with Kaila an interest in formal logic, which enabled him to develop his own philosophical approach independently of Wittgenstein (71). He is, with good reason, often considered the most original and autonomous among Wittgenstein's closest pupils.

The letters from spring and summer 1939 are also invaluable for anyone studying the emergence of von Wright's early philosophy, because he regularly reports to Kaila from Cambridge about the progress of his doctoral thesis on the problem of induction (which he defended in Helsinki in May 1941, just weeks before Finland joined Hitler's campaign against the Soviet Union). Despite his exciting work on induction and probability, von Wright seems to feel the pressure of the dark times and even occasionally expresses a wish to die (e.g., July 24, 1939), finding it somehow "wrong" ("*orätt*") to deal with philosophical issues that do not belong to the serious historical moment. Kaila (August 2, 1939) is worried about this pessimistic tone to the extent that he even quotes Wittgenstein's *Tractatus* (slightly incorrectly, though): "*Das Faktische ist für das Höhere ziemlich gleichgültig.*"

Several letters confirm the impression of Kaila as a lonely thinker working in isolation from others. In the letter giving the title to the entire collection, he tells von Wright on September 17, 1948, that he experiences a "vacuum" and even feels himself "alone in the universe" after von Wright's departure to Cambridge. On April 24, 1949, he claims to be "totally isolated". While some of the early letters are rich in von Wright's elaborations of his emerging views on induction, the late letters from Kaila are written by an aging thinker increasingly conscious of the inevitable incompleteness of his pursuit of a thoroughgoing scientific worldview. Melancholically, Kaila shares his reading experience of George Orwell's *Nineteen Eighty-Four* upon its publication in 1949: if things go as Orwell's fiction predicts, then it is better that the world will be "blown into pieces" (September 20, 1949). In contrast, von Wright is full of energy, deploring that life is too short for all the philosophical work that needs to be done (March 4, 1950).

Some of the late letters in the 1950s focus on the developments in Finnish philosophy at the time. Kaila and von Wright were both involved in delivering expert reports on the candidates for open professorships and worried that "obscurantist" currents hostile to the scientific spirit of serious philosophy they represented might prevail (e.g., Kaila to von Wright, May 7, 1949). In their final letters in 1958, they end up unanimously recommending Jaakko Hintikka to a professorship in Helsinki, to which Hintikka was appointed in 1959 (after Kaila's death); Kaila's pronouncements on some of Hintikka's rival applicants (e.g., April 20, 1958) are rather harsh. Indeed, one of the fruits born by Kaila's and von Wright's philosophical friendship was the stellar international career of Hintikka (first mentioned in von Wright's letter on July 3, 1949). Occasional references to the various philosophical and academic societies that Kaila and von Wright contributed to developing are also interesting for anyone working on the history of such organizations, including Institut International de Philosophie (Kaila to von Wright, May 25, 1947 and July 26, 1948) and the Philosophical Society of Finland (several letters in 1948–49, in particular).

Österman's and his collaborators' careful and substantial work is to be warmly recommended, both in the book format and in the more comprehensive online format, particularly to scholars interested in mid-twentieth-century philosophical developments in Northern Europe. As all the material is in Swedish (the language Kaila and von Wright used in their communication), it is presumably easily available only to readers in the Nordic countries, but with some effort a German-speaking scholar can, I suppose, study the Swedish letters, too.

Index

E. Ramharter (ed.), *The Vienna Circle and Religion*, Vienna Circle Institute Yearbook 25, https://doi.org/10.1007/978-3-030-76151-6

The manufacturer's authorised representative in the EU is Springer Nature Customer Service Centre GmbH, Europaplatz 3, 69115 Heidelberg, Germany. If you have any concerns regarding our products, please contact ProductSafety@springernature.com

Printed and bound by CPI Group (UK) Ltd, Croydon, CR0 4YY
30/09/2026
02225497-0012